Nonfermentative Gram-Negative Rods

MICROBIOLOGY SERIES

Series Editors
ALLEN I. LASKIN
Sommerset, New Jersey

RICHARD I. MATELES
Stauffer Chemical Company
Westport, Connecticut

Volume 1   Bacterial Membranes and Walls
edited by Loretta Leive

Volume 2   Eucaryotic Microbes as Model Developmental Systems
edited by Danton H. O'Day and Paul A. Horgen

Volume 3   Microorganisms and Minerals
edited by Eugene D. Weinberg

Volume 4   Bacterial Transport
edited by Barry P. Rosen

Volume 5   Microbial Testers: Probing Carcinogenesis
edited by I. Cecil Felkner

Volume 6   Virus Infections: Modern Concepts and Status
edited by Lloyd C. Olson

Volume 7   Methods in Environmental Virology
edited by Charles P. Gerba and Sagar M. Goyal

Volume 8   Microtubules in Microorganisms
edited by Piero Cappucinelli and N. Ronald Morris

Volume 9   Handbook of Indigenous Fermented Foods
edited by Keith H. Steinkraus

Volume 10   Unusual Microorganisms: Gram-Negative Fastidious Species
edited by Edward J. Bottone

Volume 11 Laboratory Manual for Medical Microbiology
edited by The Faculty of the Department of Microbiology, Schools of Medicine and Dentistry, State University of New York at Buffalo

Volume 12 Infectious Diarrheal Disease: Current Concepts and Laboratory Procedures
edited by Paul D. Ellner

Volume 13 Microbial Degradation of Organic Compounds
edited by David T. Gibson

Volume 14 The Silage Fermentation
by Michael K. Woolford

Volume 15 The Mycobacteria: A Sourcebook *(in two parts)*
edited by George P. Kubica and Lawrence G. Wayne

Volume 16 Nonfermentative Gram-Negative Rods: Laboratory Identification and Clinical Aspects
edited by Gerald L. Gilardi

Other Volumes in Preparation

Nonfermentative Gram-Negative Rods

Laboratory Identification and Clinical Aspects

Edited by
Gerald L. Gilardi
North General Hospital
New York, New York

Marcel Dekker, Inc.   New York and Basel

Library of Congress Cataloging in Publication Data
Main entry under title:

Nonfermentative gram-negative rods: laboratory
  identification and clinical aspects

  (Microbiology series ; v. 16)
  Includes index.
  1. Diagnostic microbiology. 2. Gram-negative
bacteria--Identification. 3. Fermentation tube.
4. Pseudomonas--Identification. I. Gilardi, Gerald L.
II. Series.
QR67.N66 1985   616.9'310758   85-7134
ISBN 0-8247-7370-5

COPYRIGHT © 1985 by MARCEL DEKKER, INC. ALL RIGHTS RESERVED

Neither this book nor any part may be reproduced or transmitted
in any form or by any means, electronic or mechanical, including
photocopying, microfilming, and recording, or by any information
storage and retrieval system, without permission in writing from
the publisher.

MARCEL DEKKER, INC.
270 Madison Avenue, New York, New York 10016

Current printing (last digit):
10 9 8 7 6 5 4 3 2 1

PRINTED IN THE UNITED STATES OF AMERICA

Preface

The main purpose of this latest volume in the Microbiology series is to inform the reader about recent advances in glucose-nonfermenting gram-negative rods of medical importance including the genera *Pseudomonas*, *Acinetobacter*, *Moraxella*, *Flavobacterium*, *Alcaligenes*, and *Achromobacter*. Today, most nonfermenters can be easily identified in the clinical laboratory. Infections caused by nonfermenting rods have been increasing both in number and significance among hospitalized patients. Nonfermenters are widespread in the environment, are physiologically versatile, and can cause serious infections in susceptible patients with underlying debilitating diseases. Unfortunately, nonfermenters are notable due to problems associated with antimicrobial treatment.

This volume consists of contributions by experts in various areas of microbiology and infectious diseases related to nonfermenters. Each chapter covers original investigative works by the contributors and the state of the art of the subject matter including methods and practical applications useful in clinical bacteriology. For the beginning researcher, some pertinent background information is presented on methods of isolation and identification. The cultural and biochemical characteristics of nonfermenters and the rapid and computer-assisted microsystems used for their identification are reviewed. One part of the book has been devoted to a series of chapters that survey the ecology, clinical significance, and antimicrobial susceptibilities of nonfermenters. There has been considerable emphasis recently on the possible role of various enzymes and other extracellular products in the virulence and pathogenicity of nonfermenters, and two chapters are concerned with virulence factors of *Pseudomonas*. The book concludes with newer methodologies for definitive identification of nonfermenters, including transformation assays for *Moraxella* and *Acinetobacter* using deoxyribonucleic acid samples.

This book will be useful for the bench technologist interested in identification and typing of clinical isolates, as well as those specialists in infectious diseases and hospital epidemiology concerned with the clinical role of nonfermenters, antimicrobial therapy of choice, and the epidemiological problems associated with hospital infections. Current advances in the technologies discussed in this volume will serve as a handy information source and should permit future acquisition of additional information about nonfermenters as related to effective patient care.

Gerald L. Gilardi

Contents

Preface		*iii*
Contributors		*vii*
1.	Methods of Isolation and Identification of Glucose-Nonfermenting Gram-Negative Rods *J. R. Greenwood*	1
2.	Cultural and Biochemical Aspects for Identification of Glucose-Nonfermenting Gram-Negative Rods *Gerald L. Gilardi*	17
3.	Rapid Identification of Glucose-Nonfermenting Gram-Negative Rods with Commercial Miniaturized Kits *Thomas R. Oberhofer*	85
4.	Ecology, Clinical Significance, and Antimicrobial Susceptibility of *Pseudomonas aeruginosa* *Harold C. Neu*	117
5.	Ecology, Clinical Significance, and Antimicrobial Susceptibility of *Acinetobacter* and *Moraxella* *Robert W. Lyons*	159
6.	Ecology, Clinical Significance, and Antimicrobial Susceptibility of Infrequently Encountered Glucose-Nonfermenting Gram-Negative Rods *Alexander von Graevenitz*	181

7. Pathogenic Properties and Enzyme Profiles of
   Pseudomonas and *Acinetobacter*   233
   J. Michael Janda and Edward J. Bottone

8. In Vitro and In Vivo Tests to Determine Virulence
   of *Pseudomonas aeruginosa*   255
   Ian Alan Holder

9. *Pseudomonas aeruginosa*: Serology, Phage, Pyocin
   Charles H. Zierdt   283

10. Transformation Assays for *Acinetobacter* and *Moraxella*   341
    Elliot Juni

Index   357

Contributors

EDWARD J. BOTTONE, Ph.D.   Director, Department of Microbiology, The Mount Sinai Hospital, and Professor of Clinical Microbiology, Mount Sinai School of Medicine, New York, New York

GERALD L. GILARDI, Ph.D.   Head, Bacteriology Section, Department of Laboratories, North General Hospital, and Assistant Professor, Department of Pathology, Mount Sinai School of Medicine, New York, New York

J. R. GREENWOOD, Ph.D., M.P.H.   Director, Public Health Laboratory, County of Orange, Health Care Agency, Santa Ana, California, and Adjunct Assistant Professor, Department of Epidemiology, School of Public Health, University of California, Los Angeles, California

IAN ALAN HOLDER, Ph.D.   Professor of Research Surgery and Adjunct Professor of Microbiology and Molecular Biology, Departments of Surgery and Microbiology, University of Cincinnati College of Medicine, and Director, Department of Microbiology, Shriners Burns Institute, Cincinnati, Ohio

J. MICHAEL JANDA, Ph.D.   Associate Director, Department of Microbiology, The Mount Sinai Hospital, and Associate Professor of Clinical Microbiology, Mount Sinai School of Medicine, New York, New York

ELLIOT JUNI, Ph.D.   Professor of Microbiology, Department of Microbiology and Immunology, The University of Michigan, Ann Arbor, Michigan

ROBERT W. LYONS, M.D. Director, Section of Infectious Disease and Epidemiology, Department of Internal Medicine, St. Francis Hospital and Medical Center, Hartford; Associate Professor, Department of Medicine, University of Connecticut School of Medicine, Farmington; and Associate Clinical Professor, Yale University School of Medicine, New Haven, Connecticut

HAROLD C. NEU, M.D. Professor of Medicine and Pharmacology, Columbia University College of Physicians and Surgeons, and Chief, Division of Infectious Diseases and Epidemiology, Columbia-Presbyterian Medical Center, New York, New York

THOMAS R. OBERHOFER, DrPH Chief, Microbiology Section, Department of Pathology and Area Laboratory Services, Madigan Army Medical Center, Tacoma, Washington

ALEXANDER von GRAEVENITZ, M.D. Professor and Chairman, Department of Medical Microbiology, University of Zürich, Zürich, Switzerland

CHARLES H. ZIERDT, Ph.D. Microbiologist, Department of Clinical Pathology, Clinical Center, National Institutes of Health, Bethesda, Maryland

Nonfermentative
Gram-Negative Rods

1
Methods of Isolation and Identification of Glucose-Nonfermenting Gram-Negative Rods

J. R. GREENWOOD *Public Health Laboratory, County of Orange, Health Care Agency, Santa Ana, California, and School of Public Health, University of California, Los Angeles, California*

INTRODUCTION

Nonfermentative gram-negative rods (NFGNR), as with any other group of potential pathogens encountered in clinical microbiology, should be efficiently isolated and then rapidly identified. This approach provides direct benefits to both the patient and the laboratory. Fortunately, NFGNR are not fastidious in their growth requirements, and isolation from clinical material does not require heroic measures. Additionally, NFGNR have numerous distinctive biochemical features and, if pertinent features are determined, most isolates can be identified with a minimal number of biochemical tests. This chapter will provide a practical approach to these two aspects of NFGNR microbiology.

ISOLATION PROCEDURES

Incubation Temperatures

Nonfermentative gram-negative rods represent approximately 10% of the gram-negative bacteria isolated in a clinical laboratory. Consequently, although 30° C is a commonly used temperature for NFGNR, laboratories may have only a 35–37° C incubator available for initial processing of clinical specimens. Fortunately, most of the commonly encountered NFGNR will grow at 35° C and produce visible colonies in 24—48 hrs. However, if no growth is observed after incubation (and the presence of an NFGNR is suspected from the results of a Gram stain), initial plates can be placed at room temperature and reincubated for an additional day or two. For those laboratories that are engaged in environmental studies (e.g., infection control) or offer reference services for NFGNR, a 30° C incubator provides a good compromise temperature for primary isolation.

Isolation Media: General Purpose

For the purpose of this chapter, NFGNR are members of the following genera: *Achromobacter, Acinetobacter, Alcaligenes, Bordetella, Flavobacterium, Moraxella, Pseudomonas, and Agrobacterium*. Not considered are genera such as *Eikenella* and *Brucella*, both of which contain species that are fastidious NFGNR. Thus defined, all of the species in the genera initially mentioned will grow on sheep blood agar prepared with an infusion agar base. In fact, infusion agar base, without added blood, will support the growth of virtually all NFGNR, including *Pseudomonas maltophilia, Pseudomonas diminuta,* and *Pseudomonas vesicularis*. These particular organisms are known to have rather strictly defined vitamin and/or amino acid growth requirements. Sheep blood agar has several other important advantages when used as a primary isolation medium. It can be used to determine an isolate's action on blood (e.g., hemolysis, greening), to test the ability of an isolate to produce cytochrome oxidase, and, lastly, to screen out fastidious NFGNR (i.e., an organism that requires a richer basal medium for growth).

In addition to sheep blood agar, most clinical laboratories routinely include selective or selective-differential media in their scheme for initial processing of clinical specimens. Many of these media can also be useful for isolating NFGNR, or at least they can provide additional information about an isolate. As an example, many NFGNR can grow on MacConkey agar and thus can be separated from contaminating gram-positive flora. In addition, lactose-fermenting *Enterobacteriaceae* can also be differentiated from NFGNR on this medium. Because some NFGNR, such as *Pseudomonas paucimobilis, P. vesicularis, Flavobacterium meningosepticum,* and *Moraxella osloensis*, fail to grow on MacConkey agar, initial growth of an NFGNR on blood agar, but not on MacConkey agar, is information that will be of significant help in the identification of an isolate. Other media for the isolation of gram-negative bacteria such as XLD, HEK, and Salmonella—Shigella, should be used with caution as isolation media for NFGNR, as many of these media are highly inhibitory for some species or strains of these organisms.

Isolation Media: Special Purpose

There are a number of formulations of selective media available for the isolation of NFGNR (e.g., cetrimide agar) (Grant and Holt, 1977; Simon and Ridge, 1974; vonGravenitz, 1976). While these media can be useful for environmental studies, epidemiological purposes (possible infection control programs), or industrial quality control programs, they are of little value in the routine work of a clinical laboratory. If they are used for a particular study, it is important to remember that, like other selective media formulations, not all strains of a given species will grow on them. Thus a nonselective medium should always be included when these media are used.

PRESUMPTIVE IDENTIFICATION

There are two methods available to determine the oxidative or fermentative capabilities of gram-negative isolates. The first method is the classic Hugh and Leifson (1953) approach. To use this method, two tubes of oxidative-fermentative (OF) glucose are inoculated with the unknown isolate. One of the tubes is then sealed with petrolatum and both tubes are incubated at 30–35° C. A fermentative organism will produce acid in the unsealed (aerobic) tube and the sealed (anaerobic) tube. However, NFGNR, being strictly aerobic organisms, cannot acidify glucose in the sealed tube. This method can effectively separate nonfermentative from fermentative gram-negative rods. An additional advantage of the two-tube OF method is the ability to differentiate glucose oxidizers from glucose nonoxidizers in the open OF tube. This information is necessary to use the Centers for Disease Control (CDC) identification tables for gram-negative isolates (Weaver et al., 1972).

A more practical approach to the differentiation of NFGNR from fermenters is to inoculate either a Kligler iron (KIA) or triple-sugar iron (TSI) agar slant. The medium is stabbed and the slant is streaked. The tube is incubated at 30–35° C with a loosened cap and examined at 18–24 hrs. Fermentative bacteria will acidify the butt of the tube, while NFGNR will not. When examining these tubes for acidification in the butt of the tube, it is sometimes necessary to compare the KIA/TSI reactions with an uninoculated tube. This can be particularly important with proteolytic NFGNR that alkalinize the slant, and thus, by contrast, make the butt of the tube appear acidic.

The KIA/TSI approach is also useful in separating fastidious gram-negative bacteria from NFGNR. Virtually all of the commonly encountered NFGNR can grow on the slant of a KIA/TSI, while most of the fastidious gram-negative rods (e.g., some strains of *Actinobacillus*, *Eikenella*, DF-1) cannot grow on these media.

An important point to remeber about both the OF and KIA/TSI methods used to differentiate fermenters from nonfermenters is that they do not distinguish between gram-positive oxidative organisms (i.e., *Bacillus* species) and NFGNR. Fortunately, there are now two alternative methods to the Gram stain that can be used to distinguish gram-positive from gram-negative bacteria. One of these is the lysis of gram-negative bacteria, but not gram-positive bacteria (even if they stain gram negative), in 3% KOH (Gregersen, 1978). To perform the KOH test, an opaque suspension of bacteria is prepared on a glass slide in 3% KOH. After a minute or two of mixing with a bacteriological loop, most gram-negative bacteria will lyse and the suspension will become gelatinous from released DNA. Gram-positive bacteria, however, stay in suspension. A more sensitive test than lysis by KOH is hydrolysis of L-alanine-4-nitroanilide (LANA) by gram-negative bacteria (Carlone, et al., 1982; Cerny, 1976). To perform the LANA test, an opaque suspension of bacteria is prepared in two drops of 4% reagent on a glass slide or in the lid of a Petri plate. The test is incubated at room temperature for up to 60 min before being discarded. A positive reaction, indicative of amino peptidase activity by gram-negative bacteria, is indicated by a yellow coloration.

BIOCHEMICAL METHODS

Acidification of Carbohydrates

Information derived from the oxidative acidification of carbohydrates by NFGNR can be extremely useful in the identification of isolates. It was initially recognized that there are several biological constraints in consistently demonstrating acidification of carbohydrates by NFGNR. The basic problem is that many NFGNR are proteolytic and produce alkaline by-products of metabolism which then neutralize the weak acids produced by oxidative attack of carbohydrates. Hugh and Leifson (1953) recognized this problem and formulated a oxidative—fermentative (OF) medium to detect acid production by NFGNR. They recognized not only the need to have lowered amounts of peptone, which can be alkalinized, but also an aerobic environment. An aerobic environment was created by adding agar to the OF formulation so that NFGNR could grow on the surface of this semisolid medium. While the use of OF carbohydrates provides several advantages for the processing of nonfermenters, such as a large data base and commercial availability, it also has a distinct disadvantage when used in a clinical laboratory. It can take several days for some isolates to show positive reactions on the OF medium. This is because the OF medium still contains enough peptone (200 mg %) for alkalinization to occur, thus neutralizing the weak acids produced by NFGNR. The oxidative—fermentative medium also has a limited aerobic surface area for NFGNR growth. These two factors combine to make the OF medium relatively insensitive to acid production by weakly saccharolytic NFGNR. To obviate the limitations of the OF medium, a more sensitive approach to the detection of acidification of carbohydrates by NFGNR was developed by Pickett and Pedersen (1970 a, b). This method utilizes buffered single substrates (BSS), which are prepared by adding a carbohydrate to a weakly buffered indicator solution. No peptone is present, so that neutralizing alkalinity is not produced. A heavy inoculum is added to each BSS tube and the tubes are incubated at 35° C. Preformed enzymes produced by NFGNR acidify the carbohydrate, and positive reactions are rapidly (most in 24—48 hr) detected by a pH shift in the indicator. The BSS method is probably the most sensitive approach to the detection of oxidative acidification by NFGNR. In practice, however, the method has been plagued by limited and sometimes unreliable commercial sources and incorrect use by microbiologists. Because BSS are nongrowth tests, if insufficient inoculum is added, preformed enzymes are not present in quantities sufficient to produce positive reactions. Thus false negative reactions can commonly occur.

An alternative approach to the detection of oxidative acidification by NFGNR is the use of oxidative low-peptone medium (OLP) (Greenwood and Pickett, 1978). The OLP formulation represents a compromise between the OF medium and BSS. Like the OF medium, OLP is a growth medium, but contains only 25% as much peptone. Additionally, because OLP is prepared as a slant, it has more aerobic surface area for NFGNR growth. Thus it approaches the sensitivity of BSS. Most positive reactions can be detected within 24 hr of incubation at 35° C. This speed, coupled with increased sensitivity, makes OLP a useful medium in the clinical laboratory setting. The OLP basal medium, with a pH adjustment, can also be used to detect alkalinization of organic salts and amides (e.g., citrate, acetamide). The utility of detecting the alkalinization of organic salts and amides will be discussed in a later section of this chapter.

Cytochrome Oxidase (Indophenol Oxidase, Cytochrome Oxidase)

There are numerous methods available to detect the production of cytochrome by NFGNR. Whatever method is chosen, it is necessary to remember that each method varies in its sensitivity. As an example, dimethylparaphenylenediamine monohydrochloride is less sensitive than tetramethylparaphenylenediamine dihydrochloride. Consequently, some weakly oxidase positive NFGNR, such as *P. paucimobilis,* may be oxidase negative when tested with the dimethyl reagent. Thus, in the identification process, it is important to follow the oxidase testing method used for a given set of identification charts or tables.

There are several considerations one should keep in mind when testing oxidase reactions. First of all, not only for NFGNR, but for any other organism as well, the inoculum used to test the oxidase reaction should not be too old. In most cases, 48-hr inoculum is the oldest that should be used to test oxidase. The inoculum should not be taken from selective media that contain bile or carbohydrates, such as MacConkey agar, as false negative oxidase reactions have been noted from these types of media. Inoculum taken from acid media has been shown to give false positive reactions. When working with NFGNR, I have found that paper strips saturated with the tetramethyl reagent have proven to be a sensitive and reliable oxidase reagent.

Decarboxylase and Dihydrolase Activity

Detection of decarboxylase or dihydrolase activity by NFGNR can be valuable information for identification purposes. As an example, the two commonly encountered oxidase negative isolates of NFGNR, *P. maltophilia* and *Acinetobacter anitratus,* can be separated easily with the lysine decarboxylase reaction. *Pseudomonas maltophilia* is virtually 100% positive in its ability to decarboxylate lysine, while *A. anitratus* is always negative for this feature.

Although the Moeller method has traditionally been used to detect this activity in NFGNR, this approach, like the use of OF sugars, has some limitations in a clinical laboratory setting. The Moeller method is designed for detection of decarboxylase and dihydrolase activity by fermentative bacteria. Thus it becomes readily apparent that a petrolatum overlay, while justified with fermentative bacteria, is incompatible with the aerobic growth requirements of NFGNR. The Moeller method also uses dextrose fermentation to separate weakly positive decarboxylase and dihydrolase reactions from negative reactions (i.e., acid pH shift by decarboxylase negative *Enterobacteriaceae*). Nonfermentative gram-negative rods cannot acidify dextrose under these conditions, and weakly positive reactions must be compared to a control tube of basal medium without amino acid. This approach is used to distinguish between alkalinization of the basal medium and decarboxylase or dihydrolase activity. Both of these characteristics of the Moeller method cause it to be relatively insensitive and slow when used with NFGNR. A rapid and sensitive method is the modified Carlquist (1956) chloroform—ninhydrin method. A heavy suspension of a 24-hr culture is prepared in 1 ml of amino acid broth. This is incubated for 24 hr at 35° C and then one drop of 40% KOH is added to the suspension. Next, 1 ml of chloroform—ninhydrin is added without shaking. The basic amine (produced through decarboxylation or dihydrolase activity) is extracted into the chloroform layer and turns purple in the presence of ninhydrin. This is a positive reaction.

Motility

The most rapid method available to determine motility by an NFGNR is the hanging drop or wet mount preparation. An inoculating needle is touched to an 18 to 24-hr colony growing on a nonselective medium (e.g., blood agar) and then touched to a drop of distilled water on a glass slide. A cover slip is placed on the drop and the slide is examined with low and high dry objectives. Care must be taken to avoid confusing Brownian movement with true bacterial motility.

Indirect evidence of motility by an NFGNR can be obtained by inoculating a tube of motility—nitrate medium (Pickett and Pedersen, 1970a). Motility—nitrate medium is inoculated by stabbing 5—10 mm into the medium and then the tube is incubated at 30—35° C. Because NFGNR produce little opacity in the medium, tubes are best examined for motility 3—4 hr after inoculation. At this time, a small "puffball" of motility can be seen around the original line of inoculation. If this is not observed, tubes should be reincubated and compared to an uninoculated tube after 18—24 hr of incubation.

It should be noted that enteric motility media and motility—nitrate medium are not identical. Motility—nitrate medium was formulated to allow optimal motility of NFGNR and contains 0.3% agar rather than the 0.4% found in enteric motility media. Although this appears to be an insignificant difference, numerous false negative motility reactions will be recorded when enteric motility media are used to detect NFGNR motility. Motility—nitrate medium also does not contain 2,3,5—triphenyltetrazolium chloride, a compound known to inhibit motility of some NFGNR.

As an adjunct to the determination of motility, examination of flagellar location and morphology can sometimes be useful for definitive identification of NFGNR. Several flagellar staining methods such as that of Clark (1976), Kodaka et al., (1982), or Leifson (1951) are available, and each has particular strengths, such as ease of use or extended shelf life. There is, however, little need for routine use of a flagellar stain for identification of NFGNR commonly encountered in a clinical laboratory.

Nitrate Reduction

With some important exceptions (e.g., *A. anitratus*, *Acinetobacter lwoffi*, *Moraxella urethralis*, some flavobacteria, and *Alcaligenes odorans*) most NFGNR can reduce nitrate to nitrite. A more limited number of species can also denitrify, that is, reduce nitrate or nitrite to nitrogen gas. Detection of denitrification activity, because it is limited to a relatively small number of species of NFGNR, can be useful information for identification of NFGNR isolates.

Nitrate broth with an inverted Durham tube can effectively be used to demonstrate both nitrate reduction to nitrite and dentrification.

This medium should be incubated at 30–35° C, although some NFGNR, such as *Pseudomonas pickettii*, denitrify best at room temperature. Nitrate broth should be incubated at least 4 days before being discarded as negative for denitrification. Aliquots (approximately 0.5 ml) can be tested for the presence of nitrite after 24 or 48 hr of incubation.

Motility–nitrate medium, previously described for the detection of motility, can also be used as a testing medium for both nitrate reduction and denitrification because it contains potassium nitrate. Although motility–nitrate medium is not quite as sensitive as nitrate broth for denitrification tests, it does have the advantage of giving three test results (i.e., motility, nitrate reduction, and denitrification) from one inoculated tube.

Another useful medium for detection of denitrification is fluorescence–lactose–denitrification (FLN) medium (Pickett and Pederson, 1970a). Besides the ability to detect fluorescence and oxidative acidification of lactose, each of which will be discussed separately, this medium contains both nitrate and nitrite. Fluorescence–lactose–denitrification is prepared as an agar slant with a 2- to 3-cm butt. The medium is stabbed and then streaked and incubated at 30–35° C. Nitrogen gas production is recognized by breaks in the agar. Because fluorescence–lactose–denitrification also contains nitrite, *A. odorans*, and organisms that can denitrify nitrite, but not nitrate, can also show denitrification in this medium.

Fluorescence

Demonstration of fluorescence by an NFGNR isolated from a clinical specimen virtually identifies the isolate as one of three species: *Pseudomonas aeruginosa, Pseudomonas fluorescens*, or *Pseudomonas putida*. Thus detection of fluorescence can rapidly focus identification efforts to differentiation among these three *Pseudomonas* species.

Two factors are critical in production of fluorescent pigment by NFGNR: the temperature of incubation and the medium used to demonstrate fluorescence. Temperature is critical because some strains of *P. fluorescens* and *P. putida* will only produce pyoverdin (the fluorescent pigment) at room temperature and many false negative reactions will occur when cultures are incubated at 30–35° C. Unfortunately, many clinical isolates of *P. aeruginosa* (particularly mucoid strains) fluoresce better when grown at 30–35° C. One approach to solving this problem is to inoculate two tubes of media and incubate one at room temperature and the other at 35° C. Tubes are examined under long-wave ultraviolet daily for up to 1 week before they are scored as negative for fluorescence.

Several medium formulations for detection of fluorescence are available commercially and most of these are variations of the B medium of King (Vera and Dumoff, 1974). This medium contains both magnesium and glycerol, each of which appears to enhance production of fluorescent pigment. We have also developed a fluorescent formulation which contains, in addition to the basic constituents of King's B, tryptophan and potassium nitrate (Pickett and Greenwood, 1980). Prepared as a slant, the medium is stabbed, streaked, and then incubated at room temperature. In addition to fluorescence demonstration, denitrifying strains will show nitrogen gas production in this medium. Additionally, because of the presence of tryptophan, growth from the slant can be used for the spot indole test (Vracko and Sherris, 1963). Thus the medium can also be used to screen potential flavobacteria. It should be noted that this method of detecting indole production by flavobacteria is not as sensitive as xylene extraction of growth in 2% tryptophan broth. Therefore isolates that appear to have biochemical profiles consistent with a flavobacterium yet have a negative spot indole test should be retested with the more sensitive xylene method.

Alkalinization of Organic Salts and Amides

The differential ability of Enterobacteriaceae isolates to alkalinize organic salts and amides (e.g., acetate, citrate) has always provided important diagnostic identification information. This type of information is also useful for identification of NFGNR. Besides providing results that produce distinctive biochemical profiles, alkalinization of organic salts and amides can be particularly useful in identifying the relatively inert, nonsaccharolytic NFGNR.

The OLP basal medium, previously described in the discussion of oxidative acidification of carbohydrates, can also be used to determine alkalinization of organic substrates. By adjusting the initial pH of the medium to 6.5, alkalinizing substrates such as acetamide, citrate, and urea can be detected by a color shift of the phenol red indicator from yellow to red. The OLP is incubated at 35° C and observed daily for up to 3 days before being scored as negative.

Other procedures/basal media that can be used to detect alkalinization are Simmons' citrate basal, Christensen's urea agar, the method of Arai (1970) (Nessler's reagent), and buffered single substrates. Each of these methods varies in sensitivity and results cannot always be correlated from one method to another.

CONCLUSION

In the introduction of this chapter it was suggested that most commonly encountered NFGNR could be identified with a minimal number of properly chosen tests. Table 1 lists a limited battery of suggested media selected according to the oxidase reaction of an isolate. With the exception of the fluorescence—indole—denitrification medium, which

Table 1 Primary Media for NFGNR Identification

Media for oxidase-negative NFGNR

    Fluorescence—lactose—dentrification medium

    Buffered lysine decarboxylase medium

    Motility nitrate medium

    Oxidative low-peptone sucrose

    Oxidative low-peptone arabinose

    Oxidative low-peptone urea

Media for oxidase-positive NFGNR

    Fluorescence—lactose—dentrification medium

    Motility nitrate medium

    Buffered lysine decarboxylase medium

    Fluorescence—indole—dentrification medium

    Oxidative low-peptone acetamide

    Oxidative low-peptone glucose

is incubated at room temperature, all media are incubated at 35° C and reactions read after 24 hr of incubation.

Table 2 gives the biochemical reactions of the more commonly encountered NFGNR when grown on the media listed in Table 1. As Table 2 illustrates, by testing for key features with sensitive methods, it is possible to rapidly distinguish many NFGNR on the basis of a relatively brief biochemical profile. These features also provide a good starting point in the identification of less frequently encountered NFGNR.

FORMULATIONS

L-Alanine-4-Nitroanilide Reagent

L-Alanine-4-nitroanilide reagent is prepared by adding 1.0 g of L-alanine-4-nitroanilide (EM Laboratories Inc., Elmsford, N.Y.) to 100 ml of 50 mM Tris-maleate buffer at pH 7.0. To perform the test, a suspension of bacteria is prepared in 0.5—1.0 ml of L-alanine-4-nitroanilide reagent in a 13 X 100 mm test tube. The tube is incubated at room temperature and a yellow coloration indicates a positive reaction (i.e., gram-negative bacterium). Tubes are held for up to 60 min before being scored as negative.

Oxidative Low-Peptone Medium

Ingredients	Amount (g %)
$(NH_4)_2HPO_4$	0.1
KCL	0.02
$MgSO_4\ 7H_2O$	0.02
Yeast extract, Difco	0.05
Casitone, Difco	0.05
Agar, Difco	1.5
Phenol red	0.002

Oxidative Low-Peptone "Sugar" (Carbohydrates, Glycosides, and Alcohols) Media

Adjust the basal medium to pH 7.8, steam to dissolve ingredients, dispense in 3-ml amounts in 13 X 100 mm tubes, and autoclave. While the sterilized basal medium is still molten, aseptically add 0.3 ml of a sterile 10% sugar solution to each tube, mix, and allow to solidify with tubes in a slanted position. Inoculate by streaking the surface of the medium. Yellow indicates positive; red, negative; orange, indeterminate, in which case one should reincubate.

Oxidative Low-Peptone (Organic Salts, Amides, Amino Acids, and Other Nitrogenous Compounds, Including Urea and Gelatin) Media

To each 100 ml of basal medium add 0.1 g of glucose and 1.0 g of desired salt, adjust pH to 6.5, steam to dissolve ingredients, dispense in 3-ml amounts, autoclave, and slant tubes until agar is solidified. Inoculate by streaking. Red indicates positive, yellow, negative; orange or orange-pink, indeterminate, in which case one should reincubate. (Note: The OLP is a patented basal medium of the Irvine Diagnostic Services.)

Table 2 Differential Features of Commonly Encountered Nonfermentative Bacteria[a]

Species	Oxidase	Fluorescence	Denitrification	Lysine decarboxylase
Pseudomonas aeruginosa	+	+	+	−
Pseudomonas fluorescens / Pseudomonas putida complex	+	+	−	−
Pseudomonas cepacia	+	−	−	+
Pseudomonas maltophilia	−	−	−	+
Pseudomonas pickettii	+	−	+/−	−
Flavobacterium meningosepticum	+	−	−	−
Acinetobacter anitratus	−	−	−	−
Alcaligenes denitrificans	+	−	+	−
CDC Ve-1	−	−	−	−

[a] +, positive reaction; −, negative reaction; +/−, variable reaction, (+), delayed positive reaction; NT, not tested.

Oxidase Test Strips

Oxidase test strips are prepared by saturating filter paper with 1% tetramethylparaphenylenediamine dihydrochloride in 0.2% ascorbic acid. The strips are dried at 35° C and stored in the dark. For the test, rub a bit of cell paste into the paper. Strongly oxidase-positive cultures yield a dark purple color within 5 sec, weakly positive cultures within 15 sec, and oxidase-negative cultures will remain colorless through 20 sec.

Motility	Indole	Acetamide	Lactose	Dextrose	Sucrose	Arabinose	Urea
+	−	+	−	+	NT	NT	NT
+	−	−	−	+	NT	NT	NT
+	−	+/−	+	+	NT	NT	NT
+	−	NT	−	NT	(+)	−	+/−
+	−	−	+	+	NT	NT	NT
−	+	−	−	(+)	NT	NT	NT
−	−	NT	+	NT	−	+	+/−
+	−	+	−	−	NT	NT	NT
+	−	NT	−	NT	−	+	+/−

Lysine Decarboxylase Broth and Chloroform–Ninhydrin Reagent

Ingredients	Amount (g%)
L-Lysine HCl	0.5
Glucose	0.5

Ingredients	Amount (g %)
KH_2PO_4	0.5
(to give pH 4.6)	

Incubate the inoculated medium 18–24 hr, add one drop of 40% NaOH, mix, and add 1 ml (not less than this or false negative tests may result) of ninhydrin reagent (0.1% ninhydrin in chloroform). Do not mix. A purple color appearing in the chloroform phase within 3–5 min represents a positive test. Color in the aqueous phase (upper layer) is irrelevant.

Motility—Nitrate Medium

Ingredients	Amount (g %)
Infusion broth, Difco	0.5
KNO_3	0.1
Agar, Difco	0.3

Steam to dissolve ingredients and dispense in 4-ml amounts. Autoclave. Inoculate by stabbing 5–10 mm into the medium. Detect N_2 as bubbles of gas. Detect nitrite by adding 5–10 drops of nitrite reagent.

Fluorescence—Lactose—Denitrification Medium

Ingredients	Amount (g %)
Proteose peptone #3 (Difco)	1.0
Lactose	2.0
Agar	1.5
Phenol red	0.002
$MgSO_4 \ 7H_2O$	0.15

Ingredients	Amount (g %)
K_2HPO_4	0.15
KNO_3	0.2
KNO_2	0.05
Glucose	0.05
Sodium acetate trihydrate	0.05

Steam to dissolve ingredients. Dispense in 4-ml amounts, sterilize, and solidify to give butt and slant of approximately same length. Inoculate by stabbing and streaking. Read at 24 and 48 hr for fluorescence (under ultraviolet light), acidification of lactose (yellow slant), and denitrification (gas).

REFERENCES

Arai, T. (1970). *Jpn. J. Microbiol.* 14:279.
Carlone, G. M., Valdez, M. J., and Pickett, M. J. (1982). *J. Clin. Microbiol.* 16:1157.
Carlquist, P. R. (1956). *J. Bacteriol.* 71:339.
Cerny, G. (1976). *Eur. J. Appl. Microbiol.* 3:223.
Clark, W. A. (1976). *J. Clin. Microbiol.* 3:632.
Grant, M. A., and Holt, J. G. (1977). *App. Environ. Microbiol.* 33:1222.
Greenwood, J. R., and Pickett, M. J. (1978). *Abstr. Annu. Mtg., Am. Soc. Microbiol.*, p. 290.
Gregersen, T. (1978). *Eur. J. Appl. Microbiol. Biotechnol.* 5:123.
Hugh, R., and Leifson, E. (1953). *J. Bacteriol.* 66:24.
Kodaka, H., Armfield, A. Y., Lombard, G. L., and Dowell, Jr., V. R. (1982). *J. Clin. Microbiol.* 16:948.
Leifson, E. (1951). *J. Bacteriol.* 62:377.
Pickett, M. J., and Greenwood, J. R. (1980). *J. Gen. Microbiol.* 120:439.
Pickett, M. J., and Pedersen, M. M. (1970a). *Can. J. Microbiol.* 16:351.
Pickett, M. J., and Pedersen, M. M. (1970b). *Can. J. Microbiol.* 16:401.
Simon, A., and Ridge, E. H. (1974). *J. Appl. Bacteriol.* 32:459.

Vera, H. D., and Dumoff, M. (1974). In *Manual of Clinical Microbiology*, 2nd ed. (E. H. Lennette, E. H. Spaulding, and J. P. Truant, eds.), American Society for Microbiology, Washington, D.C., pp. 270—294.

Von Gravenitz, A. (1976). *Mt. Sinai J. Med. 43*:727.

Vracko, R., and Sherris, J. C. (1963). *Am. J. Clin. Pathol. 39*: 429.

Weaver, R. E., Tatum, H. W., and Hollis, D. G. (1972). *The Identification of Unusual Pathogenic Gram Negative Bacteria*, Centers for Disease Control, Atlanta, Ga.

2
Cultural and Biochemical Aspects for Identification of Glucose-Nonfermenting Gram-Negative Rods

GERALD L. GILARDI *North General Hospital and Mount Sinai School of Medicine, New York, New York*

INTRODUCTION

The use of molecular methodology has greatly improved the resolution or taxonomic designation among glucose-nonfermenting gram-negative rods (Oyaizu and Komagata, 1983), yet problems remain with a number of nonfermenters. Bacteria comprising nonfermenters (Table 1) have the following phenotypic properties: They are gram-negative asporogenous rods. Cells vary from single, straight, or slightly curved rods to very short and plump rods, often approaching coccus shape, predominately in pairs. Motile species have one or more polar or peritrichous flagella. Some nonflagellated species demonstrate gliding movement or swarming growth on agar media. Their metabolism is respiratory, never fermentative or photosynthetic. Some produce acids without gas oxidatively from carbohydrates. Molecular oxygen is the universal electron acceptor; some can denitrify, using nitrate or nitrite as an alternate acceptor. They are obligate aerobes, except for those species which can use denitrification as a means of anaerobic respiration. Catalase is produced. Most members produce indophenol oxidase. Tests for methyl red and acetyl methyl carbinol production are negative. Indole production is limited to one phylogenetic group. None are fastidious; growth occurs on blood agar base media with or without the addition of blood at 30–35° C. The characteristics of the species and characters useful in the identification of each are recorded in Tables 2–39. The distinguishing features of each species are summarized.

Table 1 Glucose-Nonfermenting Gram-Negative Rods Encountered in Clinical Specimens

Name	Synonyms
I. Family Pseudomonadaceae	
Pseudomonas	
Fluorescent group	
Pseudomonas aeruginosa	*Bacterium aeruginosum, Bacillus pyocyaneus, Pseudomonas polycolor*
Pseudomonas fluorescens	*Pseudomonas chlororaphis, Pseudomonas aureofaciens, Pseudomonas lemonnieri*
Pseudomonas putida	*Pseudomonas ovalis*
Pseudomallei group	
Pseudomonas pseudomallei	*Bacillus pseudomallei, Malleomyces pseudomallei*
Pseudomonas mallei	*Bacillus mallei, Actinobacillus mallei*
Pseudomonas cepacia	*Pseudomonas multivorans, Pseudomonas kingae*, group EO-1
Pseudomonas gladioli	*Pseudomonas marginata, Pseudomonas alliicola*
Pseudomonas pickettii	*Pseudomonas thomasii*, groups Va-1 and Va-2
Stutzeri group	
Pseudomonas stutzeri	*Bacillus denitrificans* II, *Pseudomonas stanieri*, groups Vb-1 and Vb-3
Pseudomonas mendocina	Group Vb-2
Alcaligenes group	
Pseudomonas alcaligenes	
Pseudomonas pseudoalcaligenes	*Pseudomonas alcaligenes* biovar B
Acidovorans group	
Pseudomonas acidovorans[a]	
Pseudomonas testosteroni[a]	

Table 1 (continued)

Name	Synonyms
Dimunta group	
Pseudomonas diminuta	Group Ia
Pseudomonas vesicularis	*Corynebacterium vesiculare*
Species of uncertain affiliation	
Pseudomonas maltophilia	*Xanthomonas maltophilia*, *Pseudomonas melanogena*, group I
Pseudomonas putrefaciens	*Achromobacter putrefaciens*, *Pseudomonas rubescens*, *Alteromonas putrefaciens*, groups Ib-1 and Ib-2
Pseudomonas paucimobilis	*Flavobacterium devorans*, *Flavobacterium capsulatum*, group IIk-1
Pseudomonas extorquens	*Pseudomonas mesophilica*, *Vibrio extorquens*
Pseudomonas pertucinogena	
Pseudomonas delafieldii	
Pseudomonas species CDC group 1	*Pseudomonas denitrificans*
Genera of uncertain affiliation	
Group Ve-1	
Group Ve-2	
Alcaligenes faecalis	Group VI
Alcaligenes odorans	*Pseudomonas odorans*, *Alcaligenes odorans* var. *viridans*
Alcaligenes denitrificans	Group Vc
Achromobacter xylosoxidans	Groups IIIa and IIIb
Achromobacter species Vd-1	
Achromobacter species Vd-2	
Bordetella bronchicanis	*Bordetella bronchiseptica*, *Bacillus bronchicanis*, *Bacillus bronchisepticus*, group IVa

Table 1 (continued)

Name	Synonyms
Group IVc-2	
Group IVe	
II. Family Rhizobiaceae	
Agrobacterium tumefaciens	*Agrobacterium radiobacter*, group Vd-3
III. Family Vibrionaceae	
Genera of uncertain affiliation	
Flavobacterium meningosepticum	Group IIa
Flavobacterium indologenes	Group IIb, *Flavobacterium aureum*
Flavobacterium breve	*Bacillus brevis*, *Bacillus canalis parvus*
Flavobacterium odoratum	Group M-4f
Flavobacterium species IIe	Group IIe
Flavobacterium species IIf	Group IIf, *Flavobacterium genitale*
Flavobacterium species IIh	Group IIh
Flavobacterium species IIi	Group IIi
Flavobacterium species IIj	Group IIj
IV. Family Cytophagaceae	
Sphingobacterium multivorum	*Flavobacterium multivorum*, group IIk-2
Sphingobacterium spiritivorum	*Sphingobacterium versatilis*, *Flavobacterium spiritivorum*, group IIk-3
Sphingobacterium mizutae	
V. Family Neisseriaceae	
Moraxella lacunata	*Moraxella liquefaciens*
Morexella nonliquefaciens	
Morexella osloensis	*Moraxella duplex*, *Mima polymorpha* var. *oxidans*

Table 1 (continued)

Name	Synonyms
Moraxella phenylpyruvica	Group M-2
Moraxella atlantae	Group M-3
Moraxella urethralis	Group M-4
Acinetobacter calcoaceticus	*Bacterium anitratum, Mima polymorpha, Moraxella lwoffi, Acinetobacter anitratus, Acinetobacter calcoaceticus* biovar *anitratus, Acinetobacter calcoaceticus* biovar *lwoffi, Acinetobacter calcoaceticus* biovar *alcaligenes, Acinetobacter calcoaceticus* biovar *haemolyticus*

VI. Bacteria of uncertain affiliation

Group EO-2

[a]*Pseudomonas terrigena, Vibrio terrigenus, Vibrio percolans, Lophomonas alcaligenes, Pseudomonas indoloxidans, Pseudomonas desmolytica, Pseudomonas testosteroni,* and *Pseudomonas acidovorans* are synonyms of *Comamonas terrigena.*

PSEUDOMONAS SPECIES

Fluorescent Group

The attributes of 551 strains of the fluorescent group (*P. aeruginosa, P. fluorescens, P. putida*) are recorded in Table 2, and characters useful for the identification of most *P. aeruginosa, P. fluorescens,* and *P. putida* strains are recorded in Table 3.

Most strains of the three species in the fluorescent group produce water-soluble, yellow-green, yellow-brown, or colorless fluorescent pigments (pyoverdins) that are influenced by nutritional factors and may not develop on media supporting growth (Stanier et al., 1966). Indophenol oxidase and arginine dihydrolase are synthesized.

Pseudomonas aeruginosa

Most *P. aeruginosa* strains are identified on the basis of the characteristic grapelike odor of aminoacetophenone, the structure of the

Table 2 Characteristics of *Pseudomonas* Species in the Fluorescent Group[a]

Characteristic	P. aeruginosa (84 strains)		P. fluorescens (194 strains)		P. putida (273 strains)	
	Sign[b]	Percent positive	Sign	Percent positive	Sign	Percent positive
Acid: glucose, 1% (OFBM[c])	+	98	+	100	+	100
Fructose	+ or −	89	+	99	+	99
Xylose	+ or −	85	+	97	+	98
Lactose	−	0	− or +	12	− or +	14
Sucrose	−	0	− or +	47	−	8
Maltose	− or +	11	− or +	36	− or +	21
Mannitol	+ or −	68	+	93	− or +	18
ONPG	−	4	−	1	−	1
Hydrogen sulfide (KIA)	−	0	−	0	−	0
Indole	−	0	−	0	−	0
Nitrate to gas	+ or −	61	−	5	−	0
Indophenol oxidase	+	100	+	100	+	100

Characteristic						
Arginine dihydrolase (DBM)	+	100	+	99	+	99
Lysine decarboxylase	−	0	−	0	−	0
Ornithine decarboxylase	−	0	−	0	−	0
Phenylalanine deaminase	−	7	−	3	−	0
Hydrolysis: urea	+ or −	67	− or +	41	− or +	43
Esculin	−	0	−	0	−	0
Tween 80	+ or −	71	+ or −	63	−	0
Deoxyribonucleic acid	− or +	10	−	0	−	0
Gelatin	+ or −	50	+	100	+	0
Growth on MBM + acetate	+	96	+	100	+	100
Growth at 42° C	+	100	−	0	−	0
Pyoverdin	+ or −	71	+	95	+ or −	84
Motile	+	97	+	100	+	100
Number of flagella	1		>1		>1	

[a] Except where indicated, cultures were incubated at 30° C. *Pseudomonas aeruginosa* strains did not produce pyocyanin.
[b] Sign: +, 90% or more positive within 2 days; −, no reaction (90% or more); + or −, most strains positive; − or +, most strains negative; 1, polar monotrichous flagella; >1, polar multitrichous flagella.
[c] Abbreviations: OFBM, oxidatitive-fermentative basal medium; ONPG, ortho-nitrophenol -B-D-galactopyranoside; KIA, Kligler iron agar, DBM, decarboxylase base Moeller; MBM, mineral base medium.

Table 3 Characters Useful for Identification of *Pseudomonas aeruginosa*, *Pseudomonas fluorescens*, and *Pseudomonas putida* Strains[a]

Character	P. aeruginosa	P. fluorescens	P. putida
Gram negative, rod shaped, asporogenous	+	+	+
Indophenol oxidase	+	+	+
Motility	+	+	+
Polar monotrichous flagella	+	−	−
Polar tuft of flagella	−	+	+
Pyocyanin	+ or −	−	−
Pyoverdin	+ or −	+	+ or −
Glucose, acid (OFBM)	+	+	+
Arginine dihydrolase (DBM)	+	+	+
Gelatin hydrolysis	− or +	+	−
Growth at 42° C	+	−	−

[a] From Table 2 and text.

colonies on agar media, and pyocyanin production, a water-soluble, blue, nonfluorescent phenazine pigment. *Pseudomonas aeruginosa* is the only gram-negative rod known to excrete pyocyanin. Strains may synthesize various combinations of pyocyanin, pyoverdins, pyorubin (red water-soluble pigment), and pyomelanin (brown to black water-soluble pigment). Pyocyanin may be masked by other pigments. Strains that produce pyocyanin and the yellowish pyoverdins often impart a greenish color to culture media. A number of colonial types (smooth, coliform-type, rough, mucoid, gelatinous) may be observed. Colonial variants arising by dissociation of a single strain give the false impression that different bacterial species are present. The variant forms may demonstrate altered exoenzyme activity, antibiograms, bacteriophage pattern, serovar, and antigenicity.

Apyocyanogenic variants can be recognized as strains of *P. aeruginosa* by several uniform characters, namely, the presence of polar monotrichous flagella, growth at 42° C, and failure to produce acid from disaccharides. Strains designated unidentified fluorescent

Cultural and Biochemical Aspects 25

Pseudomonas species are similar to apyocyanogenic *P. aeruginosa* strains (in that they grow at 42° C but possess a polar tuft of flagella.

Pseudomonas fluorescens and *Pseudomonas putida*

Pseudomonas fluorescens and *P. putida* are differentiated from *P. aeruginosa* by the fact that the former do not produce pyocyanin, fail to grow at 42° C, and possess a polar tuft of flagella. Whereas *P. fluorescens* and *P. putida* are usually susceptible to kanamycin and resistant to carbenicillin, the reverse is true for most *P. aeruginosa* strains. Nonproteolytic *P. putida* strains are distinguished from proteolytic *P. fluorescens* strains.

Pseudomallei Group

The attributes of 266 strains of the pseudomallei group (*P. pseudomallei, P. mallei, P. cepacia, P. pickettii*) are recorded in Table 4, and characters useful for the identification of most *P. pseudomallei, P. mallei, P. cepacia,* and *P. pickettii* strains are recorded in Table 5.

Nutritional versatility is demonstrated by strains of the species in the pseudomallei group in the types and number of organic compounds utilized as sole sources of carbon and energy. A few strains produce a negative or slow and very weak indophenol oxidase reaction. They are not susceptible to antibiotics of the polymyxin class.

Pseudomonas pseudomallei

Colonies range from mucoid and smooth to rough with a wrinkled texture and from cream to bright orange color. Initial growth is accompanied by an odor of putrefaction followed with a distinctive, aromatic, pungent odor. Strains are motile with a polar tuft of flagella. Abundant growth occurs under anaerobic conditions in nitrate-containing media. Gelatin is hydrolyzed, lactose is utilized, and growth occurs at 42° C. Suspicious strains should be confirmed by agglutination or fluorescent-antibody reactions utilizing antisera for *P. pseudomallei.*

Pseudomonas mallei

Colonies are smooth and their colors range from white to cream. *Pseudomonas mallei* grows slowly when compared with *P. pseudomallei* or *P. aeruginosa* (Hugh, 1970). It is the only nonmotile species in the genus *Pseudomonas*. Strains do not produce nitrogen gas and fail to hydrolyze gelatin or grow at 42° C.

Pseudomonas cepacia

Some strains produce a green-yellow, water-soluble, nonfluroescent phenazine pigment. The pigment occurs in the bacterial cell and may

Table 4 Characteristics of *Pseudomonas* Species in the Pseudomallei Group[a]

Character	P. pseudomallei (7 strains) Sign[b]	P. pseudomallei (7 strains) Percent positive	P. mallei (6 strains) Sign	P. mallei (6 strains) Percent positive
Acid: glucose, 1% (OFBM[c])	+	100	+	100[d]
Fructose	+	100	+ or (+)	100
Xylose	+ or −	86	−	0
Lactose	+	100	(+) or +	100
Sucrose	+ or −	86	−	0
Maltose	+	100	(+)	100
Mannitol	+	100	(+) or −	83
ONPG	−	0		·
Hydrogen sulfide (KIA)	−	0	−	0
Indole	−	0	−	0
Nitrate to gas	+	100	−	0
Indophenol oxidase	+ or +(w)	100	+, +(w), or −	67
Arginine dihydrolase (DBM)	+	100	+	100
Lysine decarboxylase	−	0	−	0
Ornithine decarboxylase	−	0	−	0
Phenylalanine deaminase	−	0		·
Hydrolysis: urea	− or +	43	− or +	17
Esculin	+ or −	57		·
Tween 80	+ or −	86		·
Deoxyribonucleic acid	−	0		·
Gelatin	+	100	−	0
Growth on MBM + acetate	+	100	+	100
Growth at 42° C	+	100	−	0
Motile	+	100	−	0
Number of flagella	>1			

[a] Except where indicated, cultures were incubated at 30° C.
[b] Sign: +, 90% or more positive within 2 days; −, no reaction (90% or more); + or −, most strains positive; − or +, most strains negative; (+), reactions delayed 2 days or more; (w), weakly reactive; ·, not tested; 1, polar monotrichous flagella; >1, polar multitrichous flagella.

P. cepacia (201 strains)		P. pickettii biovar 1 (26 strains)		P. pickettii biovar 2 (24 strains)	
Sign	Percent positive	Sign	Percent positive	Sign	Percent positive
+	100	(+)	100	(+)	100
+	100	(+)	100	(+)	100
+	99	(+)	100	(+)	100
+	98	(+)	100	−	0
+ or −	81	−	0	−	0
+	97	(+)	100	−	0
+	100	−	0	−	0
+ or −	78	−	0	−	0
−	0	−	0	−	0
−	0	−	0	−	0
−	0	(+)	88	(+)	100
+ or +(w)	92	+	100	+	100
−	0	−	0	−	0
+	90	−	0	−	0
+ or −	65	−	0	−	0
−	2	−	4	− or +	38
− or +	43	+	100	+	100
+ or −	65	−	0	−	0
+	100	+	100	+	100
−	0	−	0	−	0
+ or −	74	+ or −	77	− or +	38
+	99	+	100	+	100
+ or −	58	− or +	27	+ or −	63
+	99	+	96	+	100
>1		1		1	

[c] Abbreviations: OFBM, oxidative−fermentative basal medium; ONPG, ortho-nitrophenyl-β-D-galactopyranoside; KIA, Kligler iron agar; DBM, decarboxylase base Moeller; MBM, mineral base medium.
[d] Data from Hugh (1970).

Table 5 Characters Useful for Identification of *Pseudomonas pseudomallei*, *P. mallei*, *P. cepacia*, and *P. pickettii* Strains[a]

Character	P. pseudomallei	P. mallei	P. cepacia	P. pickettii biovar 1	P. pickettii biovar 2
Gram negative, rod shaped, asporogenous	+	+	+	+	+
Indophenol oxidase	+ or +(w)	+, +(w), or −	+ or +(w)	+	+
Motility	+	−	+	+	+
Polar monotrichous flagella	−	−	−	−	−
Polar tuft of flagella	+	−	+	−	−
Glucose, acid (OFBM)	+	+	+	(+)	(+)
Lactose	+	(+) or +	+	(+)	−
Maltose	+	(+)	+	(+)	−
Mannitol	+	(+) or −	+	−	−
Nitrate to gas	+	−	−	(+)	(+)
Arginine dihydrolase (DBM)	+	+	−	−	−
Lysine decarboxylase	−	−	+	−	−

[a]From Table 4.

diffuse into the surrounding agar medium. Many strains are nonpigmented. Strains are motile with a polar tuft of flagella. None of the strains denitrify or synthesize arginine dihydrolase. Most strains hydrolyze ortho-nitrophenol-β-D-galactopyranoside (ONPG) and produce decarboxylases for lysine and ornithine. *Pseudomonas gladioli*, a rare isolate from clinical specimens, closely resembles but is distinguished from *P. cepacia* on the basis of carbon substrate utilization.

Pseudomonas pickettii

The growth of nonpigmented colonies on blood agar and other media characteristically is slow. Strains are motile, with polar monotrichous flagella. The development of acid reactions in carbohydrates and reduction of nitrate to nitrogen gas characteristically are slow and may require 48 hr of incubation to detect. Urease is produced. Dihydrolase and decarboxylase are not produced. *Pseudomonas pickettii* is a distinct but heterogeneous species and comprises up to seven biovars based on different oxidation patterns and other phenotypic features, that is, acid production from lactose, maltose, and mannitol and gelatinase production (King, et al., 1979; Pickett and Greenwood, 1980). Biovars 1 and 2 (see Tables 4 and 5) correspond to CDC groups Va-1 and Va-2, respectively. Phenotypic and genotypic characterization indicates that *P. pickettii*, the phytopathogenic pseudomonads *P. cepacia*, *P. gladioli (P. marginata, P. alliicola)*, and *P. caryophylli*, and the animal pathogens *P. pseudomallei* and *P. mallei* are related (Palleroni and Holmes, 1981; Whitaker et al., 1981).

Stutzeri Group

The attributes of 174 strains of the stutzeri group (*P. stutzeri, P. mendocina*) are recorded in Table 6, and characters useful for the identification of most *P. stutzeri* and *P. mendocina* strains are recorded in Table 7.

Strains of the species in the stutzeri group are salt tolerant and nonhalophilic but require sodium cation for growth. They are motile, with polar monotrichous flagella, and grow under anaerobic conditions in nitrate-containing media accompanied by nitrogen gas production. Oxidative acidity from glucose and indophenol oxidase are produced. They are susceptible to polymyxins.

Pseudomonas stutzeri

Most freshly isolated strains produce dry, wrinkled, tough, and adherent colonies, smooth colonies, and various intermediate colony types. Colonies form a buff to light-brown intracellular pigment. Smooth colony types predominate after repeated subculture. The ability to produce nitrogen gas may be lost on repeated subculture.

Table 6 Characteristics of *Pseudomonas* Species in the Stutzeri Group[a]

Character	P. stutzeri (168 strains)		P. mendocina (6 strains)	
	Sign[b]	Percent positive	Sign	Percent positive
Acid: glucose, 1% (OFBM[c])	+	100	+	100
Fructose	+	94	+	100
Xylose	+	94	+	100
Lactose	−	0	−	0
Sucrose	−	0	−	0
Maltose	+	98	−	0
Mannitol	+ or −	69	−	0
ONPG	−	0	−	0
Hydrogen suflide (KIA)	−	0	−	0
Indole	−	0	−	0
Nitrate to gas	+	100	+	100
Indophenol oxidase	+	100	+	100

Cultural and Biochemical Aspects

Arginine dihydrolase (DBM)	−	100
Lysine decarboxylase	−	0
Ornithine decarboxylase	−	0
Phenylalanine deaminase	+ or −	51
Hydrolysis: urea	− or +	15
Esculin	−	0
Tween 80	+	97
Deoxyribonucleic acid	−	0
Gelatin	−	1
Growth on MBM + acetate	+	100
Growth at 42° C	+ or −	89
Motile	+	100
Number of flagella	1	1

	+	100
	−	0
	−	0
	+ or −	50
	+ or −	50
	−	0
	+ or −	83
	−	0
	−	0
	+	100
	+	100
	+	100
	1	

[a] Except where indicated, cultures were incubated at 30° C.
[b] Sign: +, 90% or more positive within 2 days; −, no reaction (90% or more); + or −, most strains positive; − or +, most strains negative; 1, polar monotrichous flagella.
[c] Abbreviations: OFBM, oxidative–fermentative basal medium; KIA, Kliger iron agar; DBM, decarboxylase base Moeller; MBM, mineral base medium.

Table 7 Characters Useful for Identification of *Pseudomonas stutzeri* and *Pseudomonas mendocina* Strains[a]

Character	P. stutzeri	P. mendocina
Gram negative, rod shaped asporogenous	+	+
Indophenol oxidase	+	+
Motility	+	+
Polar monotrichous flagella	+	+
Glucose, acid (OFBM)	+	+
Lactose	−	−
Maltose	+	−
Nitrate to gas	+	+
Arginine dihydrolase (DBM)	−	+

[a]From Table 6.

Strains which no longer denitrify or produce wrinkled colonies will regain these features by serial passage on nitrate-containing medium. Maltose and usually starch are utilized. Acid is not produced from lactose or sucrose. Strains are nonproteolytic.

Pseudomonas mendocina

Colonies are flat, smooth, and butyrous and form a brown-yellow intracellular carotenoid pigment. Wrinkled colonies are not produced. Arginine dihydrolase is produced. Starch and maltose are not utilized. (Palleroni et al., 1970).

Pseudomonas maltophilia (Xanthomonas maltophilia)

The attributes of 527 strains of *P. maltophilia* are recorded in Table 8, and characters useful for the identification of most *P. maltophilia* strains are recorded in Table 9.

Colonies develop a characteristic lavender-green color on blood agar media. Brown by-products of metabolism may accumulate in certain agar media. Growth on nutrient agar media is usually pale yellow. Growth on blood agar is accompanied by a strong odor of ammonia. Strains are motile with a polar tuft of flagella. Oxidative acidity accumulates in oxidative−fermentative (OF) maltose and glucose media.

Table 8 Characteristics of *Pseudomonas (Xanthomonas) maltophilia*[a] (527 Strains)

Character	Sign[b]	Percent positive
Acid: glucose, 1% (OFBM[c])	+	100
Fructose	+	99
Xylose	+ or −	54
Lactose	+ or −	88
Sucrose	+	92
Maltose	+	100
Mannitol	−	0
ONPG	+	95
Hydrogen sulfide (KIA)	−	0
Indole	−	0
Nitrate to gas	−	0
Indophenol oxidase	−	2
Arginine dihydrolase (DBM)	−	0
Lysine decarboxylase	+	99
Ornithine decarboxylase	−	0
Phenylalanine deaminase	−	0
Hydrolysis: urea	−	0
Esculin	+	100
Tween 80	+	100
Deoxyribonucleic acid	+	100
Gelatin	+	100
Growth on MBM + acetate	−	2
Growth at 42° C	− or +	48
Motile	+	100
Number of flagella	>1	

[a] Except where indicated, cultures were incubated at 30° C.
[b] Sign: +, 90% or more positive within 2 days; −, no reaction (90% or more); + or −, most strains positive; − or +, most strains negative; >1, polar multitrichous flagella.
[c] Abbreviations: OFBM, oxidative−fermentative basal medium; KIA, Kliger iron agar; DBM, decarboxylase base Moeller; MBM, mineral base medium.

Table 9 Characters Useful for Identification of *Pseudomonas (Xanthomonas) maltophilia* Strains[a]

Gram negative, rod shaped, asporogenous	+
Indophenol oxidase	−
Motility	+
Polar tuft of flagella	+
Maltose, acid (OFBM)	+
Lysine decarboxylase (DBM)	+
Deoxyribonuclease	+

[a]From Table 8.

The acid reaction is more pronounced in maltose than in glucose medium. An ONPG reaction is produced by most strains. Lysine decarboxylase, esculin, deoxyribonuclease, and gelatin reactions are produced. The indophenol oxidase reaction is usually negative, apparently due to a lack of cytochrome c, which may be necessary for the indophenol oxidase reaction. Strains require either methionine or cystine plus glycine for growth. Occasional non-methionine-requiring strains (biovar 2) are recovered from clinical and environmental sources (Ikemoto et al., 1980). They differ from methionine-requiring strains (biovar 1) by the manner in which they assimilate select carbon compounds. Enzymes including hyaluronidase, esterases, lipases, mucinase, and hemolysins may be potential virulence factors.

Proposals have been made (Byng et al., 1980; DeVos and DeLey, 1983; Swings et al., 1983) to transfer *P. maltophilia* to the genus *Xanthomonas* as *Xanthomonas maltophilia* partly on the basis of the following common reactions: alkaline reactions from OF rhamnose and mannitol; negative indophenol oxidase reaction; hydrolysis of esculin and ONPG; and a requirement for amino acids as growth factors.

Pseudomonas putrefaciens

The attributes of 72 strains of *P. putrefaciens* are recorded in Table 10, and characters useful for the identification of most *P. putrefaciens* biovars are recorded in Table 11.

Colonies on agar media are slightly viscous or mucoid and usually red-brown or pink. Strains are motile, with polar monotrichous flagella. This is the only glucose nonfermenter known to produce hydrogen sulfide in Kligler iron agar (KIA). An occasional strain may fail to

Table 10 Characteristics of *Pseudomonas putrefaciens*[a]

Character	Biovar 1 (11 strains)		Biovar 2 (52 strains)		Biovar 3 (9 strains)	
	Sign[b]	Percent positive	Sign	Percent positive	Sign	Percent positive
Acid: glucose, 1% (OFBM[c])	+ or (+)	100	+ or (+)	100	+ or (+)	100
Fructose	−, +, or (+)	46	+, (+), or −	50	+, (+), or −	78
Xylose	−	9	−	0	−	0
Lactose	−, + or (+)	18	−	0	−	0
Sucrose	+	100	−	0	−	0
Maltose	+	100	−	0	−	0
Mannitol	−	0	−	0	−	0
ONPG	−	0	−	0	−	0
Hydrogen sulfide (KIA)	+	100	+	100	+	100
Indole	−	0	−	0	−	0
Nitrate to gas	−	0	−	0	−	0
Indophenol oxidase	+	100	+	100	+	100
Arginine dihydrolase (DBM)	−	0	−	0	−	0
Lysine decarboxylase	−	0	−	0	−	0
Ornithine decarboxylase	+	91	+	100	+	100

Table 10 (continued)

Character	Biovar 1 (11 strains)		Biovar 2 (52 strains)		Biovar 3 (9 strains)	
	Sign[b]	Percent positive	Sign	Percent positive	Sign	Percent positive
Phenylalanine deaminase	−	0	−	0	−	0
Hydrolysis: urea	−	0	−	0	−	0
Esculin	− or +	27	−	0	−	0
Tween 80	+	100	+	100	+	100
Deoxyribonucleic acid	+	100	+	100	+	100
Gelatin	+	100	+	100	+ or −	78
Growth on MBM + acetate	+	100	+	100	+	100
Growth at 42° C	−	0	+	90	−	0
Motile	+	100	+	100	+	100
Number of flagella	1		1		1	

[a] Except where indicated, cultures were incubated at 30° C.
[b] Sign: +, 90% or more positive within 2 days; −, no reaction (90% or more); + or −, most strains positive; − or +, most strains negative; (+), reactions delayed 2 days or more; 1, polar monotrichous flagella.
[c] Abbreviations: OFBM, OF basal medium; KIA, Kligler iron agar; DBM, decarboxylase base Moeller; MBM, mineral base medium.

Table 11 Characters Useful for Identification of *Pseudomonas putrefaciens* Biovars[a]

Character	Biovar 1[b]	Biovar 2	Biovar 3
Gram negative, rod shaped asporogenous	+	+	+
Indophenol oxidase	+	+	+
Motility	+	+	+
Polar monotrichous flagella	+	+	+
Sucrose, maltose, acid (OFBM)	+	−	−
Hydrogen sulfide (KIA)	+	+	+
Ornithine decarboxylase (DBM)	+	+	+
Deoxyribonuclease	+	+	+
Growth on SS[c] agar	−	+	−
Growth on 6.5% NaCl agar	−	+	−

[a] From Table 10.
[b] Proposed designation *Alteromonas putrefaciens*.
[c] SS, salmonella−shigella.

produce hydrogen sulfide. Ornithine decarboxylase and deoxyribonuclease tests are positive. Sodium cations are required for the growth of some strains.

Two distinct groups of strains exist within *P. putrefaciens* (Levin, 1972), and additional groups have been described (Owen et al., 1978). One group (CDC group Ib-1) has a relatively low 49 mol % guanine plus cytosine (G + C) content in their DNA and consists primarily of strains from environmental sources. These strains produce acid from OF sucrose and maltose and fail to grow in 6.5% NaCl or on Salmonella-Shigella (SS) agar. A second group (CDC group Ib-2) with a high 58 mol % G + C content includes mostly strains from human clinical specimens. These strains grow in 6.5% NaCl and on SS agar but do not produce acid from disaccharides. Biovars 1 and 2 (see Tables 10 and 11) correspond to CDC groups Ib-1 and Ib-2, respectively. The genus *Alteromonas* was created to accomodate *Pseudomonas*-like bacteria of low GC content. Proposals that this group of *P. putrefaciens* strains to be transferred to the genus *Altermonas* as *Alteromas putrefaciens* have not found uniform support (Wilkinson and Caudwell, 1980).

Table 12 Characteristics of *Pseudomonas* Species in the Alcaligenes Group[a]

Character	P. alcaligenes (51 strains)		P. pseudoalcaligenes (62 strains)	
	Sign[b]	Percent positive	Sign	Percent positive
Acid: glucose, 1% (OFBM[c])	–	0	– or +	12
Fructose	–	0	+	100
Xylose	–	0	– or +	10
Lactose	–	0	–	0
Sucrose	–	0	–	0
Maltose	–	0	– or +	13
Mannitol	–	0	–	2
ONPG	–	0	–	0
Hydrogen sulfide (KIA)	–	0	–	0
Indole	–	0	–	0
Nitrate to gas	– or +	16	–	5

Characteristic	Sign	%	Sign	%
Indophenol oxidase	+	100	+	100
Arginine dihydrolase (DBM)	− or +	10	− or +	34
Lysine decarboxylase	−	0	−	0
Ornithine decarboxylase	−	0	−	0
Phenylalanine deaminase	− or +	20	− or +	16
Hydrolysis: urea	− or +	20	−	1
Esculin	−	0	−	0
Tween 80	− or +	49	− or +	10
Deoxyribonucleic acid	−	0	−	0
Gelatin	−	2	−	2
Growth on MBM + acetate	+	90	+	99
Growth at 42° C	+ or −	59	+ or −	76
Motile	+	100	+	90
Number of flagella	1		1	

[a] Except where indicated, cultures were incubated at 30° C.
[b] Sign: +, 90% or more positive within 2 days; −, no reaction (90% or more); + or −, most strains positive; − or +, most strains negative; 1, polar monotrichous flagella.
[c] Abbreviations: OFBM, OF basal medium; DBM, decarboxylase base Moeller; MBM, mineral base medium.

Alcaligenes Group

The attributes of 113 strains of the alcaligenes group (*P. alcaligenes, P. pseudoalcaligenes*) are recroded in Table 12, and characters useful for the identification of most *P. alcaligenes* and *P. pseudoalcaligenes* strains are recorded in Table 13.

Strains of the species in the alcaligenes group are polar monotrichous when flagellated, and the wavelength of the flagellum is approximately 1.6 µm. The flagellar anatomy is similar to that of *P. aeruginosa*. The cells of most strains have no somatic curvature. Indophenol oxidase is produced. Alkali usually accumulates in OF glucose medium.

Pseudomonas alcaligenes and *Pseudomonas pseudoalcaligenes*

Strains assigned to the alcaligenes group can be accomodated in two named species, *P. alcaligenes* (*P. alcaligenes* biovar A) and *P. pseudoalcaligenes* (*P. alcaligenes* biovar B). *Pseudomonas pseudoalcaligenes* produces a weak acid reaction from fructose in OF basal medium. Fructose is the only substrate used by all strains of *P. pseudoalcaligenes* and none of the *P. alcaligenes* strains. An additional five substrates not used by any of the latter are used by most of the more nutritionally versatile *P. pseudoalcaligenes* strains (Ralston-Barrett 1976). Phenotypic characteristics do not easily distinguish the two species.

Table 13 Characters Useful for Identification of *Pseudomonas alcaligenes* and *Pseudomonas pseudoalcaligenes* Strains[a]

Character	*P. alcaligenes*	*P. pseudoalcaligenes*
Gram negative, rod shaped asporogenous	+	+
Indophenol oxidase	+	+
Motility	+	+
Polar monotrichous flagella and normal wavelength	+	+
Glucose, acid (OFBM)	−	− or +
Fructose	−	+

[a]From Table 12 and text.

Acidovorans Group (Comamonas terrigena)

The attributes of 113 strains of the acidovorans group (*P. acidovorans, P. testosteroni*) are recorded in Table 14, and characters useful for the identification of most *P. acidovorans* and *P. testosteroni* strains are recorded in Table 15.

The distinctive character of the strains of the species in the acidovorans group is the polar tuft of flagella with a mean wavelength of 3.1 µm and an amplitude of 1.08 µm. The number of flagella at one or both poles varies from one to six. Occasional cells of some strains are markedly curved. Most strains grow in a mineral base medium (MBM), without growth factors, containing ammonium ion as sole source of nitrogen and acetate as sole source of carbon and energy. Alkali accumulates in OF glucose medium and most carbohydrates are not oxidized. Indophenol oxidase is produced.

Pseudomonas acidovorans and Pseudomonas testosteroni

Strains may be divided into two species on the basis of utilization of fructose and additional phenotypic differences. Non-fructose-utilizing *P. testosteroni* strains are distinguished from fructose-utilizing *P. acidovorans* strains. Acetamide hydrolysis, mediated by an aliphatic amidase, is a practical reaction for identifying *P. acidovorans* strains. The curved soma, flagellar anatomy, and physiology suggest a relationship to spirilla. Electron microscopy shows a regularly arranged layer of subunits on the outer surface of the cell wall of *P. acidovorans* (Lapchine, 1979). This structure is not found in any other *Pseudomonas* species but does occur in *Spirillum* species. This group deserves a separate generic rank and has been designated *Comamonas terrigena* in the family Spirillaceae (Hugh, 1962).

Diminuta Group

The attributes of 74 strains of the diminuta group (*P. diminuta, P. vesicularis*) are recorded in Table 16, and characters useful for the identification of most *P. diminuta* and *P. vesicularis* strains are recorded in Table 17.

The distinctive character of the strains of the species in the diminuta group is the very tightly coiled monotrichous flagellum, with a wavelength that varies from 0.62 to 0.98 µm, compared to the flagellar wavelength of most polar monotrichous bacteria which is approximately 2 µm. Indophenol oxidase is produced. Additional characters include a requirement for specific growth factors (pantothenate, biotin, cyanocobalamin, cystine), the production of acid from primary alcohols by all strains that can utilize alcohols, and an otherwise restricted range of biochemical activities.

Table 14 Characteristics of *Pseudomonas* Species in the Acidovorans Group (*Comamonas*)[a]

Character	P. acidovorans (95 strains)		P. testosteroni (18 strains)	
	Sign[b]	Percent positive	Sign	Percent positive
Acid: glucose, 1% (OFBM[c])	−	0	−	0
Fructose	+	100	−	0
Xylose	−	0	−	0
Lactose	−	0	−	0
Sucrose	−	0	−	0
Maltose	−	0	−	0
Mannitol	+	95	−	0
ONPG	−	0	−	0
Hydrogen sulfide (KIA)	−	0	−	0
Indole	−	0	−	0
Nitrate to gas	−	0	−	0
Indophenol oxidase	+	100	+	100

Arginine dihydrolase (DBM)	−	0	0
Lysine decarboxylase	−	0	0
Ornithine decarboxylase	−	0	0
Phenylalanine deaminase	−	3	0
Hydrolysis: urea	−	0	0
Esculin	−	0	0
Tween 80	− or +	20	− or +
Deoxyribonucleic acid	−	0	0
Gelatin	−	2	0
Growth on MBM + acetate	+	100	+ or −
Growth at 42° C	− or +	10	+ or −
Motile	+	100	+
Number of flagella	>1		>1

[a] Except where indicated, cultures were incubated at 30° C.
[b] Sign: +, 90% or more positive within 2 days; −, no reaction (90% or more); + or −, most strains positive; − or +, most strains negative; >1, polar multitrichous flagella.
[c] Abbreviations: OFBM, OF basal medium; DBM, decarboxylase base Moeller; MBM, mineral base medium.

Table 15 Characters Useful for Identification of *Pseudomonas acido-vorans* and *Pseudomonas testosteroni* (*Comamonas terrigena*) Strains[a]

Character	P. acidovorans	P. testosteroni
Gram negative, rod shaped, asporogenous	+	+
Indophenol oxidase	+	+
Motility	+	+
Polar tuft of flagella and long wavelength	+	+
Glucose, acid (OFBM)	–	–
Fructose, Mannitol	+	–

[a] From Table 14 and text.

Pseudomonas diminuta and *Pseudomonas vesicularis*

Properties that differentiate the two species include pellicle formation in broth culture by *P. diminuta*, hydrolysis of esculin and utilization of certain sugars by *P. vesicularis*, and the occurrence of orange carotenoid pigment in some strains of *P. vesicularis* (Wilkinson and Galbraith, 1979). *Pseudomonas vesicularis* strains may produce acid from glucose, galactose, rhamnose, xylose, and maltose, depending on the type of basal medium used. *Pseudomonas vesicularis* and most *P. diminuta* strains produce acid from ethanol.

Pseudomonas diminuta and *P. vesicularis* constitute a distinct group that is only distantly related to other pseudomonads and more closely realted to *Acetobacter* because of the production of acid from ethanol and the multiple requirements for growth factors (Whitaker et al., 1981). This group probably deserves a separate generic rank.

Pseudomonas paucimobilis

The attributes of 110 strains of *P. paucimobilis* are recorded in Table 18, and characters useful for the identification of most *P. paucimobilis* strains are recorded in Table 19.

Colonies develop an intracellular, nondiffusible, carotenoid (nostoxanthin) pigment on most media. This yellow pigment is distinct from the brominated aryl polyenes (xanthomonadins) present in *Xanthomonas* species. The name *paucimobilis* indicates that only a few cells in a population may be motile. Motility is best demonstrated

Table 16 Characteristics of *Pseudomonas* Species in the Diminuta Group[a]

Character	P. diminuta (42 strains)		P. vesicularis (32 strains)	
	Sign[b]	Percent positive	Sign	Percent positive
Acid: glucose, 1% (OFBM[c])	−	0	− or +(w)	41
Fructose	−	0	−	0
Xylose	−	0	− or +(w)	16
Lactose	−	0	−	0
Sucrose	−	0	−	0
Maltose	−	0	− or +(w)	47
Mannitol	−	0	−	0
ONPG	−	0	+ or −	50
Hydrogen sulfide (KIA)	−	0	−	0
Indole	−	0	−	0
Nitrate to gas	−	0	−	0
Indophenol oxidase	+	100	+	100
Arginine dihydrolase (DBM)	−	0	−	0
Lysine decarboxylase	−	0	−	0
Ornithine decarboxylase	−	0	−	0

Table 16 (continued)

Character	P. diminuta (42 strains)		P. vesicularis (32 strains)	
	Sign[b]	Percent positive	Sign	Percent positive
Phenylalanine deaminase	−	7	−	0
Hydrolysis: urea	−	0	−	0
Esculin	−	0	+	100
Tween 80	− or +	12	− or +	16
Deoxyribonucleic acid	− or +	14	−	0
Gelatin	+ or −	62	+ or −	56
Growth on MBM + acetate	−	0	−	0
Growth at 42° C	− or +	21	−	0
Motile	+	100	+	97
Number of flagella	1		1	

[a] Except where indicated, cultures were incubated at 30° C.
[b] Sign: +, 90% or more positive within 2 days; −, no reaction (90% or more); + or −, most strains positive; − or +, most strains negative; (w), weakly reactive; 1, polar monotrichous flagella.
[c] Abbreviations: OFBM, OF basal medium; DBM, decarboxylase base Moeller.

Cultural and Biochemical Aspects

Table 17 Characters Useful for Identification of *Pseudomonas diminuta* and *Pseudomonas vesicularis* Strains[a]

Character	*P. diminuta*	*P. vesicularis*
Gram negative, rod shaped, asporogenous	+	+
Indophenol oxidase	+	+
Motility	+	+
Polar monotrichous flagella and short wavelength	+	+
Glucose, maltose, acid (OFBM)	−	− or +(w)
Esculin hydrolysis	−	+

[a] From Table 16 and text.

at room temperature. When motile, *P. paucimobilis* is polar monotrichous. Indophenol oxidase is produced. Esculin and ONPG are hydrolyzed. Acid is produced in OF basal medium (OFBM) from a number of carbohydrates, but not from sugar alcohols. A negative test for urease distinguishes this species from phenotypically similar strains of *Sphingobacterium (Flavobacterium) multivorum*. Deoxyribonuclease, esterases, phosphatases, and lipase activity may be involved in its virulence.

Evidence from rRNA cistron comparisons and cellular fatty acid profiles indicate this species is not an authentic member of the genus *Pseudomonas* and that *P. paucimobilis* may justify inclusion in a new genus (Owen and Jackman, 1982; Oyaizu and Komagata, 1981).

Ve Group of Pseudomonas Species

The attributes of 98 strains of *Pseudomonas* species Ve (*Pseudomonas* species Ve-1, *Pseudomonas* species Ve-2) are recorded in Table 20, and characters useful for the identification of most *Pseudomonas* species Ve-1 and *Pseudomonas* species Ve-2 strains are recorded in Table 21.

Strains in the Ve group may produce smooth colonies or wrinkled, rough, and adherent as well as smooth colonies. They produce an intracellular, nondiffusible yellow pigment. Acid is produced from polyhydric alcohols in OF basal medium. Cells are motile. Indophenol oxidase is not produced.

Table 18 Characteristics of *Pseudomonas paucimobilis*[a] (110 strains)

Character	Sign[b]	Percent positive
Acid: glucose, 1% (OFBM[c])	+	100
Fructose	+	100
Xylose	+	100
Lactose	+	100
Sucrose	+	100
Maltose	+	100
Mannitol	−	0
ONPG	+	100
Hydrogen sulfide (KIA)	−	0
Indole	−	0
Nitrate to gas	−	0
Indophenol oxidase	+	90
Arginine dihydrolase (DBM)	−	0
Lysine decarboxylase	−	0
Ornithine decarboxylase	−	0
Phenylalanine deaminase	− or +	35
Hydrolysis: urea	−	0
Esculin	+	100
Tween 80	− or +	30
Deoxyribonucleic acid	−	0
Gelatin	−	0
Growth on MBM + acetate	+	100
Growth at 42° c	−	0
Motile	+ or −	89
Number of flagella	1	

[a] Except where indicated, cultures were incubated at 30° C.
[b] Sign: +, 90% or more positive within 2 days; −, no reaction (90% or more); + or −, most strains positive; − or +, most strains negative; (+), reactions delayed 2 days or more; 1, polar monotrichous flagella.
[c] Abbreviations: OFBM, OF basal medium; DBM, decarboxylase base Moeller.

Table 19 Characters Useful for Identification of *Pseudomonas paucimobilis* Strains[a]

Gram negative rod shaped, asporogenous	+
Indophenol oxidase	+
Motility	+ or −
Polar monotrichous flagella	+ or −
Glucose, maltose, acid (OFBM)	+
Mannitol	−
ONPG reaction	+
Urease	−
Esculin hydrolysis	+
Lysine decarboxylase (DBM)	−

[a] From Table 18.

Pseudomonas Species Ve-1 and Ve-2

Pseudomonas species Ve-1 strains have a polar tuft of flagella and hydrolyze esculin. *Pseudomonas* species Ve-2 strains are polar monotrichous and do not hydrolyze esculin. Strains of the former, but not *Pseudomonas* species Ve-2 strains, may produce arginine dihydrolase and nitratase. Cellular fatty acid composition shows group Ve strains to be similar to species of *Pseudomonas* (Dees et al., 1979). *Pseudomonas* species Ve-1 strains have 56.8 mol % G + C content in their DNA, which is at the lower limit of the range for *Pseudomonas* species (Gilardi et al., 1975). The 66.9 mol % G + C content in the DNA of *Pseudomonas* species Ve-2 strains is near the G + C values of *Chromobacterium lividum*, *Chromobacterium violaceum*, and many *Pseudomonas* species. Each warrents species recognition.

Pseudomonas extorquens

The attributes of six strains of *P. extorquens* are recorded in Table 22, and characters useful for the identification of most *P. extorquens* strains are recorded in Table 23.

Colonies produce an intracellular, nondiffusible, water-insoluble, pink oxocarotenoid pigment on most media (Oyaizu and Komagata, 1983). Pink pigmentation is a characteristic feature of most methanol-utilizing bacteria. Growth develops slowly on agar media compared

Table 20 Characteristics of *Pseudomonas* Species in the Ve Group[a]

Character	*Pseudomonas* sp. Ve-1 (21 strains)		*Pseudomonas* sp. Ve-2 (77 strains)	
	Sign[b]	Percent positive	Sign	Percent positive
Acid: glucose, 1% (OFBM)	+	100	+	100
Fructose	+	100	+	100
Xylose	+	100	+	100
Lactose	−	0	−	0
Sucrose	−	0	− or +	25
Maltose	+	100	+	99
Mannitol	+	100	+	100
ONPG	+	100	−	9
Hydrogen suflide (KIA)	−	0	−	0
Indole	−	0	−	0
Nitrate to gas	−	0	−	0
Indophenol oxidase	−	0	−	0

Arginine dihydrolase (DBM[c])	+ or −	76	0
Lysine decarboxylase	−	0	0
Ornithine decarboxylase	−	0	0
Phenylalanine deaminase	−	5	31
Hydrolysis: urea	− or +	10	30
Esculin	+	100	0
Tween 80	−	0	8
Deoxyribonucleic acid	−	0	0
Gelatin	+ or −	52	7
Growth on MBM + acetate	+	100	100
Growth at 42° C	+ or −	81	17
Motile	+	100	100
Number of flagella	>1	1	

[a] Except where indicated, cultures were incubated at 30° C.
[b] Sign: +, 90% or more positive within 2 days; −, no reaction (90% or more); + or −, most strains positive; − or +, most strains negative; 1, polar monotrichous flagella; >1, polar multitrichous flagella.
[c] DBM, decarboxylase base Moeller.

Table 21 Characters Useful for Identification of *Pseudomonas* Species Ve-1 and *Pseudomonas* Species Ve-2 Strains[a]

Character	*Pseudomonas* sp. Ve-1	*Pseudomonas* sp. Ve-2
Gram negative, rod shaped, asporogenous	+	+
Indophenol oxidase	−	−
Motility	+	+
Polar monotrichous flagella	−	+
Polar tuft of flagella	+	−
Glucose, mannitol, acid (OFBM)	+	+
Esculin hydrolysis	+	−

[a]From Table 20.

to other pseudomonads. Strains are motile, with polar monotrichous flagella. Acid is produced slowly, within 2−4 days, from fructose, xylose, and methanol in OF basal medium. Indophenol oxidase, urease, and amylase reactions are produced.

NON-PSEUDOMONAS SPECIES

Acinetobacter

The attributes of 1,700 strains of *A. calcoaceticus* are recorded in Table 24, and characters useful for the identification of most *A. calcoaceticus* biovars are recorded in Table 25.

Acinetobacter calcoaceticus

The genus *Acinetobacter* consists of a single species, *A. calcoaceticus*, which includes nonmotile rods that fail to produce indophenol oxidase. The rods are very short and plump, approaching coccus shape in stationary phase, predominately in pairs. Flagella are not present, but some strains show twitching movement on solid surfaces. Most strains are resistant to penicillin. A number of biovars exist (Baumann et al., 1968). Biovar *anitratus* produces acid from 1% glucose of OF basal medium and a number of other carbohydrates. Biovar *lwoffi* does not produce acid from OF glucose. Strains designated biovar *haemolyticus* or *alcaligenes* are done so on the basis of

Cultural and Biochemical Aspects

Table 22 Characteristics of *Pseudomonas extorquens*[a] (6 Strains)

Character	Sign[b]	Percent positive
Acid: glucose, 1% (OFBM)	−	0
Fructose	+	100
Xylose	+	100
Lactose	−	0
Sucrose	−	0
Maltose	−	0
Mannitol	−	0
ONPG	−	0
Hydrogen sulfide (KIA)	−	0
Indole	−	0
Nitrate to gas	−	0
Indophenol oxidase	+	100
Arginine dihydrolase (DBM[c])	−	0
Lysine decarboxylase	−	0
Ornithine decarboxylase	−	0
Phenylalanine deaminase	−	0
Hydrolysis: urea	+	100
Esculin	−	0
Tween 80	−	0
Deoxyribonucleic acid	−	0
Gelatin	−	0
Growth on MBM + acetate	+	100
Growth at 42° C	− or +	33
Motile	+	100
Number of flagella	1	

[a] Except where indicated, cultures were incubated at 30° C.
[b] Sign: +, 90% or more positive within 2 days; −, no reaction (90% or more); − or +, most strains negative; 1, polar monotrichous flagella.
[c] DBM, decarboxylase base Moeller.

Table 23 Characters Useful for Identification of *Pseudomonas extorquens* Strains[a]

Gram negative, rod shaped, asporogenous	+
Indophenol oxidase	+
Motility	+
Polar monotrichous flagella	+
Xylose, acid (OFBM)	+
Maltose	−
Urease	+
Nitrate to gas	−
Intracellular pink pigment	+

[a]From Table 22 and text.

their beta-hemolytic and proteolytic properties. Biovar *haemolyticus*, but not biovar *alcaligenes*, produces acid from glucose. *Acinetobacter* shares some characters with *Moraxella*, but such phenotypic attributes as cellular morphology, the oxidase reaction, and susceptibility to antibiotics separate the two genera. The genetics and physiology of *Acinetobacter* are reviewed in detail (Juni, 1978).

Group EO-2

Unnamed group EO-2 strains of uncertain affiliation are similar to *Acinetobacter* strains but differ from the latter species in several aspects, namely, the production of indophenol oxidase and oxidative acidity from fructose (Weaver et al., 1983).

Moraxella

The attributes of 117 strains of *Moraxella* species encountered in human clinical specimens are recorded in Table 26, and characters useful for the identification of most *Moraxella* species strains are recorded in Table 27.

The genus *Moraxella* includes nonmotile rods that synthesize indophenol oxidase and fail to produce acid in carbohydrates. Cellular morphology is variable; strains usually appear as short rods occurring characteristically in pairs, singly, and sometimes in short chains; some strains are filamentous. A tendency to retain the crystal violet

Table 24 Characteristics of *Acinetobacter calcoaceticus*[a]

Character	Biovar *anitratus* (1271 strains)		Biovar *haemolyticus* (166 strains)		Biovar *alcaligenes* (37 strains)		Biovar *lwoffi* (226 strains)	
	Sign[b]	Percent positive	Sign	Percent positive	Sign	Percent positive	Sign	Percent positive
Acid: glucose, 1% (OFBM)	+	100	+	100	−	0	−	0
Fructose	−	0	−	0	−	0	−	0
Xylose	+	100	+	100	−	0	−	0
Lactose	+	100	+	93	−	0	−	0
Sucrose	−	0	−	0	−	0	−	0
Maltose	+ or −	67	+ or −	63	−	0	−	0
Mannitol	−	0	−	0	−	0	−	0
ONPG	−	0	−	0	−	0	−	0
Hydrogen sulfide (KIA)	−	0	−	0	−	0	−	0
Indole	−	0	−	0	−	0	−	0
Nitrate to nitrite	−	0	−	0	−	0	−	0
Nitrate to gas	−	0	−	0	−	0	−	0
Indophenol oxidase	−	0	−	0	−	0	−	0
Arginine dihydrolase (DBM[c])	−	0	−	0	−	0	−	0

Table 24 (continued)

Character	Biovar anitratus (1271 strains)		Biovar haemolyticus (166 strains)		Biovar alcaligenes (37 strains)		Biovar lwoffi (226 strains)	
	Sign[b]	Percent positive	Sign	Percent positive	Sign	Percent positive	Sign	Percent positive
Lysine decarboxylase	−	0	−	0	−	0	−	0
Ornithine decarboxylase	−	0	−	0	−	0	−	0
Phenylalanine deaminase	−	0	−	0	−	3	−	3
Hydrolysis: urea	+ or −	72	− or +	18	−	6	− or +	24
Esculin	−	0	−	0	−	0	−	0
Tween 80	+	100	+	99	+	100	+	96
Deoxyribonucleic acid	−	0	−	0	−	0	−	0
Gelatin	−	0	+	98	+	100	−	0
Beta-hemolysis	−	0	+	100	+	100	−	1
Growth on MBM + acetate	+	99	+	100	+	100	+	100
Growth at 42° C	+	98	+ or −	65	+ or −	53	+ or −	53
Motile	−	0	−	0	−	0	−	0

[a] Except where indicated, cultures were incubated at 30° C.
[b] Sign: +, 90% or more positive within 2 days; −, no reaction (90% or more); + or −, most strains positive; − or +, most strains negative.
[c] DBM, decarboxylase base Moeller.

Table 25 Characters Useful for Identification of *Acinetobacter calcoaceticus* Biovars

Character	Biovar anitratus	Biovar haemolyticus	Biovar alcaligenes	Biovar lwoffi
Gram negative, rod shaped asporogenous	+	+	+	+
Indophenol oxidase	−	−	−	−
Motility	−	−	−	−
Glucose, acid (OFBM)	+	+	−	−
Beta-hemolysis	−	+	+	−
Gelatin hydrolysis	−	+	+	−
Nitrate to nitrite	−	−	−	−

[a]From Table 24.

Table 26 Characteristics of *Moraxella* Species[a]

Character	*M. lacunata* (3 strains) Sign[b]	Percent positive	*M. nonliquefaciens* (56 strains) Sign	Percent Positive
ONPG	−	0	−	0
Hydrogen sulfide (KIA)	−	0	−	0
Indole	−	0	−	0
Nitrate to nitrite	+	100	+	100
Indophenol oxidase	+	100	+	100
Arginine dihydrolase (DBM[c])	−	0	−	0
Lysine decarboxylase	−	0	−	0
Ornithine decarboxylase	−	0	−	0
Phenylalanine deaminase	−	0	−	0
Hydrolysis: urea	−	0	−	7
Esculin	−	0	−	0
Tween 80	−	0	−	4
Deoxyribonucleic acid	−	0	−	0
Gelatin	+	100	−	0
Growth on MBM + acetate	−	0	−	0
Growth at 42° C	−	0	− or +	21
Motile	−	0	−	0

[a] Except where indicated, cultures were incubated at 30° C. All strains gave negative results for acid production from carbohydrates.
[b] Sign: +, 90% or more positive within 2 days; −, no reaction (90% or more); + or −, most strains positive; − or +, most strains negative.
[c] DBM, decarboxylase base Moeller.

M. oslo-ensis (36 strains)		M. phenyl-pyruvica (7 strains)		M. atlan-tae (1 strain)		M. ureth-ralis (14 strains)	
Sign	Percent positive	Sign	Percent positive	Sign	Percent positive	Sign	Percent positive
−	0	−	0	−	0	−	0
−	0	−	0	−	0	−	0
−	0	−	0	−	0	−	0
− or +	14	+ or −	86	−	0	−	0
+	100	+	100	+	100	+	100
−	0	−	0	−	0	−	0
−	0	−	0	−	0	−	0
−	0	−	0	−	0	−	0
−	0	+	100	−	0	+	100
−	0	+	100	−	0	−	0
−	0	−	0	−	0	−	0
− or +	31	+	100	−	0	−	0
−	0	−	0	−	0	−	0
−	0	−	0	−	0	−	0
+	100	−	0	−	0	+	100
− or +	36	− or +	14	−	0	+	100
−	0	−	0	−	0	−	0

Table 27 Characters Useful for Identification of *Moraxella* Species Strains[a]

Character	M. lacunata	M. nonlique-faciens	M. osloensis	M. phenyl-pyruvica	M. atlantae	M. urethralis
Gram negative, rod shaped, asporogenous	+	+	+	+	+	+
Motility	−	−	−	−	−	−
Indophenol oxidase	+	+	+	+	+	+
Glucose, acid (OFBM)	−	−	−	−	−	−
Serum required	+	d[b]	−	−	+	−
Growth in MBM + acetate	−	−	+	−	−	+
Proteolytic	+	−	−	−	−	−
Phenylalanine deaminase	−	−	−	+	−	+
Urease	−	−	−	+	−	−
Nitrate to nitrite	+	+	− or +	+ or −	−	−

[a]From Table 26 and text.
[b]Different reactions.

Cultural and Biochemical Aspects

stain is noted. Indole is not produced. Some isolates produce urease and phenylalanine deaminase and are proteolytic, but the majority do not elaborate extracellular hydrolytic enzymes. Strains are usually susceptible to penicillin and most other antibiotics. With two exceptions, the species have complex organic growth factor requirements and will not grow in a mineral base medium (MBM) containing acetate as the sole source of carbon and energy. *Moraxella*, as well as *Acinetobacter*, may be distinguished from the genus *Neisseria*, since cellular elongation occurs in the former genera, but not in *Neisseria*, during growth in the presence of penicillin.

Moraxella lacunata

This species is more fastidious than other *Moraxella* species. Most strains do not grow on peptone media, owing to a toxic effect of certain components of peptone. The toxic factor is counteracted by the addition of cholesterol, oleic acid, or rabbit serum. Proteolysis is detected in coagulated serum or nutrient gelatin supplemented with serum. No growth occurs in OF basal medium. Nitrate is reduced to nitrite. *Moraxella liquefaciens* may be considered a biovar or subspecies of *M. lacunata*. The majority of isolates are from inflamed human eyes.

Moraxella nonliquefaciens

Some strains require the addition of serum or oleic acid to peptone media. Occasional strains are encapsulated and grow large mucoid colonies on blood agar. No growth occurs in OF basal medium or on MacConkey agar. Nitrate is reduced to nitrite. Strains are nonproteolytic. They are isolated primarily form the nasopharynx, throat, and sputum of humans.

Moraxella osloensis

Growth occurs in MBM, without organic growth factors, containing acetate as sole carbon source. Most strains grow in OF basal medium. Growth on MacConkey agar and reduction of nitrate to nitrite are strain variable.

Moraxella phenylpyruvica

Good growth occurs on MacConkey agar and in OF basal medium. Most strains deaminate phenylalanine and hydrolyze urea. Nitrate reduction is strain variable.

Moraxella atlantae

Some strains require the addition of serum to peptone media. Growth occurs on MacConkey agar but not in OF basal medium. Nitrate is not reduced.

Moraxella urethralis

Growth occurs in MBM containing acetate, in OF basal medium, and on MacConkey agar. Phenylalanine is deaminated. Nitrate is not reduced, but some strains reduce nitrite with the accumulation of gas. Most isolates are from the genitourinary tract.

Moraxella-like CDC groups M5 and M6 are considered to be true *Neisseria* species, *Neisseria elongata* and *Neisseria parelongata*, respectively (Bøvre and Holten, 1970). *Moraxella* species are described in greater detail (Bøvre et al., 1976; Henriksen, 1973; Riley et al., 1974).

Flavobacterium

The attributes of 250 strains of acid-producing and 62 strains of non-acid-producing *Flavobacterium* species are recorded in Tables 28 and 29, respectively, and characters useful for the identification of most acid-producing and non-acid-producing *Flavobacterium* species strains are recorded in Tables 30 and 31, respectively.

The genus *Flavobacterium* includes yellow-pigmented rods that produce indophenol oxidase and indole. Motility is not observed in hanging drop. Spreading colonies are formed by some strains. In many species the water-insoluble, yellow carotenoid pigment is very pale and not easily detectable; the pigment does not occur on all media that support growth. Indole is formed by all but one species, but the amount is small, not always detected by routine means, and requires xylene extraction before testing with Ehrlich's reagent. Those species of flavobacteria which produce acid from maltose and other carbohydrates do so oxidatively. Generally actively proteolytic, the majority of species liquefy gelatin. Specified amino acids and vitamins are required as organic growth factors. With specific exceptions, strains of most species are resistant to penicillin and polymyxin. Extensive chemotaxonomic and phenotypic characterization show that some strains designated flavobacteria, i.e., *F. meningosepticum*, *F. odoratum*, and *F. breve*, should be transferred to the genus *Cytophaga*, family Cytophagaceae (Callies and Mannheim, 1980; Collins and Jones, 1981; Oyaizu and Komagata, 1981).

Characters of Acid-Producing Flavobacteria

Flavobacterium meningosepticum: Colonies are convex, smooth or mottled, glistening, and butyrous and may produce an intracellular pale yellow pigment. Esculin, ONPG, deoxyribonucleic acid, and gelatin, but not starch or Tween 80, are hydrolyzed. Indole is produced. Nitrate is not reduced and urease is usually negative.

Flavobacterium indologenes: Strains are similar in phenotypic characters to strains of *F. meningosepticum* but differ from the latter species in several aspects. Colonies of *F. indologenes* produce an intracellular bright yellow pigment. Lipase and amylase are synthesized. Tests for ONPG and extracellular deoxyribonuclease are usually negative. Occasional strains reduce nitrate and produce urease. A rare isolate will have a phenotypic pattern that merges between *F. meningosepticum* and *F. indologenes*.

Flavobacterium breve: Colonies have an entire edge and are circular, smooth, and shiny with yellow centers and greenish edges. A fruity odor is associated with growth. Indole is produced and gelatin is liquified. Urea, ONPG, esculin, starch, and Tween 80 are not hydrolyzed. Acid is not produced from carbohydrates other than maltose and glucose.

Flavobacterium species IIh: Indole is produced and gelatin and esculin, but not ONPG, are hydrolyzed. Strains are susceptible to penicillin but not polymyxin. No distinct pigment is produced. Strains are isolated primarily from the respiratory tract of animals.

Flavobacterium species IIi: Strains of *Flavobacterium* species IIi are similar to but distinguished from strains of *Sphingobacterium multivorum* on the basis of indole production and urease activity. Colonies are pale to light yellow.

Characters of Non-Acid-Producing Flavobacteria

Flavobacterium odoratum: Deoxyribonucleic acid, gelatin, and urea are hydrolyzed. Variable features include deamination of phenylalanine and reduction of nitrite, but not nitrate, to gas. Indole is not produced. Growth occurs in OF basal medium. Strains produce yellow-green-pigmented colonies with different morphological types. Most strains produce both large, rough, spreading colonies with a raised center and smaller, smooth, convex, nonspreading colonies. Some strains produce only the latter colony type. A strong fruity odor is produced similar to that associated with colonies of *Alcaligenes odorans*. Most isolates are from urine specimens.

Flavobacterium species IIe: Indole is produced. Gelatin and urea are not hydrolyzed. Growth fails on Mueller-Hinton agar without the addition of blood and in OF basal medium. Acid may be produced in a modified oxidative–fermentative base medium. Colonies develop a pale yellow pigment on milk agar plates. Strains are susceptible to penicillin, polymyxin, and most other antibiotics.

Flavobacterium species IIf: Indole is produced and gelatin, but not urea, is hydrolyzed. Growth occurs in OF basal medium. Strains produce luxuriant growth on blood agar, giving rise to colonies that are very mucoid and moist, with a pale yellow intracellular pigment. Colonies have a tendency to stick to the agar surface. A brown discoloration of the surrounding agar medium and a green discoloration

Table 28 Characteristics of Acid-Producing *Flavobacterium* Species[a]

Character	F. meningosepticum (29 strains) Sign[b]	F. meningosepticum (29 strains) Percent positive	F. breve (8 strains) Sign	F. breve (8 strains) Percent positive
Acid: glucose, 1% (OFBM)	+	100	+ or −	88
Fructose	+	100	−	0
Xylose	−	7	−	0
Lactose	+ or −	86	−	0
Sucrose	−	0	−	0
Maltose	+	100	+	100
Mannitol	+	100	−	0
ONPG	+	100	−	0
Hydrogen sulfide (KIA)	−	0	−	0
Indole	+	100	+	100
Nitrate to gas	−	0	−	0
Indophenol oxidase	+	100	+	100
Arginine dihydrolase (DBM[c])	−	0	−	0
Lysine decarboxylase	−	0	−	0
Ornithine decarboxylase	−	0	−	0
Phenylalanine deaminase	−	0	−	0
Hydrolysis: urea	−	3	− or +	25
Esculin	+	100	−	0
Tween 80	−	0	−	0
Starch	− or +	10	−	0
Deoxyribonucleic acid	+	100	+	100
Gelatin	+	100	+	100
Growth on MBM + acetate	−	0	−	0
Growth at 42° C	− or +	34	−	0
Motile	−	0	−	0

[a] Except where indicated, cultures were incubated at 30° C.
[b] Sign: +, 90% or more positive within 2 days; −, no reaction (90% or more); + or −, most strains positive; − or +, most strains negative.
[c] DBM, decarboxylase base Moeller.

F. indologenes (206 strains)		Flavobacterium sp. IIh (2 strains)		Flavobacterium sp. IIi (3 strains)	
Sign	Percent positive	Sign	Percent positive	Sign	Percent positive
+	100	+	100	+	100
+	94	−	0	+	100
− or +	45	−	0	+	100
−	1	−	0	+	100
− or +	27	−	0	+	100
+	100	+	100	+	100
− or +	11	−	0	−	0
− or +	43	+ or −	50	+	100
−	0	−	0	−	0
+	100	+	100	+	100
−	0	−	0	−	0
+	100	+	100	+	100
−	0	−	0	−	0
−	0	−	0	−	0
−	0	−	0	−	0
− or +	22	−	0	−	0
− or +	13	−	0	−	0
+	100	+	100	+	100
+	95	−	0	− or +	33
+	100	+	100	− or +	33
−	4	+	100	−	0
+	100	+	100	−	0
−	0	−	0	−	0
− or +	20	+ or −	50	−	0
−	0	−	0	−	0

Table 29 Characteristics of Non-Acid-Producing *Flavobacterium* Species[a]

Character	*F. odoratum* (19 strains)		*Flavobacterium* sp. IIe[b] (2 strains)		*Flavobacterium* sp. IIf (32 strains)		*Flavobacterium* sp. IIj (9 strains)	
	Sign[c]	Percent positive	Sign	Percent positive	Sign	Percent positive	Sign	Percent positive
ONPG	−	0	−	0	−	0	−	0
Hydrogen sulfide (KIA)	−	0	−	0	−	0	−	0
Indole	−	0	+	100	+	100	+	100
Nitrate to gas	−	0	−	0	−	0	−	0
Indophenol oxidase	+	100	+	100	+	100	+	100
Arginine dihydrolase (DBM[d])	−	0	−	0	−	0	−	0
Lysine decarboxylase	−	0	−	0	−	0	−	0
Ornithine decarboxylase	−	0	−	0	−	0	−	0
Phenylalanine deaminase	− or +	16	+ or −	50	−	9	−	0

Cultural and Biochemical Aspects

Hydrolysis: urea	+	100	—	0	—	0	+(r)	100
Esculin	— or +	0	—	0	—	0	—	0
Tween 80	— or +	16	+	100	—	0	—	0
Starch	—	0	+	100	—	0	—	0
Deoxyribonucleic acid	+	95	—	0	—	3	—	0
Gelatin	+	100	—	0	+	100	+	100
Growth on MBM + acetate	—	0	—	0	—	0	—	0
Growth at 42° C	—	0	—	0	+ or —	75	—	0
Motile	—	0	—	0	—	0	—	0

[a] Except where indicated, cultures were incubated at 30° C. All strains gave negative results for acid production from carbohydrates.
[b] Strains may produce acid in a modified oxidative-fermentative base medium (Weaver et al., 1983).
[c] Sign: +, 90% or more positive within 2 days; —, no reaction (90% mor more); + or —, most strains positive; — or +, most strains negative; (r), rapid reaction.
[d] DBM, decarboxylase base Moeller.

Table 30 Characters Useful for Identification of Acid-Producing *Flavobacterium* Species Strains[a]

Character	*F. meningosepticum*	*F. breve*	*F indologenes*	*Flavobacterium* sp. IIh	*Flavobacterium* sp. IIi
Gram negative, rod shaped, asporogenous	+	+	+	+	+
Indophenol oxidase	+	+	+	+	+
Motility	−	−	−	−	−
Maltose, acid (OFBM)	+	+	+	+	+
Xylose	−	−	− or +	−	+
ONPG reaction	+	−	− or +	+ or −	+
Indole	+	+	+	−	+
Urease	−	− or +	− or +	+	−
Esculin hydrolysis	+	−	+	+	+
Tween 80 hydrolysis	−	−	+	−	− or +
Starch hydrolysis	− or +	−	+	+	− or +
Deoxyribonucleic acid hydrolysis	+	+	−	+	−
Gelatin hydrolysis	+	+	+	+	−

[a]From Table 28.

Table 31 Characters Useful for Identification of Non-Acid-Producing *Flavobacterium* Species Strains[a]

Character	*F. odoratum*	*Flavobacterium* sp. IIe	*Flavobacterium* sp. IIf	*Flavobacterium* sp. IIj
Gram negative, rod shaped, asporogenous	+	+	+	+
Indophenol oxidase	+	+	+	+
Motility	−	−	−	−
Maltose, acid (OFBM)	−	−	−	+
Indole	−	+	+	+
Urease	+	−	−	+(r)
Esculin hydrolysis	−	−	−	−
Starch hydrolysis	−	+	−	−
Deoxyribonucleic acid hydrolysis	+	−	−	−
Gelatin hydrolysis	+	−	+	+
Penicillin susceptible	−	+	+	+
Polymyxin susceptible	−	+	+	−

[a]From Table 29 and text.

of erythrocytes in blood agar medium may be noted. Strains are susceptible to penicillin, polymyxin, and most other antibiotics. Most isolates are from urine and vaginal specimens from human females.

Flavobacterium species IIj: Indole is produced and gelatin and urea (rapid) are hydrolyzed. Growth fails in OF basal medium. Growth is not as luxuriant as that observed with IIf strains. Smaller colonies develop which are butyrous and sticky and difficult to remove from agar surfaces. No distinct pigment is produced. A green discoloration of erythrocytes in blood agar medium may occur. Strains are susceptible to penicillin but not polymyxin. They are isolated from the respiratory tract of dogs and cats. *Flavobacterium* species are described in greater detail (Holmes et al., 1983; Homes and Owen, 1982; Homes et al., 1982; Owen and Holmes, 1978; Weaver et al., 1983).

Sphingobacterium

The attributes of 52 strains of *S. multivorum* are recorded in Table 32, and characters useful for the identification of most *S. multivorum* strains are recorded in Table 33.

The genus *Sphingobacterium*, family Cytophagaceae, includes pale to light yellow pigmented rods that produce indophenol oxidase, but, unlike members of the genus *Flavobacterium*, are nonproteolytic and fail to produce indole. The cells of species in this genus have no flagella, but may exhibit gliding motility. Acid is produced from many carbohydrates. Esculin, ONPG, and urea are hydrolyzed. Virulence factors may include phosphatases, amidase, and esterase lipase.

Strains originally designated *Flavobacterium multivorum* and *Flavobacterium spiritivorum* and a new species, *S. mizutae*, have been proposed (Yabuuchi et al., 1983) to be included in the genus *Sphingobacterium* based on the presence of sphingophospholipids as cellular lipid components. *Sphingobacterium spiritivorum* is distinguished from *Sphingobacterium multivorum* by reactions in mannitol. *Sphingobacterium mizutae* is distinguished from *F. multivorum* by reactions in rhamnose and urea.

Alcaligenes

The attributes of 252 strains of *Alcaligenes* species are recorded in Table 34, and characters useful for the identification of most *Alcaligenes* species are recorded in Table 35.

The genus *Alcaligenes* includes peritrichously flagellated rods that synthesize indophenol oxidase and produce alkali from carbohydrates. Three species are recognized, but the properties of *A. denitrificans*, *A. odorans*, and *A. faecalis* are so similar that it has been suggested that the species be reduced to two or even one species (Kiredjian et al., 1981; Ruger and Tan, 1983; Yamasato et al., 1982). Species differ-

Cultural and Biochemical Aspects

Table 32 Characteristics of *Sphingobacterium multivorum*[a] (52 Strains)

Character	Sign[b]	Percent positive
Acid: glucose, 1% (OFBM)	+	100
Fructose	+	100
Xylose	+	100
Lactose	+	100
Sucrose	+	100
Maltose	+	100
Mannitol	−	0
ONPG	+	100
Hydrogen sulfide (KIA)	−	0
Indole	−	0
Nitrate to gas	−	0
Indophenol oxidase	+	100
Arginine dihydrolase (DBM[c])	−	0
Lysine decarboxylase	−	0
Ornithine decarboxylase	−	0
Phenylalanine deaminase	−	7
Hydrolysis: urea	+	100
Esculin	+	100
Tween 80	− or +	48
Starch	+ or −	52
Deoxyribonucleic acid	−	0
Gelatin	−	0
Growth on MBM + acetate	−	0
Growth at 42° C	−	0
Motile	−	0

[a] Except where indicated, cultures were incubated at 30° C.
[b] Sign: +, 90% or more positive within 2 days; −, no reaction (90% or more); + or −, most strains positive; − or +, most strains negative.
[c] DBM, decarboxylase base Moeller.

Table 33 Characters Useful for Identification of *Sphingobacterium multivorum* Strains[a]

Gram negative, rod shaped, asporogenous	+
Indophenol oxidase	+
Motility	−
Glucose, acid (OFBM)	+
Mannitol	−
ONPG reaction	+
Indole	−
Urease	+
Esculin hydrolysis	+
Gelatin hydrolysis	−

[a]From Table 28.

entiation is based on carbon substrate utilization, the manner in which they attack nitrate and nitrite, and colonial morphology (Rarick et al., 1978). Strains of *A. odorans* reduce nitrite, but not nitrate, to gas. *Alcaligenes denitrificans* respires anaerobically in the presence of both nitrate and nitrite, with the accumulation of gas. *Alcaligenes faecalis* does not produce nitrogen gas. Approximately half of the *A. faecalis* strains reduce nitrate to nitrite.

Alcaligenes faecalis biovar 1 strains produce colonies which are small, low convex, glistening, and with an entire edge. *Alcaligenes faecalis* biovar 2 strains produce colonies which are larger, umbonate with a spreading periphery, and granular. The morphology of *A. denitrificans* colonies are similar to *A. faecalis* biovar 1. The morphology of *A. odorans* colonies is similar to that of *A. faecalis* biovar 2. In addition, colonies of *A. odorans* produce a distinct dark green discoloration of erythrocytes in blood agar media, and growth is accompanied by a characteristic fruity odor described as similar to valerian tincture.

Bordetella bronchicanis and Closely Related Unnamed Groups IVc-2 and IVe

The attributes of 53 strains of *B. bronchicanis* and unnamed groups IVc-2 and IVe are recorded in Table 36, and characters useful for

Table 34 Characteristics of Alcaligenes Species[a]

Character	A. faecalis (97 strains) Sign[b]	A. faecalis Percent positive	A. denitrificans (48 strains) Sign	A. denitrificans Percent positive	A. odorans (107 strains) Sign	A. odorans Percent positive
ONPG	−	0	−	0	−	0
Hydrogen sulfide (KIA)	−	0	−	0	−	0
Indole	−	0	−	0	−	0
Nitrate to gas	−	0	+	100	−	0
Nitrite to gas	−	0	+	100	+	100
Indophenol oxidase	+	100	+	100	+	100
Arginine dihydrolase (DBM[c])	−	0	−	0	−	0
Lysine decarboxylase	−	0	−	0	−	0
Ornithine decarboxylase	−	0	−	0	−	0
Phenylalanine deaminase	− or +	10	−	4	−	0
Hydrolysis; urea	−	6	− or +	33	−	0
Esculin	−	0	−	0	−	0
Tween 80	− or +	14	− or +	25	−	0
Deoxyribonucleic acid	−	2	−	0	−	0
Gelatin	−	0	−	0	−	5

Table 34 (continued)

Character	A. faecalis (97 strains)		A. denitrificans (48 strains)		A. odorans (107 strains)	
	Sign[b]	Percent positive	Sign	Percent positive	Sign	Percent positive
Growth on MBM + acetate	+ or −	81	+ or −	75	+	100
Growth at 42° C	+ or −	53	− or +	25	+ or −	63
Motile	+	100	+	100	+	100
Number of flagella	p		p		p	

[a]Except where indicated, cultures were incubated at 30° C. All strains gave negative results for acid production from carbohydrates.
[b]Sign: +, 90% or more positive within 2 days; −, no reaction (90% or more); + or −, most strains positive; − or +, most strains negative; p, peritrichous flagella.
[c]DBM, decarboxylase base Moeller.

Table 35 Characters Useful for Identification of *Alcaligenes* Species Strains[a]

Character	A. faecalis	A. denitrificans	A. odorans
Gram negative, rod shaped, asporogenous	+	+	+
Indophenol oxidase	+	+	+
Motility	+	+	+
Peritrichous flagella	+	+	+
Glucose, xylose, acid (OFBM)	−	−	−
Nitrate to gas	−	+	−
Nitrite to gas	−	+	+
Fruity odor	−	−	+

[a] From Table 34 and text.

the identification of most *B. bronchicanis*, IVc-2, and IVe strains are recorded in Table 37. The three recognized *Bordetella* species (*pertussis, parapertussis, bronchicanis*) may be considered members of the same species, but for practical reasons a distinction can be made on phenotypic properties (Goodnow, 1980; Kloos et al., 1981). *Bordetella bronchicanis* is the only motile species.

These bacteria are phenotypically similar to *Alcaligenes*, except that they domonstrate immediate urea hydrolysis in Christensen's medium when the medium is warmed to incubator temperature prior to use. Rapid urease activity is the most practical test for distinguishing these bacteria from *Alcaligenes*. Some *Alcaligenes* strains produce urease slowly. *Bordetella bronchicanis* and group IVe can be separated from *Alcaligenes* species on the basis of carbon substrate utilization and cellular fatty acid composition (Dees et al., 1980; Martin et al., 1981).

Strains of *B. bronchicanis* reduce nitrate to nitrite and grow on SS agar, while strains of group IVc-2 do not. Reduction of nitrate or nitrite is variable in group IVe strains. Group IVe strains deaminate phenylalanine and demonstrate both polar and long lateral flagella; an occasional strain is nonmotile. *Bordetella bronchicanis* is a pathogen of mammalian respiratory tracts. IVc-2 strains are environmental isolates primarily from water. IVe isolates are found in urine specimens.

Table 36 Characteristics of *Bordetella bronchicanis*, Group IVc-2 and Group IVe[a]

Character	B. bronchicanis (18 strains)		Group IVc-2 (23 strains)		Group IVe (12 strains)	
	Sign[b]	Percent positive	Sign	Percent positive	Sign	Percent positive
ONPG	−	0	−	0	−	0
Hydrogen sulfide (KIA)	−	0	−	0	−	0
Indole	−	0	−	0	−	0
Nitrate to nitrite	+	100	−	0	+	100
Nitrate to gas	−	0	−	0	+ or −	83
Indophenol oxidase	+	100	+	100	+	100
Arginine dihydrolase (DBM[c])	−	0	−	0	−	0
Lysine decarboxylase	−	0	−	0	−	0
Ornithine decarboxylase	−	0	−	0	−	0

Phenylalanine deaminase	− or +	22	−	9	+	100
Hydrolysis: urea	+(r)	100	+(r)	100	+(r)	100
Esculin	−	0	−	0	−	0
Tween 80	−	0	+ or −	57	−	0
Deoxyribonucleic acid	−	0	−	0	−	0
Gelatin	−	0	−	0	−	0
Growth on MBM + acetate	+	100	+	100	+	100
Growth on SS agar	+	100	−	9	− or +	33
Growth at 42° C	− or +	39	+ or −	61	−	8
Motile	+	100	+	100	+	92
Number of flagella	p		p		pl	

[a] Except where indicated, cultures were incubated at 30° C. All strains gave negative results for acid production from carbohydrates.
[b] Sign: +, 90% or more positive within 2 days; −, no reaction (90% or more); + or −, most strains positive; − or +, most strains negative; (r), rapid reaction; p, peritrichous flagella; pl, polar and lateral flagella.
[c] DBM, decarboxylase base Moeller.

Table 37 Characters Useful for Identification of *Bordetella bronchicanis*, Group IVc-2 and IVe Strains[a]

Character	B. bronchicanis	Group IVc-2	Group IVe
Gram negative, rod shaped, asporogenous	+	+	+
Indophenol oxidase	+	+	+
Motility	+	+	+
Peritrichous flagella	+	+	−
Polar and lateral flagella	−	−	+
Glucose, xylose, acid (OFBM)	−	−	−
Urease	+(r)	+(r)	+(r)
Nitrate to nitrite	+	−	+
Nitrate to gas	−	−	+ or −
Growth on SS agar	+	−	− or +
Phenylalanine deaminase	+ or −	−	+

[a] From Table 36.

Achromobacter

The attributes of 129 strains of *Achromobacter xylosoxidans*, *Achromobacter* species Vd-1 and *Achromobacter* species Vd-2 are recorded in Table 38, and characters useful for the identification of most *A. xylosoxidans*, *Achromobacter* sp. Vd-1, and *Achromobacter* sp. Vd-2 strains are recorded in Table 39.

The genus *Achromobacter* includes nonpigmented, peritrichously flagellated rods that synthesize indophenol oxidase and produce acid from xylose and other carbohydrates, excluding lactose. The name *Achromobacter* is widely used in clinical bacteriology, but the designation may be rejected.

Achromobacter xylosoxidans

Achromobacter xylosoxidans biovar IIIa reduces nitrate to nitrite only; biovar IIIb reduces nitrate to gas. Urease is negative. At least seven serotypes (A through G) are recognized (Shigeta et al., 1983).

Table 38 Characteristics of Achromobacter xylosoxidans, Achromobacter sp. Vd-1, Achromobacter Species Vd-2, and Agrobacterium tumefaciens[a]

Character	A. xylosoxidans (85 strains)		Achromobacter sp. Vd-1 (24 strains)		Achromobacter sp. Vd-2 (20 strains)		A. tumefaciens (15 strains)	
	Sign[b]	Percent positive	Sign	Percent positive	Sign	Percent positive	Sign	Percent positive
Acid: glucose, 1% (OFBM)	+	99	+	96	+	100	+	100
Fructose	−	7	+	96	+	100	+	100
Xylose	+	100	+	100	+	100	+	100
Lactose	−	0	−	0	−	0	+	100
Sucrose	−	0	−	0	+	100	+	100
Maltose	−	0	−	0	+	100	+	100
Mannitol	−	0	−	0	+ or −	85	+	100
ONPG	−	0	−	0	−	0	+	100
Hydrogen sulfide (KIA)	−	0	−	0	−	0	−	0
Indole	−	0	−	0	−	0	−	0
Nitrate to gas	+ or −	62	+	100	+	100	−	7
Indophenol oxidase	+	100	+	100	+	100	+	100
Arginine dihydrolase (DBM[c])	−	0	− or +	42	− or +	25	−	0
Lysine decarboxylase	−	0	−	0	−	0	−	0

Table 38 (continued)

Character	A. xylosoxidans (85 Strains)		Achromobacter sp. Vd-1 (24 strains)		Achromobacter sp. Vd-2 (20 strains)		A. tumefaciens (15 strains)	
	Sign[b]	Percent positive	Sign	Percent positive	Sign	Percent positive	Sign	Percent positive
Ornithine decarboxylase	−	0	−	0	−	0	−	0
Phenylalanine deaminase	−	1	+	100	+	100	+	100
Hydrolysis: urea	−	0	+(r)	100	+(r)	100	+(r)	100
Esculin	−	0	− or +	29	+ or −	50	+	100
Tween 80	−	0	−	0	−	0	−	0
Deoxyribonucleic acid	−	0	−	0	−	0	−	0
Gelatin	−	0	−	0	−	0	−	0
Growth on MBM + acetate	+	100	+	100	+	100	+	100
Growth at 42° C	+ or −	84	− or +	17	−	5	− or +	20
Motile	+	100	+	100	+	100	+	100
Number of flagella	p		p		p		p	

[a] Except where indicated, cultures were incubated at 30° C.
[b] Sign: +, 90% or more positive within 2 days; −, no reaction (90% or more); + or −, most strains positive; − or +, most strains negative; (r), rapid reaction; p, peritrichous flagella.
[c] DBM, decarboxylase base Moeller.

Table 39 Characters Useful for Identification of *Achromobacter xylosoxidans*, *Achromobacter* species Vd-1, *Achromobacter* Species Vd-2, and *Agrobacterium tumefaciens*[a]

Character	*A. xylosoxidans*	*Achromobacter* sp. Vd-1	*Achromobacter* sp. Vd-2	*A. tumefaciens*
Gram negative, rod shaped, asporogenous	+	+	+	+
Indophenol oxidase	+	+	+	+
Motility	+	+	+	+
Peritrichous flagella	+	+	+	+
Xylose, acid (OFBM)	+	+	+	+
Lactose	−	−	−	+
Sucrose, maltose	−	−	+	+
Mannitol	−	−	+ or −	+
Nitrate to gas	+ or −	+	+	−
Urease	−	+(r)	+(r)	+(r)
3-Ketolactose	−	−	−	+

[a]From Table 38 and text.

Achromobacter Species Vd-1 and Vd-2

Achromobacter sp. Vd strains reduce nitrate to gas and produce rapid hydrolysis of urea. This is an ill-defined group; two and possibly more biovars (Vd-1, Vd-2) are recognized by means of their oxidative metabolism of carbohydrates. Strains of Vd-2, but not Vd-1, oxidize sucrose, maltose, and mannitol. Achromobacter species Vd strains are essentially identical but differ markedly from A. xylosoxidans both in cellular fatty acid composition and production of keto acids (Dees and Moss, 1978).

Agrobacterium

The attributes of 15 strains of Agrobacterium tumefaciens are recorded in Table 38, and characters useful for the identification of most A. tumefaciens strains are recorded in Table 39.

Colonies of A. tumefaciens are nonpigmented. The peritrichously flagellated rods produce positive reactions when tested for indophenol oxidase, phenylalanine deaminase, and hydrolysis of urea (rapid), esculin, and ONPG. The production of 3-ketolactose separates this species from the genus Achromobacter (Holmes and Roberts, 1981). Additonal species of agrobacteria are recognized, including Achromobacter rhizogenes and Achromobacter rubi. These species do not grow at 37° C, are nonmotile, and fail to produce 3-ketolactose.

REFERENCES

Baumann, P., Doudoroff, M., and Stanier, R. Y. (1968). J. Bacteriol. 95:1520.
Bøvre, K., and Holten, E. (1970). J. Gen. Microbiol. 60:67.
Bøvre, K., Fuglesang, J. E., Hagen, N., Jantzen, E., and Frøholm, L. O. (1976). Int. J. Syst. Bacteriol. 26:511.
Byng, G. S., Whitaker, R. J., Gherna, R. L., and Jensen, R. A. (1980). J. Bacteriol. 144:247.
Callies, E., and Mannheim, W. (1980). Antonie van Leeuwenhoek J. Microbiol. Serol. 46:41.
Collins, M. D., and Jones, D. (1981). Microbiol. Rev. 45:316.
Dees, S. B., and Moss, C. W. (1978). J. Clin. Microbiol. 8:61.
Dees, S. B., Moss, C. W., Weaver, R. E., and Hollis, D. (1979) J. Clin. Microbiol. 10:206.
Dees, S., Thanabalasundrum, S., Moss, C. W., Hollis, D. G., and Weaver, R. E. (1980). J. Clin. Microbiol. 11:664.
De Vos, P., and De Ley, J. (1983). Int. J. Syst. Bacteriol. 33:487.
Gilardi, G. L., Hirschel, S., and Mandel, M. (1975). J. Clin. Microbiol. 1:384

Goodnow, R. A. (1980). *Microbiol. Rev.* 44:722.
Henriksen, S.D. (1973). *Bacteriol. Rev.* 37:522.
Holmes, B., and Roberts, P. (1981). *J. Appl. Bacteriol.* 50:443.
Holmes, B., and Owen, R. J. (1982). *Int. J. Syst. Bacteriol.* 32:233.
Holmes, B., Owen, R. J., and Hollis, D. G. (1982). *Int. J. Syst. Bacteriol.* 32:157.
Holmes, B., Hollis, D. G., Steigerwalt, A. G., Pickett, M. J., and Brenner, D. J. (1983). *Int. J. Syst. Bacteriol.* 33:677.
Hugh, R. (1962). *Int. Bull. Bacteriol. Nomencl. Taxon.* 12:33.
Hugh, R. (1970). Pseudomonas and Aeromonas, in *Manual of Clinical Microbiology*, 1st ed. (J. E. Blair, E. H. Lennete, and J. P. Truant, eds.), American Society for Microbiology, Washington, D.C., p. 175.
Ikemoto, S., Suzuki, K. Kaneko, T., and Komagata, K. (1980). *Int. J. Syst. Bacteriol.* 30:437.
Juni, E. (1978). *Annu. Rev. Microbiol.* 32:349.
King, A., Holmes, B., Phillips, I., and Lapage, S. P. (1979). *J. Gen. Microbiol.* 114:137.
Kiredjian, M., Popoff, M. Coynault, C., Lefevre, M., and Lemelin, M. (1981). *Ann. Microbiol.* 132B:337.
Kloos, W. E., Mohapatra, N., Dobrogosz, W. J., Ezzell, J. W., and Manclark, C. R. (1981). *Int. J. Syst. Bacteriol.* 31:173.
Lapchine, L. (1979). *FEMS Microbiol. Lett.* 5:223.
Levin, R. E. (1972). *Antonie van Leeuwenhoek J. Microbiol. Serol.* 38:121.
Martin, R., Riley, P. S., Hollis, D. G., Weaver, R. E., Krichevsky, M. I. (1981). *J. Clin. Microbiol.* 14:39.
Owen, R. J., and Holmes, B. (1978). *FEMS Microbiol. Lett.* 4:41.
Owen, R. J., and Jackman, P. J. H. (1982). *J. Gen. Microbiol.* 128:2945.
Owen, R. J., Legros, R. M., and Lapage, S. P. (1978). *J. Gen. Microbiol.* 104:127.
Oyaizu, H., and Komagata, K. (1981). *J. Gen. Appl. Microbiol.* 27:57.
Oyaizu, H., and Komagata, K. (1983). *J. Gen. Appl. Microbiol.* 29:17.
Palleroni, N. J., and Holmes, B. (1981). *Int. J. Syst. Bacteriol.* 31:479.
Palleroni, N. J., Doudoroff, M., Stanier, R. Y., Solánes, R. E., and Mandel, M. (1970). *J. Gen. Microbiol.* 60:215.
Pickett, M. J., and Greenwood, J. R. (1980). *J. Gen. Microbiol.* 120:439.
Ralston-Barrett, E., Palleroni, N. J., and Doudoroff, M. (1976). *Int. J. Syst. Bacteriol.* 26:421.
Rarick, H. R., Riley, P. S., and Martin, R. (1978). *J. Clin. Microbiol.* 8:313.

Riley, P. S., Hollis, D. G., and Weaver, R. E. (1974). *Appl. Microbiol.* 28:355.
Ruger, H. -J., and Tan, T. L. (1983). *Int. J. Syst. Bacteriol.* 33:85.
Shigeta, S., Hyodo, S., Chonan, E., and Yabuuchi, E. (1983). *J. Clin. Microbiol.* 17:181.
Stanier, R. Y., Palleroni, N. J., and Doudoroff, M. (1966). *J. Gen. Microbiol.* 43:159.
Swings, J., De Vos, P., Van den Mooter, M., and De Ley, J. (1983). *Int. J. Syst. Bacteriol.* 33:409.
Weaver, R. E., Hollis, D. G., Clark, W. A., and Riley, P. (1983). *Revised Tables from the Identification of Unusual Pathogenic Gram Negative Bacteria,* U.S. Department of Health and Human Services, Centers for Disease Control, Atlanta, Ga.
Whitaker, R. J., Byng, G. S., Gherna, R. L., and Jensen, R. A. (1981). *J. Bacteriol.* 145:752.
Wilkson, S. G., and Caudwell, P. F. (1980). *J. Gen. Microbiol.* 118:329.
Wilkinson, S. G., and Galbraith, L. (1979). *Biochem. Biophys. Acta* 575:244.
Yabuuchi, E., Kaneko, T., Yano, I., Moss, C. W., and Miyoshi, N. (1983). *Int. J. Syst. Bacteriol.* 33:580.
Yamasato, K., Akagawa, M., Oishi, N., and Kuraishi, H. (1982). *J. Gen. Appl. Microbiol.* 28:195.

3
Rapid Identification of Glucose-Nonfermenting Gram-Negative Rods with Commercial Miniaturized Kits

THOMAS R. OBERHOFER *Madigan Army Medical Center, Tacoma, Washington*

INTRODUCTION

Conventional biochemical testing of nonfermentative bacteria (NFB) is time-consuming, expensive, and often confusing to the laboratorian, and this mode of identification is beyond the capabilities of many clinical laboratories lacking the necessary expertise. To satisfy the need for a simplified means of identification of the NFB, microsystems both in kit form and as part of automated identification systems have been introduced to the microbiology community. These systems offer an attractive alternative to conventional methodologies through such features as rapid (24—48 hr) and computer-assisted identification, convenience, ease of use, and overall cost effectiveness.

Use of rapid methods and automation in clinical microbiology for the identification of significant bacterial pathogens is firmly established. Whereas conventional methods used to identify the NFB require 48 hr and sometimes 72 or 96 hr of incubation, newer techniques can reduce this time to 48 hr and in many cases to 24 hr. It should be noted, however, that some NFB do not possess striking distinguishing characteristics and yield few specific positive reactions of the high probability necessary for accurate species characterization. As a result, the rapid and automated systems may not include in their substrates or in their computer data bases adequate discriminants to distinguish species composed of similar phenotypic characteristics. This inevitably leads to a need for a wide array of supplemental tests which must be selected for definitive identification. Fortunately, many of these organisms are seen infrequently in the laboratory, and as a rule pose no identification problem.

When choosing a microsystem to identify the NFB, certain factors must be addressed in order to make a wise choice of "systems." The guidelines suggested by Sherris and Ryan (1981) are worth adopting and form, in part, the basis for the following considerations.

Test Accuracy

1. Test results should be consistently accurate at the genus level. Where a firm identification is not given in the code profiles, the system should provide acceptable choices for selection of supplemental tests to give genus identification.
2. Species of particular clinical and epidemiological significance must be identified consistently. Furthermore, frequently enocuntered organisms and those with high recognition values (e.g., *Acinetobacter*) must be correctly identified. Although data vary between institutions, it is generally agreed that *Pseudomonas putida*, *Pseudomonas fluorescens*, *Pseudomonas maltophilia*, *Acinetobacter anitratus*, and *Acinetobacter lwoffi* account for at least 80% of all nonpyocyanogenic isolates of NFB recovered from clinical sources. It is therefore reasonable to expect that these organisms should be identified correctly.
3. Overall correct species identification should approximate 95% or more for the commonly encountered species. For the frequent and some infrequent isolates, supplemental tests should be cited in the system to achieve this accuracy. It is also preferable, and indeed, desirable that the manufacturer of the system selected for use have the courage to state that closely related species cannot be identified with reliability, rather than to invite laborious and unusual testing that can easily lead to misidentification.

Quality Control

A standard panel of NFB representing those strains that will be tested in practice must be selected for each system to ensure that the performance of the system is satisfactory. Each system will have recommended strains, usually American Type Culture Collection designates, that must be used to determine the reactivity of each substrate in the test system.

Effect of the System on the Quality of Results

The accuracy and reproducibility of results obtained with the system must be at least comparable to the system currently in use. Ease of use cannot be substituted for accuracy, and elaborate code profiles are of little value if they cannot be reproduced.

Effect on Laboratory Operation and Personnel

The system must be fully integrated into the laboratory, personnel must be well trained with the system and must accept the system, and storage (space) of the components or substrates must be adequate and comply with the manufacturer's directions.

Cost Effectiveness

Cost of the system relative to other procedures should be evaluated, as well as the cost in time that the system may save when in operation. This evaluation must include cost of reagents, disposable supplies, quality control time and costs, calibration if necessary, and, of course, the cost of labor. It would not be wise to purchase an automated instrument to use just with the NFB; all possible organisms should be tested with the system to justify the cost. Also, if one test system (e.g., the API test strip) can be used for all nonfastidious, gram-negative bacteria, then the system may be an attractive one. If, on the other hand, the system may be slightly more expensive than a competing system or the conventional procedure but provides accurate data 24 or 48 hr earlier than these, then that system may be deemed the more desirable despite the slight additional cost.

Rapidity of Identification

The concept of rapidity is only a relative term. But rapid systems, as applied to identification of the NFB, refer to systems that can identify the organisms quicker than conventional systems without sacrificing accuracy. Whereas 8 to 12 hr for identification may not be rapid for the *Enterobacteriaceae*, since it is no longer a same-day test, 24–36 hr is considered rapid when testing the NFB when contrasted to the 48 hr or more that is usually required for identification. A high degree of accuracy is achieved after 5 days of incubation with conventional systems, but the information is at the expense of speed and may not be relevant to the patient's condition. But rapid identification must be consistent with accurate identification, or else the identification obtained will have little meaning.

Summary

The rapid and automated test systems that are available to the clinical microbiology laboratory have given the laboratory the opportunity to identify the NFB with comparative ease, and have at least standardized the approach to their identification. The miniaturized multimedium kits all include computer-generated interpretation manuals for convenience and standardization of the respective identification systems. Any laboratory, regardless of size, is now capable of identifying most of the NFB by simply selecting the system that most closely fits the needs of the particular laboratory. The systems bring an increase in diagnostic ability to the laboratory and permit less well-trained personnel to provide high quality of test results. This latter is not implied to be a substitute for well-trained workers, however.

Of the many factors to be considered prior to the selection of a microsystem, only accuracy of identification and to some extent rapidity of identification will be addressed. During the ensuing discussion, attention should be given to the species of organisms tested with each system, to the number of strains of a species tested, and to the total number of organisms tested, since each component of the total will influence the overall accuracy of the test system.

AUTOMATED METHODS

Automicrobic System

The Automicrobic System (AMS, Vitek Systems, Inc., Hazelwood, Mo.) provides automated identification of commonly encountered glucose-nonfermenting (NFB) and oxidase-positive fermentative gram-negative bacteria (OPFB) within 8–13 hr. The system utilizes the EBC-plus (Enterobacteriaceae Biochemical Card), a modification of the EBC. The EBC-plus is a sealed, disposal plastic template that contains 28 biochemical substrates. The EBC-plus contains three additional tests beyond the EBC: acetamide, cetrimide, and glucose oxidation tests. A fourth test, a cytochrome oxidase test, must be done separately and the results are recorded at the time of inoculation by blackening a specific circular depression on the card. An inoculum is prepared in the specially designed AMS tubes, and the tubes are inserted into the apparatus. After the manual steps of labeling and inoculum preparation, the AMS proceeds with automated inoculation of the biochemical tests, with reading of the tests, interpretation of the biochemical patterns, and printing of a hard copy containing the final organism identification. The programmed computer determines when a biochemical reaction has taken place, and the results are printed out after the completion of the incubation cycle. This is 8 hr for *P. aeruginosa* and 13 hr for other NFB. If there is insufficient growth in the control well, the tests are not interpreted and "no growth" or "nonviable organism" is recorded.

The AMS system includes two categories of NFB-OPFB: organisms that are included in the AMS data base and organisms that are not included in the data base. The latter represent species which are less frequently isolated in most clinical laboratories and are the most diffucult to identify by conventional means. Failure to include these organisms by species into the data base, however, limits the utility of the system.

Identification

Of the organisms which are included in the data base (Table 1), accuracy of identification is high and has ranged from 89% (Smith et al., 1982), to 99% (Marso et al., 1981). Of those organisms not included in the data base (Table 2), correct identification was scored by the investigators if the organisms were correctly called "unidentified organisms." Correct reporting as "unidentified organisms" varied from 67% (Smith et al., 1982) to 95% (Johnson and Brinkley, 1982). It is evident that the larger the selection of organisms tested, the lesser are the correct responses obtained. The most common problem using the AMS was misidentification of various organisms as *Pseudomonas cepacia* (Johnson and Brinkley, 1982; Smith et al., 1982; Barry et al., 1982a).

Rapid reporting of the NFB-OPFB using the AMS system is 8–13 hr of incubation. Johnson and Brinkely (1982) found that 96% of the test organisms were correctly identified in this time. In another study (Barry and Badal, 1982) attempts to identify 98 of 234 isolates at 4 hr gave an accuracy of only 64%. By the end of 13 hr, all but three strains were identified with an accuracy of 94%. Many NFB simply fail to grow or to produce identifiable test patterns in the first 4–8 hr of incubation and cannot be identified during this time period.

Advantages

Identification of the more common pathogens is available within 13 hr, which is, on the average, 48 hr earlier than with conventional tube tests or with some of the commercially available miniaturized test systems. The automated system gives constancy of test measurement by eliminating operator variations that are common when reading weak reactions. The AMS system requires few if any supplemental tests for definitive identification and can accurately identify the commonly encountered NFB and OPFB that are within the manufacturer's claim for identification. The system is completely automated after the EBC-plus has been inserted into the apparatus and has the least hands-on requirement of the automated systems.

Disadvantages

The AMS system does not attempt to identify, and has excluded from its data base, those organisms which prove difficult to identify using conventional or other commercial systems. The biochemically inactive NFB provide few specific positive reactions of the high probability necessary for accurate species characterization. This obliges the computer to make an identification based on limited information. But it is problematic whether it is more valuable to obtain a questionable identification or an "unidentified organism" result. The uncommon

Table 1 Identification of NFB and OPFB That Are Included in the AMS Data Base

Organism	Johnson and Brinkley (1982)		Smith et al. (1982)		Barry et al. (1982a)		Marso et al. (1981)	
	Number	Percent	Number	Percent	Number	Percent	Number	Percent
Acinetobacter anitratus	20	95	38	71	59	98	48	100
Acinetobacter lwoffi	13	100	12	100	NT[a]		10	100
Pseudomonas aeruginosa	58	100	211	96	231	98	55	100
Pseudomonas fluorescens/ putida	10	70	26	85	16	88	22	86
Pseudomonas maltophilia	10	100	69	77	41	93	53	100
Pseudomonas cepacia	NT		31	94	6	80	6	100
Aeromonas hydrophila	5	100	NT		17	100	25	100
P. shigelloides	2	100	NT		NT		1	100
Vibrio cholerae	2	100	NT		NT		1	100
Vibrio parahaemolyticus	1	100	NT		NT		2	100
Total	121	97	410	89	370	97	223	99

[a]NT, not tested.

Table 2 Identification of NFB and OPFB That Are Not Included in the AMS Data Base

Organism	Johnson and Brinkley (1982)		Smith et al. (1982)		Barry et al. (1982a)		Marso et al. (1981)	
	Number	Percent	Number	Percent	Number	Percent	Number	Percent
Alcaligenes faecalis	1	0	5	80	NT[a]		2	50
Alcaligenes denitrificans	NT		4	75	NT		NT	
Alcaligenes odorans	NT		3	33	NT		1	0
Alcaligenes sp.	NT		NT		17	65	2	100
Bordetella bronchiseptica	2	100	NT		1	100	3	100
Achromobacter xylosoxidans	NT		6	0	6	67	NT	
Achromobacter sp.	4	100	NT		NT		3	0
Pseudomonas alcaligenes	NT		3	33	NT		1	100
Pseudomonas diminuta	NT		6	83	NT		NT	
Pseudomonas paucimobilis	NT		4	50	NT		1	100
Pseudomonas pseudoalcaligenes	NT		5	60	1	100	1	100
Pseudomonas putrefaciens	NT		2	100	NT		1	100
Pseudomonas stutzeri	NT		7	86	6	100	7	100
Pseudomonas vesicularis	NT		NT		NT		1	100

Table 2 (continued)

Organism	Johnson and Brinkley (1979)		Smith et al. (1982)		Barry et al. (1982a)		Marso et al. (1981)	
	Number	Percent	Number	Percent	Number	Percent	Number	Percent
Pseudomonas sp.	9	100	NT		6	100	1	100
CDC Va-1	NT		1	100	NT		NT	
Flavobacterium breve	NT		1	100	NT		NT	
Flavobacterium indologenes	NT		10	70	NT		1	0
Flavobacterium meningosepticum	NT		1	100	NT		NT	
Flavobacterium odoratum	NT		NT		NT		5	80
Flavobacterium sp.	NT		NT		8	100	3	100
Moraxella sp.	2	100	1	100	6	33	3	100
CDC groups	10	90	NT		NT		NT	
Pasteurella sp.	5	100	NT		2	100	NT	
Comamonas terrigena	5	100	3	100	2	100	2	100
Total	38	95	62	67	55	78	38	84

[a]NT, not tested.

isolates were most often misidentified as *P. cepacia*. Since *P. cepacia* is resistant to polymyxin/colistin, the correlation of the AMS identification of *P. cepacia* with susceptibility tests would aid in preventing misidentifications.

Autobac ID System

The Autobac ID System (General Diagnostics, Morris Plains, N.J.) utilizes inhibitory compounds, including antimicrobial agents, to predict the identity of bacterial strains as a result of growth pattern recognition. The compounds are used in the form of impregnated disks. The system consists of five main components: a light-scattering photometer, and incubator—shaker, a data terminal, a disk dispenser, and a 19-chamber cuvette. Eighteen different antimicrobial or substrate-containing disks are dispensed into the cuvette, leaving one chamber free to serve as a control. A standardized inoculum is prepared using the photometer, a sample is placed in special broth, and a prescribed aliquot of the broth inoculum is delivered to each of 19 chambers of the cuvette. The cuvette is placed in the incubator-shaker and incubated for 3 hr, after which the cuvette is placed in the photometer, which reads the cuvette and computes a light-scatter index value for each chamber. If sufficient growth is not initiated in the control chamber, the cuvettes are reincubated and read at hourly intervals for the next 3 hr. The values of each reading are utilized in a two-stage quadratic discriminant analysis to arrive at an identification. The two best identifications are printed along with the probability (P value) that can be given to each identification. Five additional tests are performed and the results are entered into the computerized identification system before the cuvettes are read. These include growth on MacConkey agar, lactose fermentation on MacConkey agar, presence of precipitated bile around colonies on MacConkey agar, spot oxidase test, and swarming on blood agar.

Identification

Tables 3 and 4 show the level of agreement between the Autobac method and standard reference methods for identification of the NFB and OPFB. Of the organisms included in the data base, accuracy of identification is quite high. The Autobac assembles related species into broad groups (e.g., *Flavobacterium* sp.) for which accuracy is established for the group (genus) rather than for individual species. Sielaff et al. (1982) correctly identified 89% of the isolates, Barry et al., (1982c) identified 96%, and Barry et al. (1982b) identified 97%. Of the organisms not included in the data base (Table 4), none of 10 isolates tested by Barry et al. (1982c) were identified and only 13 of 37 (35%) tested by Barry et al. (1982b) were identified.

Table 3 Identification of NFB and OPFB That Are Included in the Autobac Data Base

Organism	Sielaff et al. (1982)		Barry et al. (1982c)		Barry et al. (1982b)	
	Number	Percent	Number	Percent	Number	Percent
Pseudomonas aeruginosa	180	96	308	97	152	97
Pseudomonas putida/fluorescens	173	80	30	93	12	83
Pseudomonas maltophilia	111	93	57	95	32	100
Pseudomonas cepacia	102	81	9	78	4	50
Pseudomonas stutzeri	99	92	6	83	NT[a]	
Pseudomonas sp.	NT		10	70	NT	
Acinetobacter anitratus	197	95	101	97	NT	
Acinetobacter sp.	NT		NT		57	98
Alcaligenes sp.	108	66	17	88	NT	
Aeromonas sp.	123	94	38	100	14	100
Flavobacterium sp.	182	98	13	100	NT	
Moraxella sp.	68	85	4	100	NT	
Total	1343	89	593	96	271	97

[a]NT, not tested.

Table 4 Identification of NFB and OPFB That Are Not Included in the Autobac Data Base

Organism	Barry et al. (1982c)		Barry et al. (1982b)	
	Number	Percent	Number	Percent
Achromobacter xylosoxidans	6	0	6	0
Bordetella bronchiseptica	1	0	1	100
Pseudomonas stutzeri	NT[a]		3	100
Pseudomonas pseudoalcaligenes	NT		1	100
Pseudomonas acidovorans	NT		1	0
Alcaligenes sp.	NT		15	0
Flavobacterium sp.	NT		6	100
Moraxella sp.	NT		2	100
Pasteurella multocida	3	0	2	0
Total	10	0	37	35

[a]NT, not tested.

Advantages

The Autobac ID system is both accurate and reliable for identifying the NFB. The sytem is as sensitive, specific, and precise as the standardized reference methods. The mechanized system, with computer-assisted interpretation, requires a minimum amount of technologist time and provides reliable results within 3–6 hr of incubation.

Disadvantages

The Autobac system demands manual manipulation during the identification cycles as well as the dispensing of the compound disks into the cuvettes. The inability to identify the uncommon strains detracts from the system's utility.

Automated Systems: Summary

The automated systems are very much rapid systems, and the subjectivity of the test result interpretation that is characteristic of the traditional biochemical test procedures is eliminated. The Autobac system identifies more genera of NFB than does the AMS computer program. The Autobac system, in contrast to AMS, can identify *Alcaligenes* species, *Flavobacterium* species, *Moraxella* species, *Pseudomonas stutzeri*, and many other *Pseudomonas* species. Most Autobac identifications are completed after 3 hr, with none requiring more than 6 hr. With the AMS system, the NFB require 13 hr of incubation for proper identification. The AMS, on the other hand, is more reproducible than the Autobac system. Also, the AMS is almost completely automated, whereas the Autobac system involves more manual operations. Neither system is capable of identifying all species of nonfastidious NFB, and both require supplementary tests to confirm the identity of select isolates.

MINIATURIZED NONAUTOMATED METHODS

Oxi/Ferm System

The Oxi/Ferm System (Roche Diagnostics, Nutley, N.J.) is a self-contained assembly of eight compartments in a plastic tube, each containing a specific medium to generate nine test results. The media are agar substrates familiar to the user in the conventional tube systems. An inoculum rod traverses the media, and after an inoculum is gathered at the tip of the rod, the inoculum is pulled through the tube and all eight compartments. The rod is reinserted into the last three compartments and broken off to produce an atmosphere of reduced aeration in these compartments. The tube in incubated at 35° C for 48 hr, at which time Kovacs reagent is introduced to one of the compartments for the indole test. The results of the nine tests plus the oxidase test are coded as a four-digit number called the ID value. The identification manual lists the organism corresponding to the ID value. Where two or more organisms are listed for one value, up to eight supplemental tests are given to differentiate among the different species. The coding manual does not provide data for species identification of *Alcaligenes* sp., *Flavobacterium* sp., and *Moraxella* sp., but identification to the genus level for these organisms is considered correct.

Identification

When considering only those organisms that are included in the manufacturer's list, accuracy of identification to species varies widely between the more common and less common organisms. Table 5 shows the organisms that are considered to be common isolates. The accuracy of identification is high and is close in range, from 88% (Koestenblatt et al., 1982) to 98% (Isenberg and Sampson-Scherer, 1977). Experiences differ using strains of the same organism, since identification of *P. putida* and *P. fluorescens* lies between 0 and 100% identification. Of the less common isolates (Table 6), there is a vide disparity of findings overall, varying from 59% accuracy of identification (Oberhofer, 1983) to 96% (Dowda, 1977). Differences in identification are seen with *Pseudomonas diminuta, Pseudomonas testosteroni, Pseudomonas acidovorans,* and *Alcaligenes* species. Whereas most workers accepted genus identification only, Oberhofer (1983) required species identification for accuracy. When a few species are selected for testing (Dowda, 1977), accuracy is high, but as the number of species increases (Shayegani et al., 1978a; Oberhofer, 1983) a larger proportion of the more difficult to identify organisms become included in the study population. The Oxi/Ferm system shows considerable accuracy in identification at the species level as compared to standard biochemical testing; however, the identification usually depends on supplemental tests as suggested in the numerical coding manual.

The Oxi/Ferm system is a 48 hr test system, although attempts have been made to identify the test strains at 24 hr. Identification accuracy and efficacy at 24 hr ranges from 15% (Koestenblatt et al., 1982) to 22% (Dowda, 1977), and almost all of these isolates are oxidase-negative bacteria. *Pseudomonas aeruginosa* and *A. anitratus* can be identified at 24 hr when the user becomes familiar with the system, but other organisms almost always require 48 hr of incubation. Nadler et al. (1978) showed that 47% of the test organisms were more rapidly identified by the Oxi/Ferm system, with a mean identification time of 2.6 days, in contrast to 3.3 days for the conventional system used.

Requirements for supplemental tests for definitive identification have been reported as 19% (Dowda, 1977), 40% (Koestenblatt et al., 1982), 41% (Oberhofer, 1983), 84% (Isenberg and Sampson-Scherer, 1977), and 85% (Shayegani et al., 1978a). This means that a minimum of 72 hr after isolation is necessary to accurately identify these organisms. The number of supplemental tests per organism has been reported to be 1.7 tests for 42% of the strains, and 1.8 tests for 73% to provide a final identification (Koestenblatt et al., 1982).

Advantages

The advantages of the Oxi/Ferm system are the rapid inoculation (does not require an initial suspension of the test organism), the ease of handling, the reduction of the coding system to common and recurring codes for most organisms, and the small number of tests necessary to provide the minimum number of characteristics required for identification. The system employs conventional media as substrates, and the test reactions are similar to conventional techniques and require little user training. The Oxi/Ferm system is suitable for identification of the commonly encountered NFB such as *P. aeruginosa* and the *Acinetobacter* species.

Table 5 Identification of Commonly Encountered NFB and OPFB Using the Oxi/Ferm System

Organism	Shayegani et al. (1978a)		Nord et al. (1977)	
	Number	Percent	Number	Percent
Pseudomonas aeruginosa	9	100	20	90
Pseudomonas putida	7	0	18	94
Pseudomonas fluorescens	3	0	11	100
Pseudomonas maltophilia	8	88	16	100
Pseudomonas cepacia	3	0	11	73
Acinetobacter anitratus	10	100	NT	
Acinetobacter lwoffi	10	100	NT	
Achromobacter xylosoxidans	17	82	NT	
Aeromonas hydrophila	NT		9	89
Total	67	94	85	92

[a] NT, not tested.

Disadvantages

The disadvantages of the Oxi/Ferm system include the length of incubation (48 hr in most cases, without supplemental tests), and the high frequency of supplemental tests required for accurate identification. There is difficulty in discerning subtle changes in the xylose and citrate tests (Oberhofer et al., 1977). The Oxi/Ferm tubes require refrigeration for storage and they easily dry out if the container is left uncovered. The system fails to meet the needs of the reference laboratory because it fails to identify the organisms that are usually received by these laboratories (Shayegani et al., 1978a). Overall, the code manual must be updated to reflect current methodologies.

API 20E

The API 20E (Analytab Products, Inc., Plainview, N.Y.) system is a standardized, ready-to-use microtube system designed to identify the fermentative bacteria, but adapted to identify the NFB as well.

Koestenblatt et al. (1982)		Dowda (1977)		Isenberg and Sampson-Scherer (1977)		Oberhofer (1983)	
Number	Percent	Number	Percent	Number	Percent	Number	Percent
5	100	44	98	65	100	151	99
8	88	NT		12	100	63	64
NT		14	100	10	100	42	95
15	87	23	100	6	100	71	72
3	67	6	67	1	100	27	100
5	100	25	100	40	100	144	99
8	100	7	100	42	100	65	100
3	33	9	56	8	63	29	86
2	100	6	100	6	83	30	97
49	88	134	95	190	98	622	95

The system consists of a plastic strip with microtubes containing dehydrated substances for the performance of 23 standard biochemical tests. The oxidase test is performed separately. A light inoculum is prepared in saline and the substrates are reconstituted by adding the bacterial suspension to the microtubes. The system is incubated at 35°C for 24 to 48 hr, as necessary. After 24 hr of incubation, the API strip is examined for reactivity in three of the microtubes. These reactions plus the oxidase test are sufficient for an identification to be attempted. Systems failing to give three reactions are reincubated for an additional 24 hr. The reactions are encoded and identification is obtained from the appropriate sections of the profile code book. The API system is the only one that contains a profile for results obtained at 24 hr as well as at 48 hr.

Identification

The accuracy of identification of the most common organisms using the API system (Table 7) ranges from 65% (Shayegani et al., 1978b) to 99% (Appelbaum et al., 1980). Discrepancies between different

Table 6 Identification of Infrequently Encountered NFB and OPFB Using the Oxi/Ferm System

Organism	Shayegani et al. (1978a)		Nord et al. (1977)	
	Number	Percent	Number	Percent
Alcaligenes denitrificans	5	60	4	75
Alcaligenes faecalis	5	40	26	100
Alcaligenes odorans	4	50	NT	
Bordetella bronchiseptica	7	100	NT	
CDC Vd group	NT		NT	
Pseudomonas diminuta	7	100	6	100
Pseudomonas pseudoalcaligenes	9	0	3	100
Pseudomonas stutzeri	4	75	18	83
Pseudomonas testosteroni	7	14	3	100
Pseudomonas paucimobilis	7	57	NT	
Pseudomonas acidovorans	5	20	3	100
Pseudomonas putrefaciens	NT		6	83
Pseudomonas pickettii	NT		NT	
Pseudomonas alcaligenes	NT		NT	
Pseudomonas vesicularis	NT		NT	
CDC Va group	NT		NT	
CDC Ve group	7	43	NT	
Moraxella sp.	4	100	NT	
Flavobacterium indologenes	4	25	NT	
Flavobacterium odoratum	8	38	NT	
Flavobacterium meningosepticum	NT		NT	
Flavobacterium multivorum	NT		NT	
Flavobacterium IIf	NT		NT	
Pasteurella multocida	NT		NT	
Total	83	60	69	93

[a]NT, not tested.

Koestenblatt et al. (1982)		Dowda (1977)		Isenberg and Sampson-Scherer (1977)		Oberhofer (1983)	
Number	Percent	Number	Percent	Number	Percent	Number	Percent
NT[a]		NT		3	67	7	0
5	80	4	100	7	71	12	75
3	100	NT		9	89	13	0
NT		2	100	NT		8	100
NT		NT		NT		7	29
NT		NT		11	100	8	0
NT		NT		3	100	19	0
6	67	NT		13	93	27	56
1	100	NT		6	100	9	0
2	100	NT		NT		11	91
2	100	NT		6	100	13	0
3	100	NT		2	100	5	100
1	100	NT		NT		6	0
NT		NT		7	100	4	0
NT		NT		NT		2	0
2	100	NT		NT		5	60
3	100	3	100	NT		17	100
2	100	11	100	3	100	34	100
4	100	4	100	5	100	37	84
2	50	NT		NT		4	0
1	100	NT		NT		7	14
NT		NT		NT		8	75
1	0	3	100	NT		12	100
NT		4	50	NT		43	74
38	87	31	94	75	93	318	59

Table 7 Identification of Commonly Encountered NFB and OPFB Using the API 20E System

Organism	Nord et al. (1977)		Dowda (1977)		Shayegani et al. (1978a)		Appelbaum et al. (1980)		Warwood et al. (1979)		Burdash et al. (1980)		Oberhofer (1983)	
	Number	Percent	Number	Percent	Number	Percent	Number	Percent	Number	Percent	Number	Percent	Number	Percent
Pseudomonas aeruginosa	20	85	44	91	9	100	53	100	52	77	73	99	112	73
Pseudomonas fluorescens	11	91	14	93	3	0	11	91	4	50	5	80	28	36
Pseudomonas putida	18	62	NT[a]		6	83	7	100	17	77	13	100	39	85
Pseudomonas maltophilia	16	100	23	100	8	88	34	100	28	93	18	94	58	86
Pseudomonas cepacia	11	73	6	50	2	100	12	100	13	46	9	100	23	65
Acinetobacter anitratus	NT		25	92	10	90	26	100	31	90	28	100	152	84
Acinetobacter lwoffi	NT		7	100	10	90	16	100	NT		4	50	45	89
Achromobacter xylosoxidans	NT		9	100	17	6	NT		8	88	6	83	22	50
Aeromonas hydrophila	9	89	6	100	NT		NT		NT		NT		13	54
Total	85	82	134	93	65	65	158	99	153	80	156	96	492	76

[a]NT, not tested

reports are apparent, especially with *P. fluorescens* and *Achromobacter xylosoxidans*. The effect of the population of organisms selected for testing on the outcome of testing is clear with the less common strains (Table 8). Accuracy of identification ranges from very good (99%, Appelbaum et al., 1980) to poor (19%, Nord et al., 1977; 33%, Oberhofer, 1983). Wide-ranging differences are also seen with different groups of organisms, such as the *Alcaligenes* species, *P. acidovorans*, *P. diminuta,* and *Pseudomonas pseudoalcaligenes,* where identification varies from 0 to 100%. Moreover, some organisms such as the Va/*Pseudomonas pickettii* group cannot be identified at all.

The need for supplemental tests also is all embracing. For all organisms tested, extra tests were required in 23% of instances (Warwood et al., 1979), in 49% (Burdash et al., 1980), and in 54% (Shayegani et al., 1978b). One study showed the requirement to be mostly with the oxidase-positive organisms (28%) rather than with the oxidase-negative organisms (4%, Oberhofer, 1983).

Advantages

The main advantage of the API system is that the same system that is used to identify the Enterobacteriaceae also can be used for the NFB. This adds to supply economy. Furthermore, additional tests are not necessary to determine the glucose fermentation status of the oxidase-negative organisms before identification attempts are made. Of all of the microsystems, the API system requires the fewest supplemental tests for definitive identification of the NFB.

Disadvantages

Most isolates require 48 hr of incubation when using the API system. Indeed, only 13% (Dowda, 1977) and 29% (Appelbaum et al., 1980) of the strains tested could be identified at 24 hr. Burdash et al. (1980) found that API identified 26% of the isolates within 24 hr, and 58% of these were oxidase-negative strains. Whereas oxidase-negative organisms are reliably identified, problems are encountered with the various oxidase-positive organisms, especially the nonglucolytic strains. From a technical standpoint, the zinc test in the system gives erroneous results (Oberhofer, 1979) and must be performed using zinc pellets rather than zinc dust to confirm the nitrite reactions (Appelbaum et al., 1980; Oberhofer, 1983).

Minitek System

The Minitek System (BBL Microbiology System, Cockeysville, Md.) consists of a plastic plate containing 20 wells to hold dried disks impregnated with the appropriate biochemicals and substrates selected

Table 8 Identification of Infrequently Encountered NFB and OPFB Using the API 20E System

Organism	Nord et al. (1977)		Dowda (1977)		Shayegani et al. (1978b)	
	Number	Percent	Number	Percent	Number	Percent
Alcaligenes denitrificans	4	0	NT[a]		5	20
Alcaligenes faecalis	26	0	4	100	5	80
Alcaligenes odorans	NT		NT		4	25
Bordetella bronchiseptica	NT		2	100	7	100
CDC IVc	NT		NT		6	17
CDC Vd group	NT		8	38	NT	
Pseudomonas acidovorans	3	0	NT		6	17
Pseudomonas diminuta	6	0	NT		7	29
Pseudomonas putrefaciens	6	67	NT		3	100
Pseudomonas pseudoalcaligenes	3	0	NT		9	11
Pseudomonas stutzeri	18	50	NT		3	100
Pseudomonas testosteroni	3	0	NT		7	43
Pseudomonas alcaligenes	NT		NT		3	67
Pseudomonas paucimobilis	NT		NT		5	40
Pseudomonas pickettii	NT		NT		1	0
Pseudomonas vesicularis	NT		NT		NT	
CDC Va-1	NT		NT		2	0
CDC Ve group	NT		3	0	5	80
Moraxella sp.	NT		11	45	21	71
Flavobacterium indologenes	NT		4	100	3	0
Flavobacterium odoratum	NT		NT		NT	
Flavobacterium meningosepticum	NT		NT		3	33
Flavobacterium multivorum	NT		NT		NT	
Flavobacterium IIf	NT		3	67	4	0
Pasteurella multocida	NT		4	75	NT	
Total	69	19	39	59	109	47

[a]NT, not tested.

Appelbaum et al. (1980)		Warwood et al. (1979)		Burdash et al. (1980)		Oberhofer (1983)	
Number	Percent	Number	Percent	Number	Percent	Number	Percent
12	92	1	0	NT		7	0
12	100	1	0	NT		5	20
14	100	6	17	NT		7	0
NT		6	100	6	100	2	100
NT		2	100	2	0	4	25
NT		NT		NT		7	0
NT		6	0	6	100	11	9
NT		2	0	6	100	1	0
NT		1	100	3	100	3	100
NT		NT		5	100	9	0
NT		6	83	8	100	12	58
NT		NT		NT		4	0
NT		2	0	6	0	2	0
NT		NT		2	50	14	79
NT		NT		NT		7	0
NT		NT		NT		2	0
NT		NT		10	0	5	0
NT		1	0	3	100	8	50
24	100	18	44	7	43	23	65
NT		8	63	NT		23	22
NT		NT		3	67	3	33
9	100	NT		2	100	3	0
NT		NT		2	100	4	0
NT		1	0	NT		13	23
NT		NT		NT		32	56
71	99	61	46	71	66	211	33

to identify the NFB. A heavy suspension of the test organism is made in the manufacturer-formulated inoculum broth, which is then dispensed into the disk-containing wells with a pipette. Two drops of the inoculum broth are dispensed into an empty well for performance of the indole test, or the vial of the remaining inoculum broth is incubated and used for the indole test. The test system is incubated for 48 hr, after which the reactions are recorded and interpreted using the profile code book and accompanying tables.

Identification

The common isolates encountered in the laboratory are identified with a high degree of accuracy (Table 9). Correctness of identifications range from 89% (Wellstood-Nuesse, 1979) to 100% (Appelbaum et al., 1980). There is a great disparity, however, in identification of the uncommon organisms (Table 10), and this is due to the selection of test organisms, as well as the demand for accuracy by the various workers. Appelbaum et al. (1980) identified 100% of the strains tested, although only five species made up the battery of organisms and the *Alcaligenes* were not speciated but were identified to genus only. On the other hand, Oberhofer (1983) and Chester and Cleary (1980) tested a comprehensive collection of organisms. Oberhofer (1983) chose identification to species as the criterion for identification of the *Alcaligenes*, whereas Chester and Cleary (1980) chose genus identification as their criterion.

Similar to other test systems, the Minitek System required supplemental tests for final identification. One or more supplemental biochemical tests were needed for 15% (Appelbaum et al., 1980), 22% (Oberhofer, 1983), 37% (Burdash et al., 1980), 49% (Chester and Cleary, 1980), and 61% of the isolates (Wellstood-Nuesse, 1979). In almost all of the studies, at least 48 hr of incubation, excluding supplemental tests, were required for identifications to be complete.

Advantages

The Minitek disks have a shelf life of at least 2 years when stored at 4°C. The system itself is flexible whereby the disk substrates may be selected and varied to meet the needs of the individual laboratory. Moreover, supplemental tests such as esculin can be anticipated to aid in the identification of *P. maltophilia* or the flavobacteria and can be preselected for inclusion into the original setup. In contrast to tubed media, the Minitek system takes less time to set up, and the system is at least as rapid as conventional test methods.

Disadvantages

Difficulty is often experienced with the nitrate test, specifically false negative tests for nitrogen gas production (Appelbaum et al.,

Table 9 Identification of Frequently Encountered NFB and OPFB Using the Minitek System

Organism	Appelbaum et al. (1980)		Wellstood-Nuesse (1979)		Chester and Cleary (1980)		Oberhofer (1983)	
	Number	Percent	Number	Percent	Number	Percent	Number	Percent
Pseudomonas aeruginosa	52	100	20	80	29	90	111	82
Pseudomonas fluorescens	11	100	3	100	24	100	46	74
Pseudomonas putida	7	100	8	75	33	100	60	92
Pseudomonas maltophilia	34	100	21	86	33	100	90	89
Pseudomonas cepacia	12	100	5	40	10	100	15	80
Acinetobacter anitratus	41	100	36	100	51	100	148	99
Acinetobacter lwoffi	30	100	14	100	14	100	50	82
Achromobacter xylosoxidans	NT[a]		13	85	21	100	30	97
Aeromonas hydrophila	NT		13	92	15	93	22	73
Total	187	100	133	89	230	98	572	90

[a]NT, not tested.

Table 10 Identification of Infrequently Encountered NFB and OPFB Using the Minitek System

Organism	Appelbaum et al. (1980)		Wellstood-Nuesse (1979)		Chester and Cleary (1980)		Oberhofer (1983)	
	Number	Percent	Number	Percent	Number	Percent	Number	Percent
Alcaligenes faecalis	12	100	18	100	11	100	6	0
Alcaligenes odorans	14	100	5	100	22	100	12	0
Alcaligenes denitrificans	12	100	NT[a]		2	0	7	0
Bordetella bronchiseptica	NT		2	100	10	100	4	100
CDC IVc	NT		5	80	6	100	6	100
CDC Vd	NT		NT		34	91	9	44
Pseudomonas stutzeri	NT		3	33	23	91	20	80
Pseudomonas acidovorans	NT		3	33	12	100	12	0
Pseudomonas testosteroni	NT		3	100	8	100	5	0
Pseudomonas pseudoalcaligenes	NT		6	100	20	95	9	89
Pseudomonas putrefaciens	NT		3	100	16	100	3	100

Organism								
Pseudomonas alcaligenes	NT		5	100	7	100	3	0
Pseudomonas pickettii	NT		NT		6	17	9	0
Pseudomonas paucimobilis	NT		1	0	8	100	12	0
Pseudomonas diminuta	NT		NT		11	100	2	0
Pseudomonas vesicularis	NT		NT		10	100	6	33
CDC Ve group	NT		NT		20	100	5	100
CDC Va group	NT		3	100	8	58	7	0
Moraxella sp.	24	100	5	100	18	67	28	96
Flavobacterium indologenes	NT		4	75	12	100	35	74
Flavobacterium meningosepticum	9	100	2	100	13	92	3	0
Flavobacterium odoratum	NT		NT		26	85	7	86
Flavobacterium multivorum	NT		NT		11	100	8	75
Flavobacterium IIf	NT		NT		10	60	30	43
Pasteurella multocida	NT		2	100	12	92	48	21
Total	71	100	70	90	336	90	296	46

[a] NT, not tested.

1980). Also isolates of *Flavobacterium indologenes* (*Flavobacterium* IIb) do not give consistent results for indole production or dextrose utilization. Carbohydrate oxidation reactions for *Flavobacterium meningosepticum* and *Pasteurella multocida* are difficult to interpret (Wellstood-Nuesse, 1979; Oberhofer, 1983) because the color changes in the disks are very subtle and often undistinguishable from the nonoxidized substrates.

Flow N/F System

The Flow N/F System (Flow Laboratories, Inc., Roslyn, N.Y.) consists of two screening tubes (GNF and 42P) and a circular plate (Uni-N/F-Tek) that is divided into 12 sections. The GNF tube is constricted to detect glucose fermentation or denitrification in the base of the agar below the constriction, and fluorescein production on the slant of the agar. The second tube, 42P, detects growth at 42°C and pyocyanin production at that emperature. The two-tube set is inoculated only with oxidase-positive organisms and is used as a screen for the expedient identification of *P. aeruginosa*. The Uni-N/F-Tek plate consists of 11 independently sealed peripheral wells and a central unsealed well. Each well contains an agar medium for the determination of 12 biochemical reactions. The plate is used to identify oxidase-negative organisms and the less common oxidase-positive organisms that are not identified by the two-tube screen. Each well is inoculated with a heavy saline suspension of the test organism. The GNF and 42P tubes are incubated overnight at 35 and 42°C, respectively, and the Uni-N/F-Tek plate at 35°C for 24 or 48 hr. The results of the N/F screen are used to construct a two-digit profile, and the profile screen manual is used for identification. If the Uni-N/F-Tek is used, a profile number is generated for use with the profile screen manual.

Identification

Tables 11 and 12 show the results of the N/F tests using the common and less common organisms, respectively. Identification accuracy of the N/F system with the common isolates ranges from 88% (Warwood et al., 1979) to 100% (Appelbaum et al., 1980). Ranges of accuracy with the less common organisms are 66% (Warwood et al., 1979) to 98% (Koestenblatt et al., 1982).

A total of 90–98% of *P. aeruginosa* are identified in 24 hr with the two-tube screen (Appelbaum et al., 1980; Barnishan and Ayers, 1979; Burdash et al., 1980) and 97% are identified at 48 hr, whereas only 18–19 and 35–45% of *P. fluorescens* and *P. putida,* respectively, are identified during this time (Appelbaum et al., 1980; Barnishan and Ayers, 1979). Of the oxidative organisms tested, 95% are identified at 24 hr, whereas only 26% of the weakly oxidative or nonoxidative

Table 11 Identification of Commonly Encountered NFB Using the Flow N/F System

Organism	Warwood et al. (1979)		Koestenblatt et al. (1982)		Burdash et al. (1980)		Appelbaum et al. (1980)	
	Number	Percent	Number	Percent	Number	Percent	Number	Percent
Pseudomonas aeruginosa	52	90	5	100	73	99	52	100
Pseudomonas putida	17	100	8	100	13	77	7	100
Pseudomonas fluorescens	4	0	NT[a]		5	100	11	100
Pseudomonas maltophilia	28	96	15	100	18	100	34	100
Pseudomonas cepacia	13	54	3	100	9	100	12	100
Acinetobacter anitratus	31	97	5	100	28	100	41	100
Acinetobacter lwoffi	NT		8	100	4	75	30	100
Achromobacter xylosoxi-dans	8	75	3	67	6	83	NT	
Total	153	88	47	98	156	96	187	100

[a]NT, not tested.

Table 12 Identification of Infrequently Encountered NFB and OPFB Using the Flow N/F System

Organism	Warwood et al. (1979)		Koestenblatt et al. (1982)		Burdash et al. (1980)	
	Number	Percent	Number	Percent	Number	Percent
Alcaligenes denitrificans	1	100	NT[a]		NT	
Alcaligenes faecalis	1	100	5	100	NT	
Alcaligenes odorans	6	100	3	100	NT	
Alcaligenes sp.	5	80	NT		8	75
Bordetella bronchiseptica	6	100	NT		6	100
CDC IVc	2	0	NT		2	50
Pseudomonas paucimobilis	1	100	2	100	2	100
Pseudomonas acidovorans	6	33	2	100	6	100
Pseudomonas alcaligenes	2	0	NT		6	0
Pseudomonas diminuta	2	100	NT		6	100
Pseudomonas putrefaciens	1	100	3	100	3	100

Organism					
Pseudomonas stutzeri	6	83	6	100	8	88
Pseudomonas pickettii	NT		2	100	NT	
Pseudomonas testosteroni	NT		1	100	NT	
Pseudomonas pseudoalcaligenes	NT		NT		5	40
Pseudomonas sp.	5	100	NT		NT	
CDC Va group	NT		1	100	10	100
CDC Ve group	1	0	3	100	3	100
Moraxella sp.	18	44	2	100	7	71
Flavobacterium indologenes	8	75	4	75	5	80
Flavobacterium odoratum	1	100	2	100	3	100
Flavobacterium meningosepticum	NT		1	100	2	100
Flavobacterium multivorum	NT		NT		2	100
Flavobacterium IIf	1	0	1	100	NT	
Aeromonas hydrophila	NT		2	100	NT	
Total	73	66	40	98	84	81

[a]NT, not tested.

organisms are so identified (Barnishan and Ayers, 1979). It should be pointed out that this is 24 hr after the results of the two-tube system are obtained. Koestenblatt et al. (1982) correctly identified 85% of the isolates within 48 hr and 98% in 72 hr of incubation.

The N/F system requires supplemental tests for definitive identification with 18% (Koestenblatt et al., 1982) to 25% of the organisms tested (Warwood et al., 1979). In a comparative study Burdash et al. (1980) found the N/F system to require the fewest supplemental tests (18%) versus 37% for Minitek, 49% for API and 67% for Oxi/Ferm.

Advantages

One advantage of the N/F system is that the two-tube test allows rapid and economical identification of P. aeruginosa within 24 hr. Also, oxidase-negative organisms are identified rapidly with the plate system, and nonfluorescent oxidase-positive organisms are reliably identified after 24 hr following the initial two-tube screen. The Uni-N/F-Tek system employs conventional media. Reactions are similar to conventional techniques and require little user training. Few supplemental tests are needed for definitive identification.

Disadvantages

The two-tube system is designed to identify a select group of organisms. In their absence, additional tests or the Uni-N/F-Tek plate must be used, adding to the cost and to a delay in identification. The Uni-N/F-Tek plate requires refrigeration for storage and may dry out if left uncovered. Problems have been encountered in reading gas production in the GNF tube (Warwood et al., 1979), because the manufacturer recommends that any bubble be scored as a positive test (Koestenblatt et al., 1982), which leads to misidentification if only one or two bubbles are present as a result of nonspecific causes.

Other Systems

The MicroScan system (MicroScan Inc., Campbell, Calif.) is a relatively new introduction to the field of nonfermentative bacteriology, allowing for only one evaluation at this time. The MicroScan system contains a set of 24 miniaturized identification tests combined with a susceptibility system in a microtiter plate. The system comes frozen from the manufacturer and is thawed for use. The system is inoculated with approximately 10^5 colony-forming units per well and is incubated at 35°C. Positive reactions are recorded at 24 hr, and at 48 hr reagents are added to the tests requiring them. Positive tests are used to code a profile number for each isolate and, using the MicroScan data base, the identity of the isolate is determined. Mathewson et al. (1983) tested 182 veterinary isolates and found that the MicroScan Urinary Combo Panel correctly identified 72% of these. Of 118 strains of the three most common species of NFB isolated from animals (*P. aeruginosa, A. anitratus* and *Bordetella bronchiseptica*), the system correctly identified 86%. The system had the most difficulty identifying the biochemically inert NFB.

The most recent introduction is the Rapid NFT test by DMS Laboratories, Inc., Flemington, N.J. The Rapid NFT kit provides a standardized micromethod combining 20 biochemical and assimilation tests for the identification of the NFB and OPFB. The test system is in a strip form similar to the API system. The manufacturer claims 24-hr identification with many isolates of the NFB. Preliminary evaluations have begun, but no definitive report on the usefulness or accuracy of the Rapid NFT has appeared in the literature.

REFERENCES

Appelbaum, P. C., Stavitz, J., Bentz, M. S., and von Kuster, L. C. (1980). *J. Clin. Microbiol.* 12:271.
Barnishan, J., and Ayers, L. W. (1979). *J. Clin. Microbiol.* 9:239.
Barry, A. L., and Badal, R. E. (1982). *J. Clin. Microbiol.* 9:239.
Barry, A. L., Gavan, T. L., Badal, R. E., and Telenson, M. J. (1982a). *J. Clin. Microbiol.* 15:582.
Barry, A. L., Gavan, T. L., Badal, R. E., and Telenson, M. J. (1982b). *Am. J. Clin. Pathol.* 78:462.
Barry, A. L., Gavan, T. L., Smith, P. B., Matsen, J. M., Morrello, J. A., and Sielaff, B. H. (1982c). *J. Clin. Microbiol.* 15:1111.
Burdash, N. M., Bannister, E. R., Manos, J. P., and West, M. E. (1980). *Am. J. Clin. Pathol.* 73:564.
Chester, B., and Cleary, T. J. (1980). *J. Clin. Microbiol.* 12:509.
Dowda, H. (1977). *J. Clin. Microbiol.* 6:605.
Isenberg, H. D., and Sampson-Scherer, J. (1977). *J. Clin Microbiol.* 5:336.
Johnson, J. E., and Brinkley, A. W. (1982). *J. Clin. Microbiol.* 15:25.
Koestenblatt, E. K., Larone, D. H., and Pavletich, K. J. (1982) *J. Clin. Microbiol.* 15:384.
Marso, E., Hamada, S. S., and Martin, W. J. (1981). In *Rapid Methods and Automation in Microbiology* (R. C. Tilton, ed.), American Society for Microbiology, Washington, D.C., p. 90.
Mathewson, J. J., Simpson, R. B., and Brooks, F. L. (1983). *J. Clin. Microbiol.* 17:139.
Nadler, H., George, H., and Barr, J. (1978). *J. Clin. Microbiol.* 9:180.
Nord, C., Wretlind, B., and Dahlback, A. (1977). *Med. Microbiol. Immunol.* 163:93.
Oberhofer, T. R. (1979). *J. Clin. Microbiol.* 9:220.

Oberhofer, T. R. (1983). *Diagn. Microbiol. Infect. Dis.* 1:241.
Oberhofer, T. R., Rowen, J. W., Cunningham, G. F., and Higbee, J. W. (1977). *J. Clin. Microbiol.* 6:559.
Shayegani, M., Lee, A. M., and McGlynn, D. M. (1978a). *J. Clin. Microbiol.* 7:533.
Shayegani, M., Maupin, P. S., and McGlynn, D. M. (1978b). *J. Clin. Microbiol.* 7:539.
Sherris, J. C., and Ryan, K. J. (1981). In *Rapid Methods and Automation in Microbiology* (R. C. Tilton, ed.), American Society for Microbiology, Washington, D.C., p. 1.
Sielaff, B. H., Matsen, J. M., and McKie, J. E. (1982). *J. Clin. Microbiol.* 15:1103.
Smith, S. M., Cundy, K. R., Gilardi, G. L., and Wong, W. (1982). *J. Clin. Microbiol.* 15:302.
Warwood, N. M., Blazevic, D. J., and Hofherr, L. (1979). *J. Clin. Microbiol.* 10:175.
Wellstood-Nuesse, S. (1979). *J. Clin. Microbiol.* 9:511.

4
Ecology, Clinical Significance, and Antimicrobial Susceptibility of *Pseudomonas aeruginosa*

HAROLD C. NEU *Columbia University College of Physicians and Surgeons and Columbia-Presbyterian Medical Center, New York, New York*

INTRODUCTION

Pseudomonas aeruginosa is an organism that is ubiquitous in nature. It has been the cause of serious illness in man for several hundred years, but only in the past two decades have physicians appreciated the serious nature of infections caused by this organism. *Pseudomonas aeruginosa* infrequently causes infection in normal individuals unless they have suffered major trauma or burns. The organism is an important cause of respiratory, cutaneous, and disseminated infections in individuals who have defective host defenses of cutaneous barriers, granulocytes, or immunoglobulins. In the past decade progress has been made in our understanding of how *Pseudomonas* causes infection in man and in understanding its pathogenic properties. Although a large number of antimicrobial agents have been developed to treat infections due to this organism, it has become increasingly apparent that *P. aeruginosa* has the potential to become resistant to antimicrobial agents of every class, be they aminoglycoside antibiotics, anti-*Pseudomonas* penicillins, or drugs of the cephalosporin or carbapenem groups (Neu, 1982d). Attempts to develop efficient vaccine programs to protect susceptible individuals against life-threatening infections due to this organism have not, thus far, been highly successful. Furthermore, *P. aeruginosa* has become an increasingly serious problem in patients with cystic fibrosis, as many of these individuals live to older age. This chapter will discuss the distribution of this organism in nature, its pathogenic properties, antimicrobial susceptibility, and the clinical infections due to *Pseudomonas* currently being seen.

DISTRIBUTION IN NATURE

Pseudomonas aeruginosa has been found in many surface waters and soil and it infects a number of common plants (Schroth et. al., 1977). It is particularly frequent in warm waters in which there has been human or animal activity. Domestic waters act as a source of dissemination of the organism to man (Verstrate et al., 1975). *Pseudomonas* will contaminate chlorinated water even though chlorine normally retards the growth of *Pseudomonas*. Because of the unique biochemical ability of *Pseudomonas*, this organism is able to grow in distilled water, even though it has originally grown in an enriched medium (Favero et al., 1971). Fruits, vegetables and ornamental plants can be a source of *P. aeruginosa*. Indeed, ornamental plants and flowers are a much more common source of *Pseudomonas* than are foodstuffs (Kominos, et al., 1972). Leafy vegatables and carrots frequently are contaminated with *Pseudomonas*, and salads may contain 10^3-10^4 *P. aeruginosa* per gram, depending upon methods of preparation (Kominos et al., 1977).

There are many sources of *Pseudomonas* in the hospital (Petras and Bognar, 1981). Solutions, body lotion, faucets, sinks, inhalation and resuscitation equipment, and body secretions have all been shown to be transitory vectors of *Pseudomonas* (Table 1). Ophthalmological solutions, soaps, hand creams, and aqueous quaternary ammonium disinfectant solutions have been the source of serious hospital infections. Nebulizers, including the ultrasonic forms, and

Table 1 Sources of *Pseudomonas aeruginosa* in Hospitals

Soaps, solutions	Catheters
Antiseptic creams	Sponges, mops
Ophthalmological solutions	Iodine solutions
Sinks: traps	Tracheotomies
Water faucets: aerators	Biliary drainage fluid
Incubators	Catheter bags
Inhalation equipment	Personnel: hands
Resuscitation equipment	Fecal material
Flowers	Burn tissue from patients
Hospital food	

humidifiers have been a means of disseminating *Pseudomonas* in the hospital environment. Indeed, hospitals appear to be the major reservoir of *Pseudomonas*, contributing to the spread of this organism into the environment and the persistence of the organism in nature. Streams or rivers into which hospital waste has been dischared have much higher counts of *Pseudomonas* than of common coliforms.

Certain patient groups are a major source of *Pseudomonas* in the hsopital. For example, intensive therapy units for respiratory surgical patients often contain large amounts of *Pseudomonas* on the sinks and on the hands of personnel. Tracheotomized patients treated with various antibiotics will become colonized with *Pseudomonas* and be a source of contamination to other patients during suctioning procedures (Lowbury et al., 1970; Noone et al., 1983). Sputum from patients with cystic fibrosis may be a source of contamination within the hospital environment (Zimakoff et al., 1983). But most studies have failed to show that these highly mucoid forms of *Pseudomonas* cause infection in other patients. Solutions used for rinsing endotracheal suction equipment can spread *Pseudomonas* to other individuals, as can urinary catheter bags from patients who have *Pseudomonas* urinary tract infections or the devices used to measure urine (Marrie et al., 1978). Since *Pseudomonas* frequently colonizes biliary tract fluid in patients treated with cephalosporin antibiotics, this fluid also may be a source of contamination of the hands of attendants and thereby be transferred to other patients.

The intestinal carriage of *Pseudomonas* ranges from 1% to as much as 75% of hospitalized patients (Stoodley and Thom, 1970; Sutter et al., 1967; Grogan, 1966; Buck and Cooke, 1969). Normal individuals rarely have the organism as part of their fecal flora unless they have been treated with antimicrobial agents (Table 2), however, patients who have been hospitalized for gastrointestinal surgery, leukemia, or burns are rapidly colonized with *Pseudomonas*, and the gastrointestinal prevalence of *Pseudomonas* in these populations ranges from 25 to 45%. Schimpff et al. (1970) cultured *P. aeruginosa* from at least one site in 30% of patients with cancer. Colonization was greatest in patients with leukemia, and 21% of the patients colonized developed bacteremia, compared to only 7% who were not colonized. Bodey (1970) found that colonization with *Pseudomonas* increased in patients with leukemia from 25% on admission to 47% after 4 weeks of hospitalization. Schimpff et al. (1973) also found that 46% of patients with acute leukemia became colonized. The colons of cadavers have been found to be colonized with *Pseudomonas* in 23% of examined individuals. Individuals receiving chronic suppressive antimicrobial drugs, such as young women with recurrent urinary tract infections, will have *P. aeruginosa* as part of their fecal flora.

Table 2 Gastrointestinal Levels of *Pseudomonas aeruginosa*

	Percent carrying *Pseudomonas*
Normal individuals	3—12
Burn patients on admission to hospital	5
Burn patients after 3 weeks' hospitalization	25—50
Surgical patients receiving antibiotics in hospital	12—38
Leukemia patients on admission to hospital	17—25
Leukemia patients after 2—4 weeks' hospitalization	41—47
Autopsy cases	36
Newborn	0
Ileostomy patients	73

RISK OF INFECTION

It is difficult to determine whether there has been a major increase in *Pseudomonas* infections in the last decade, since it was not until 1975 that *P. aeruginosa* was reported as a separate species in the national nosocomial infection study (Allen et al., 1981; Stamm et al., 1981). This is a voluntary nationwide surveillance system for nosocomial infection that was initiated by the Centers for Disease Control in 1970. There are approximately 80 hospitals of various sizes, geographical locations, and categories that participate in the study, providing data on over 1 million hospitalized patients annually (Allen et al., 1981). It is important to note that during 1976—1980 *P. aeruginosa* increased as a cause of hospital infections, so that by 1978 it was second only to *Escherichia coli* as a cause of nosocomial infection and accounted for 8—9% of nosocomial infections at all sites (Cross et al., 1983). Among the various forms of infection produced by different bacteria, 9.5% of lower respiratory tract infections were caused by *P. aeruginosa*, nearly as many as the 10.7% due to *Klebsiella*. Our group (Gross et al., 1980) has demonstrated that nosocomially acquired respiratory infections are a major cause of death due to nosocomial infections. *Pseudomonas aeruginosa* also was the second most common cause of nosocomial urinary infections in 1980, 11.5%, compared to 31.7% due to *E. coli*. What was of particular interest was that 5.8% of primary bacteremias were caused by *Pseudomonas*, in comparison to 10.6% due to *Klebsiella* and 13.6% due to *E. coli*. Analysis of infection rates due to *Pseudomonas* showed that it is not possible to accurately assess the impact of *P. aeruginosa* in a given institution on the basis of literature reports, since wide variations may occur within an individual institution. For example, they noted that in a referral center in the western United States the total number of isolates of *P. aeruginosa* from nosocomial infections increased from 8% in 1971 to 18% in 1979. *Pseudomonas aeruginosa* is an extremely prominent pathogen in patients with malignancy and/or granulocytopenia (Winston et al., 1979). It has uniformly been the second or third most common organism found to produce bacteremia and disseminated infection in patients with hematological malignancy. *Pseudomonas aeruginosa* is almost as common as *Staphylococcus aureus* in recent years as a cause of bacteremia in burn patients. Indeed, about one-fourth of all burn patients admitted to hospitals develop *Pseudomonas* infections at burn sites and, in addition, develop infections of the urinary tract or at the sites of surgery of infection of the lower respiratory tract due to this organism. *Pseudomonas* has also been shown to be extremely common as a pathogen in patients who aspirate in the hospital setting. Stamm et al. (1981) reviewed the data of the comprehensive hospital infection project of the Centers for Disease Control and found that *Pseudomonas* was the fourth most common cause of endemic and epidemic infection. Thus it is apparent the *Pseudomonas* infections are relatively uncommon in institutions dealing with patients with normal host defenses, such as a community hospital, but burn centers, institutions dealing with hematological malignancy, and institutions in which there is use of long-term indwelling urethral catheters, such as neurological services dealing with paraplegic patients or the elderly, will have *Pseudomonas* infection in up to 25% of their patient population (Young, 1981; Noone et al., 1983; Feeman and McPeake, 1983).

PATHOGENIC PROPERTIES

Although *P. aeruginosa* produces a number of toxic substances, the pathogenic significance of each of these toxins remains unclear (Wretlind and Pavlorskis, 1981). The surface slimes of *Pseudomonas* are polysaccharides which function similarly to the capsules of organisms such as *Streptococcus pneumoniae*. Thus these capsules can interfere with normal phagocytosis. Lipopolysaccharides from

P. aeruginosa have the same general architecture as those from the Enterobacteriaceae. There are differences in the fatty acid composition and there is an unusually high degree of phosphorylation. Furthermore, the outer region of the core oligosaccharides is also distinctive, since it contains D-glucose, L-rhamnose, D-glucosamine, and L-alanine. It is probable that lipopolysaccharides are partially responsible for the high intrinsic resistance of P. aeruginosa to many antibacterial agents.

Costerton et al. (1983) have proposed that there is a microcolony mode of growth of Pseudomonas in certain clinical situations that promotes its ability to cause serious infection. Some Pseudomonas produce an exopolysaccharide, alginate, which encloses bacterial microcolonies and protects them from attack by host defenses. The alginate surrounds Pseudomonas, from which microcolonies spread and multiply, producing a chronic albeit nonlethal infection. Thus the slime material produced by Pseudomonas is itself not toxic but acts as a means of protection for the organisms.

Although Pseudomonas produces hydrocyanic acid, it is doubtful that sufficient HCN is produced to be toxic (Golfarb and Margraf, 1967). It is conceivable that in the lung it could be locally toxic to alveolar cells, particularly macrophages. Most strains of P. aeruginosa produce proteases with broad substrate specificity. One of the three proteases has elastolytic activity and is an elastase (Liu, 1966; Leake et al., 1978). This particular enzyme can degrade various plasma proteins, such as immunoglobulins, coagulation and complement factors and α-proteinase inhibitor (Schultz and Miller, 1974). Wretlind and Pavlorskis (1983) have shown that elastase is important in localized infections, such as Pseudomonas keratitis, pneumonia, and burn infection. Conversely, there appears to be a small role for the proteases in septicemia, once colonization and invasion with Pseudomonas have occurred. Only in localized infection where the protease inhibitor, plasma, and α-macroglobulin will be present in insufficient amounts would the proteases exert their damaging activity and facilitate the spread of the organism to other tissues and into the bloodstream. It is clear, however, that elastase and other exoproteins are produced by Pseudomonas during infection in humans, since antibodies to these enzymes can be found in cancer patients (Crow et al., 1982). Studies of strains isolated from cases of Pseudomonas bacteremia have not shown any differences, however, in the frequency or amounts of protease or elastase produced, as compared with strains from other sources such as urinary tract or pulmonary infections (Balch et al., 1979). A number of other factors such as hemolysins and leukocidins are produced by Pseudomonas, but their role in the infective process has not been established.

Most strains of *Pseudomonas* produce exotoxin A (Pollack, 1983). This is an extracellular enzyme that is a single-chain polypeptide of molecular weight 71,000 with A and B fragments that mediate enzymatic and cell-binding functions (Vasil et al., 1977). Exotoxin A catalyzes the transfer of the adenosine diphosphate—ribosyl moiety from the nocotinaminde—adenine dinucleotide to elongation factor-2, which results in the inactivation of the latter factor and the inhibition of protein biosynthesis (Iglweski and Kabat, 1975). It is clear that exotoxin A is produced during *Pseudomonas* infection in vivo (Pollack et al., 1976) and that it causes disease by inhibiting protein synthesis, its direct cytopathic effects, and interference with cellular immune functions in the host. Further evidence that exotoxin A plays a pathogenic role is that the presence of serum antibodies to exotoxin A in bacteremic human *Pseudomonas* infection results in greater survival (Pollack and Young, 1979; Pollack et al., 1983). Animal experiments also demonstrate that there is less tissue invasion in bacteremia in animals that have received antibodies against exotoxin A (Cross et al., 1980). It is possible that exotoxin A may also play a role in allowing *Pseudomonas* to penetrate the host physical defenses, which is suggested by its necrotizing effects on the skin and the eye. It is also possible that exotoxin A released by *Pseudomonas* during infection interferes with normal bacterial clearance mechanisms, partially because of its effect on phagocytic function as well as on spleen cells and mononuclear cells.

It is clear that there are a number of factors that contribute to the pathogenicity of *P. aeruginosa* (Table 3). The characteristic

Table 3 Factors Contributing to *Pseudomonas aeruginosa* Pathogenicity

Surface pili	Adherence to intestinal, respiratory, or bladder epithelium leading to colonization
Slime polysaccharide	Antiphagocytic effect; protection of microcolonies
Extoxin A	Inhibition of protein synthesis; cytotoxic to macrophages; inhibition of spleen and T-cell function
Proteases	Degradation of plasma proteins and complement; damage to the eye, skin, or lung

lesions of *Pseudomonas* are a diffuse, acute vasculitis of small vessels with countless organisms located in the walls of the vessels. The interaction of exotoxin A, proteolytic enzymes, phospholipase, and lipopolysaccharide, as well as the outer polysaccharide mucin material, may be responsible for the total pathological picture. It is clear, however, that there are major pathogenic differences between infections due to *Pseudomonas* and those due to other enteric organisms.

HOST DEFENSES

The first defense against *P. aerginsoa* is prevention of colonization. Gastric acidity, intestinal motility, and secretory immunoglobulins protect humans from colonization by many bacteria. Whether such mechanisms are operative for *Pseudomonas* has been less clearly established. However, it has been shown that normal colonic flora and short-chain fatty acids in a dissociated state inhibit the growth of *Pseudomonas*. *Pseudomonas* normally does not enter skin pores on the surface and we know from outbreaks of infection occurring in hot tubs that opening skin pores allows *Pseudomonas* to penetrate. The most important host protection against *Pseudomonas* is the granulocyte (Peterson, 1980). Wheter granulocytopenia itself permits increased colonization is unclear, however. There are a variety of humoral resistance factors to *Pseudomonas* which include classic IgG and IgM antibodies as well as other opsonic substances (Pollack et al., 1983). Complement has been shown to be necessary for phagocytosis and intracellular killings of *Pseudomonas*. It is important, however, to note that presence of antibody does not prevent colonization, since cystic fibrosis and burn patients have high antibody titers and yet are readily colonized. A role for T-cell function in the prevention of infection has not been demonstrated. Generally, patients with agammaglobulinemia do not develop infection due to *Pseudomonas* unless they have simultaneously depressed white cell function.

The pulmonary host defenses of *Pseudomonas* are not fully characterized. In contrast to the rapid clearance of *E. coli* from the lungs of normal mice, *Pseudomonas* is not readily removed from the mouse lung after aerosol deposition (Reynolds, 1974). There has been a correlation between the adherence of *P. aeruginosa* to the upper respiratory tract epithelium of seriously ill patients and subsequent infection. *Pseudomonas aeruginosa* adherence in vitro can be correlated with the loss of a protease-sensitive glycoprotein, fibronectin, from the cell surface (Woods et al., 1983). Thus loss of fibronectin may be associated with colonization of the oral pharynx with this organism. Fibronectin normally covers receptors on respiratory epithelium which may be exposed by salivary protease activity (Woods et al.,

1980, 1981). Under normal circumstances man can control aspirated *Pseudomonas* through pulmonary macrophages, but in the presence of defective pulmonary clearance the organisms will proliferate. Secretory IgA has not been shown to improve pulmonary clearance or to be protective, whereas IgG will improve pulmonary clearance (Southern et al., 1970). Clearly, local factors such as mucus secretion and ciliary function are an important protection against *Pseudomonas* in the lung (Holby, 1977). Table 4 lists host defects associated with *Pseudomonas* infection.

ANTIMICROBIAL SUSCEPTIBILITY AND RESISTANCE

Although the number of antimicrobial agents that inhibit *P. aeruginosa* is fairly large, resistance has been a major problem with this organism (Neu, 1982d, 1984). Agents which interfere with cell wall synthesis of *Pseudomonas* are the penicillins, particularly the anti-*Pseudomonas* penicillins for which there are many agents available: carbenicillin, ticarcillin, azlocillin, mezlocillin, sulbenicillin, piperacillin and apalcillin (Neu, 1982a, 1983; Neu and Labthavikul, 1982c). There are in addition a large number of cephalosporins of the third generation

Table 4 Host Defects Associated with *Pseudomonas aeruginosa* Infection

Condition	Host defects
Burns	Loss of integument, loss of immunoglobulin, depressed white cell function, altered complement activity
Cystic fibrosis	Defective pulmonary clearance mechanisms, altered macrophage activity
Hematological malignancy	Granulocyte deficiency
Chemotherapy	Altered gastrointestinal mucosa
Transplantation patients	Granulocyte deficiency, immunoglobulin deficiency
Patients requiring ventilation	Loss of pulmonary clearance mechanisms, loss of fibronectin in lung

which have various degrees of activity against *Pseudomonas* (Neu, 1982a). These will be discussed in greater detail subsequently. Fosfomycin and fosmidomycin also interfere with cell was biosynthesis in *Pseudomonas* (Neu and Kamimura, 1981) by interfering with a phosphoenolpyruvate necessary for the production of UDP-GlcNAc-enolpyruvate.

There are a number of aminocyclitol antibiotics more commonly referred to as aminoglycosides which inhibit *Pseudomonas*, and recently there have been a number of fluorinated carboxy quinolone compounds which inhibit DNA gyrase. Although other antimicrobial agents occasionally inhibit *Pseudomonas*, for practical purposes the other agents are not used clinically. The mechanism of action of these antibiotics against *Pseudomonas* is similar to their activity against other gram-negative organisms.

Activity of penicillins against *P. aeruginosa* can be explained on the basis of affinity for penicillin-binding proteins (PBP), ability to penetrate the outer wall of *Pseudomonas*, and stability against attack by β-lactamases (Neu, 1983). Carbenicillin and ticarcillin are carboxy penicillins differing by the presence of a benzyl and a thienyl group on the acyl side chain. Sulbenacillin contains a $-SO_3$ group replacing the $-COOH$ of the other drugs. These acidic moieties allow the agents to penetrate the outer cell wall and to resist attack by the inducible β-lactamase present in all *Pseudomonas*.

The new ureidopenicillins, namely, azlocillin, mezlocillin, apalcillin, and piperacillin, are derivatives of ampicillin with a ureido configuration on the acyl side chain. This molecular modification results in better entry into the periplasmic space of bacteria and better binding to the PBP of *Pseudomonas*. This explains the higher activity of these agents. They are not more β-lactamase stable, and this explains why there are differences in the minimum inhibitory concentration (MIC) values of ticarcillin azlocillin, and piperacillin when inocula of 10^6 or 10^7 colony-forming units (CFU) are used in susceptibility tests. bility tests.

The new cephalosporins are able to pass the outer barrier of *Pseudomonas* and to bind to PBP, but those which contain an oxime ether group do not have excellent anti-*Pseudomonas* activity. Use of a carboxy propyl group as in ceftazidime causes a major increase in activity against *Pseudomonas* and a loss of gram-positive activity. Cefoperazone is a ureidocephalosporin which has a 2,3-dioxopiperazine group which gives excellent activity against *Pseudomonas* but does not give very good β-lactamase stability. Moxalactam is a com pound in which the sulfur of the dihydrothiazine ring has been replaced by an oxygen. The increased *Pseudomonas* activity probably lies more in the $-COOH$ group of its acyl side chain.

Aztreoman is a monobactam compound with an aminothiazoly side chain and a carboxy propyl group similar to ceftazidime which provides its excellent anti-*Pseudomonas* activity. Imipenem, n-formimidoylthienamycin, is a novel structure in which there is neither a sulfur nor an oxygen in the ring fused to the β-lactam ring. This compound has a 6-(R-hydroxyethyl) side chain which provides excellent activity against *Pseudomonas* and also β-lactamase stability.

The activities of these compounds have ranged from 100% of isolates being susceptible to only 50% of isolates being susceptible to concentrations that could be achieved in humans. Azlocillin, apalcillin, and piperacillin will inhibit many isolates resistant to carbenicillin. Cefsulodin and cefoperazone have similar activities against *P. aeruginosa* and susceptibility depends to a great extent on the type of β-lactamases present in the strains tested. Cefotaxime ceftizoxime, ceftriaxone, and cefmenoxime differ among themselves by two-to fourfold dilutions in their activity against *Pseudomonas*. Ceftazidime and aztreonam are of similar activity, with ceftazidime slightly more active. Both agents inhibit carbenicillin-, piperacillin-, and aminoglycoside-resistant isolates. Imipenem is the most active β-lactam tested against *P. aeruginosa*. The activity of the β-lactams is shown in Table 5.

Among the aminoglycosides the most active agents available are tobramycin and sisomicin which are 2- to 8-fold more active than gentamicin and 4- to 16-fold more active than netilmicin and amikacin. Tobramycin will inhibit some isolates resistant to gentamicin, but much of the susceptibility is an artifact of testing due to the greater effect of the cations Mg^{2+} and Ca^{2+} on the activity of the other aminoglycosides. Amikacin inhibits isolates resistant to all of the other agents. The mean MIC values are shown in Table 5.

Resistance of *P. aeruginosa* to antimicrobial agents is a complex phenomena (Table 6). Some *P. aeruginosa* strains have modified target enzymes, penicillin-binding proteins, so that β-lactams do not inhibit the organisms (Mirelman et al., 1981). More frequently *Pseudomonas* relies on the mechanism of prevention of access to the target by altering the porin channels through which the anionic penicillins and cephalosporins or the cationic aminoglycosides pass to reach their receptor sites (Zimmerman, 1980). Finally, *Pseudomonas* has the ability to either inactivate various substances, such as penicillins, cephalosporins, and chloramphenicol, through β-lactamases or chloramphenicol acetyltransferase, or to modify aminoglycosides by acetylation, phosphorylation, or adenylation.

Compounds such as ampicillin and the antistaphylococcal penicillins are not effective against *Pseudomonas* because they do not readily cross the outer wall of the organism and hence do not reach the penicillin-binding proteins (PBP). Furthermore, these compounds do not have a high affinity for penicillin-binding proteins; conversely the anti-*Pseudomonas* penicillins such as carbenicillin and the ureidopencillins are active against *Pseudomonas* because of their

Table 5 Activities of Various Antibiotics Against *Pseudomonas aeruginosa*[a]

	Range	MIC_{50}	MIC_{90}
Penicillins			
Apalcillin	1->128	16	64
Azlocillin	1->128	16	64
Carbenicillin	1->128	64	>128
Mezlocillin	1->128	32	128
Piperacillin	1->128	16	64
Sulbenacillin	1->128	64	>128
Ticarcillin	1->128	32	128
Cephalosporins			
Cefotaxime	2->128	16	128
Ceftizoxime	2->128	16	128
Ceftriaxone	2->128	16	128
Cefmenoxime	2->128	16	128
Ceftazidime	<1-32	2	8
Cefoperaxone	<1->128	4	32
Cefsulodin	<1->128	4	16
Moxalactam	1->128	16	32
HR 810	<1->128	4	8
BMY 28,142	<1->128	4	8
Other agents			
Imipenem	<1-16	2	4
Aztreoman	<1-64	4	16
Aminoglycosides			
Kanamycin	8->128	64	>128
Gentamicin	<1->128	2	8
Sisomicin	<1->128	1	4

Table 5 (continued)

	Range	MIC$_{50}$	MIC$_{90}$
Tobramycin	<1->128	1	4
Netilmicin	<1->128	2	16
Amikacin	1->128	4	32
Dibekacin	1-32	2	8
O-Demethyfortimicin	4-128	8	>128
Phosphonic acids			
Fosfomycin	2->128	16	>128
Fosmidomycin	1->128	8	>128
Quinolones			
Norfloxacin	0.25-16	2	4
Ofloxacin	0.25-16	4	8
Enoxacin	0.25-8	2	4
Pefloxacin	0.25-8	2	4
Ciprofloxacin	0.01-2	0.25	1
Amifloxacin	0.25-8	2	4

[a]Based on literature cited in this article.

ability to reach adequate concentrations within the periplasmic space and to bind to PBP. Plasmid-mediated β-lactamases, for example, the Richmond-Sykes IIIa enzyme, also referred to as the TEM-1 enzyme, when present in *Pseudomonas* will make these organisms resistant to compounds such as carbenicillin, ticarcillin, azlocillin, mezlocilin, piperacillin, and apalcillin (Furth, 1979; Livermore, 1983). There are other plasmid-mediated β-lactamases, PSE-1 through 4 that are found in *Pseudomonas* (Furth, 1979; Livermore, 1983). These enzymes also contribute to the resistance of *Pseudomonas* against various penicillins, particularly PSE-4, which is also referred to as the Dalgleish enzyme (Table 7).

All *P. aeruginosa* organisms contain an inducible, chromosomally mediated β-lactamase which is an inducible cephalosporinase (Sabath et al., 1965). This inducible cephalosporinase explains why first- and second-generation cephalosporins such as cefazolin and cefaman-

Table 6 Various Mechanisms of Resistance to Antimicrobial Agents Found in *Pseudomonas aeruginosa*

Agent	Mechanism
Carbenicillin	β-Lactamase, TEM-1, decreased permeability
Ticarcillin	β-Lactamase, decreased permeability
Azlocillin	β-Lactamase
Piperacillin	β-Lactamase
Cefoperazone	β-Lactamase, PSE-4
Cefsulodin	β-Lactamase, PSE-1, PSE-4
Aztreoman	Decreased permeability
Ceftazidime	Decreased permeability
Imipenem	Decreased permeability
Aminoglycosides	Aminoglycoside-modifying enzymes, decreased uptake of drugs
Chloramphenicol	Modified drug (acetylated) enters less well
Tetracyclines	Altered permeability
Fluorinated carboxy quinolones	Altered DNA gyrase
Fosfomycin	Altered transport system
Fosmidomycin	Altered transport system

dole do not inhibit these bacteria. An agent such as cefoxitin does not inhibit *Pseudomonas* even though it is not destroyed by the β-lactamases of *Pseudomonas*. The reason for this is that the presence of a methoxy group on the β-lactam ring decreases the anti-*Pseudomonas* activity of these compounds. Temocillin is a methoxy derivative of ticarcillin. This compound is β-lactamase stable, but the presence of the methoxy group has caused it to lose all *Pseudomonas* activity. Thus it is clear that resistance of *Pseudomonas* to β-lactam compounds is a combination of both β-lactamase activity and the ability of the compound to reach a receptor site (Sykes and Bush, 1982). Compounds such as azlocillin and pipercillin are more active against *Pseudomonas* than earlier anti-*Pseudomonas* agents because of their ability to pass more readily through the outer membrane of the organism and their high affinity for critical PBP in *Pseudomonas*.

Table 7 β-Lactamases Present in *Pseudomonas*

Genetic basis	Trivial name	Richmond-Sykes class	Substrates hydrolyzed
Plasmid	TEM-1	IIIa	Carbenicillin[a]
			Cefoperazone
			Cefsulodin
Plasmid	OXA-2	V	Cefoperazone
Plasmid	PSE-1	V	Carbenicillin[a]
Plasmid	PSE-2	V	Carbenicillin[a]
			Cefotaxime
			Cefoperazone
Plasmid	PSE-3	V	Carbenicillin[a]
			Cefsulodin
			Cefoperazone
Plasmid	PSE-4 (Dalgleish)	V	Carbenicillin[a]
			Cefoperazone
			Cefsulodin
Chromosomal	Sabath-Abraham	Id	First-generation cephalosporins

[a]Azlocillin, apalcillin, mezlocillin, piperacillin, and ticarcillin are also hydrolyzed.

Although a number of third-generation cephalosporins, cefotaxime, ceftizoxime, cefmenoxime, and ceftriaxone, have activity against *Pseudomonas,* for practical purposes these compounds should not be considered anti-*Pseudomonas* agents (Neu, 1982a, 1983). Conversely, agents such as cefoperazone, cefsulodin, and ceftazidime show fairly good anti-*Pseudomonas* antibacterial activity (Neu et al., 1979; Neu and Labthavikul, 1982a). Indeed, ceftazidime is not hydrolyzed by any of the *Pseudomonas* β-lactamases, whereas cefoperazone and cefsulodin are hydrolyzed by TEM-1, TEM-2, PSE-1, and PSE-4 (Neu and Fu, 1979; Neu et al., 1979; Neu and Labthavikul, 1982a). A new class of compounds which has high activity against *Pseudomonas* is the monobactams. Aztreonam is the first of these compounds and it has activity comparable to that of ceftazidime (Neu and Labthavikul, 1981). A similar monobactam drug has been developed by Takeda pharmaceuticals and it is also active against *Pseudomonas*. The oxacephem moxalactam has activity against many *Pseudomonas* strains; however, it should not be considered a major anti-*Pseudomonas* drug, in spite of the fact that it shows excellent stability against attack by *Pseudomonas* β-lactamases.

Several new cephalosporin antimicrobial agents active against *Pseudomonas* are under study. One of these, HR 810, shows excellent activity against many *Pseudomonas* agents, as does the compound BMY 28,142. Another excellent *Pseudomonas* β-lactam compound is the thienamycin-class drug imipenem (Neu and Labthavikul, 1982b). Impenem will inhibit many *Pseudomonas* strains that are resistant to all the other β-lactam compounds. In addition to these drugs, there are several other carbapenems which will inhibit *Pseudomonas*, but these agents are still in laboratory stages of investigation. We have isolated *P. aeruginosa* strains from patients with cystic fibrosis or from certain skin structure infections which are resistant to all β-lactam compounds, including imipenem (Prince and Neu, 1981). The type of β-lactamases found in *Pseudomonas* varies from one country to another (Jouvenot et al., 1983).

The precise mechanism of aminoglycoside resistance in *Pseudomonas* is complex. Porin channels exist for the cationic aminoglycosides. Once these agents cross the outer cell wall, the positively charged aminoglycosides are moved across the cell membrane by the proton-motive force after they bind to some outer cytoplasmic binding sites (Bryan and Van Den Elzen, 1977). The initial uptake of aminoglycosides into the bacterial cell is energy dependent (Bryan et al., 1976). When the aminoglycosides enter the cytoplasm, they bind to 30S ribosomes and to a degree to one protein in the 50S ribosomes and cause faster uptake of the drug through induction of a protein that increases the drug transport. Many bacteria possess plasmid-mediated enzymes that can acetylate, phosphorylate, or adenylate the amino or hydroxyl groups of the aminoglycoside (Moellering, 1983; Mitsuhashi and Kawabe, 1982; see Table 8). Modified aminoglycosides do not bind as well to ribosomes and do not induce the uptake of the drug, thus resulting in resistance. When streptomycin and kanamycin were first introduced, there were *Pseudomonas* isolates in the community susceptible to these agents. Today none of these organisms would be found owing to the widespread use of such agents and the fact that virtually all *Pseudomonas* strains possess a 6-phosphotransferase enzyme that will modify streptomycin and a 3-phosphotransferase which will modify neomycin and kanamycin (Shannon and Phillips, 1982). Resistance of *Pseudomonas* to gentamicin and to tobramycin varies from one institution to another and from one community to another. A great deal depends upon the presence of a 3-acetyltransferase en-

Table 8 Aminoglycoside-Modifying Enzymes Found in *Pseudomonas*

	Compounds modified
Acetylating	
3-Acetyltransferase (AAC-3)	Gentamicin, tobramycin, kanamycin, neomycin, sisomicin, fortimicin
6'-Acetyltransferase (AAC-6')	Kanamycin, tobramycin, amikacin, sisomicin
Phosphorylating	
3'-Phosphotransferase (APH-3'-1)	Neomycin, kanamycin
6-phosphotransferase (APH-6)	Streptomycin
3"-Phosphotransferase (APH-3")	Streptomycin
Adenylating	
2"-Adenyltransferase (ANT-2')	Gentamicin, tobramycin, kanamycin, netilmicin

zyme in the organisms. The other aminoglycoside-modifying enzymes are shown in Table 8. Resistance of *Pseudomonas* to amikacin is fairly uncommon, and in general the resistance in the United States of *Pseudomonas* to amikacin has been due to a failure of accumulation of the drug because of altered cell wall properties making the organism resistant to all aminoglycoside antibiotics (Miller et al., 1980). However, some strains of *Pseudomonas* do possess 6'-acetyltransferase (AAC-6') which will make them resistant to amikacin. Ironically, such strains are susceptible to gentamicin, provided that they do not have any of the other acetyltransferase enzymes.

Chloramphenicol resistance of *Pseudomonas* appears to be due to poor entry of the antibiotic (Mitsuhashi et al., 1975), and a similar mechanism is operative for trimethoprim. In the case of chloramphenicol most *Pseudomonas* strains possess a chloramphenicol transacetylase enzyme, which causes the antibiotic to be taken into the cell much less efficiently. Ironically, some *Pseudomonas* isolates are susceptible to sulfonamides but resistant to trimethoprim. Resistance to sulfonamides is due to an altered dihydropteroate synthetase enzyme. Resistance of *Pseudomonas* to fosfomycin and fosmidomycin is due to lack of the glycerol-phosphate transport system in the resistant isolates (Neu and Kamimura, 1981). Most *Pseudomonas* strains are resistant to naladixic acid and to cinoxicin either because of poor entry of the compounds or because of a resistant DNA gyrase. There are currently a number of fluorinated carboxy quinolones such as enoxacin, ofloxacin, perfloxacin, and ciprofloxacin which inhibit *Pseudomonas,* including isolates resistant to the anti *Pseudomonas* penicillins, cephalosporins, and to aminoglycosides (Neu and Labthavikul, 1982d; Chin and Neu, 1983, 1984; Kumada and Neu, 1983). Ciprofloxacin is an extremely potent anti-*Pseudomonas* drug. High-level resistance to quinolones can be produced in the laboratory. Such resistance has not been reported thus far in early clinical trials. Resistance does not appear to be due to permeability, since it is not altered by treatment of the bacteria with EDTA but, rather, is due to a chromosomal change in the DNA gyrase.

Resistance to tetracycline appears to be related to a decrease in tetracycline accumulation similar to that found in Enterobacteriaceae. There appear to be both R-factor-containing strains and strains in which impermeability is expressed constitutitively through a chromosomal mechanism. Resistance to macrolides and lincosamides is related to impermeability, since mutant strains of *Pseudomonas* are susceptible to these compounds.

Pseudomonas also possesses resistance to various antiseptics and preservatives, including mercury (Table 9). The organisms are able to reduce mercury and volatize Hg^0 from the bacterial cell. There are also organisms that are resistant to silver, but the mechanism is not fully clarified. Finally, resistance to quaternary ammonium compounds may be due to reduced permeability but is also apparently related to the ability of the organisms to grow on these compounds (Hague and Russell, 1974).

Table 9 Germicides That Can Be Contaminated with *Pseudomonas*

Benzalkonium chloride

Benzalkonium chloride/picloxydine

Cetrimide

Chlorhexidine

Hexachlorophene

Poloxamer-iodine

Quaternary ammonium

Phenolic

Pine tar

A number of antibiotics show synergistic activity against *Pseudomonas*. The most common synergy is shown between the anti-*Pseudomonas* penicillins and the aminoglycosides. This was first demonstrated for the combination of carbenicillin and gentamicin. However, virtually all aminoglycosides and anti-*Pseudomonas* penicillins show some degree of synergy (Fu and Neu, 1978). Synergy of aminoglycosides is also shown with the anti-*Pseudomonas* cephalosporins and with the monobactam aztreonam. Synergy can also be demonstrated for the combonation of fosmidomycin with aminoglycosides and with fosmidomycin and anti-*Pseudomonas* penicillins. This snynergy is not merely a laboratory phenomena, since synergistic combinations of antimicrobial agents result in increased protection of animals infected with *Pseudomonas*. Indeed, in neutropenic animal models of infection combination antimicrobial agents such as ticarcillin and tobramycin proved the highest degree of protection (Andriole, 1971; Scott and Robson, 1976).

CLINICAL INFECTIONS

Thermal Injuries

In the 1960s *P. aeruginosa* was recognized as a major problem in burn patients. A total of 60% of patients at the Army Burn Institute in the beginning of the 1960s died from *Pseudomonas* sepsis (Pruitt, 1974). One might initially think that burn injury produces only skin damage and necrosis, but burns cause a major effect on many of the organ systems within the body that are protective against bacterial invasion. There is impairment of humoral and cellular defenses following burn injury, and there is postburn depression of serum immunoglobulin levels and of serum opsonic activity and particularly functional impairment of the complement system (Alexander and Fisher, 1970; Alexander, et al. 1971). There is a decrease in granulocyte number and suppression of stimulated granulocyte activity, including granulocyte chemotaxis, phagocytosis, and bacterial capacity. There is alteration of T- and B- lymphocyte ratios and depressed skin test reactivity with supressor lymphocyte activity increased. Furthermore, the reticuloendothelial system phagocytic function is depressed after thermal injury.

The burn vitiates the barrier function of the skin and the thermal injury provides an avascular area which is the burn eschar. The avascular eschar is a culture medium which supports the growth of microorgansims that are seeded onto the burn wound. The lack of an effective blood supply precludes the delivery of protective mechanisms. Sutter and Hurst (1966) found that only 15–25% of *Pseudomonas* on burn wounds came from the patient themselves. It appears that most *Pseudomonas* present in burn wounds are derived from other burn patients present in the hospital. The organisms prevalent on burn wounds in a burn unit change with time. It is unclear, however, why *Pseudomonas*, of all organisms, appears to be most able to persist in this burn wound environment.

Pseudomonas tends to colonize the burn patient so that by the end of the third week 27% of burn wounds are colonized with *Pseudomonas*. They proliferate in vascular eschar, and when the bacterial density exceeds 10^5 organisms per gram of tissue, bacteremia usually ensues as the organisms pass from the subeschar space and enter viable tissue. Development of topical agents effective against *Pseudomonas* such as mafenide, silver sulfadizine, and silver nitrate provided agents that would significantly reduce the number of organisms within the eschar. No topical agent sterilizes the burn wound, but by reducing the bacterial load it permits healing to occur and thus avoids burn wound sepsis.

The likelihood of infection due to *Pseudomonas* increases with increasing size of the burn. Burn patients must be examined daily to look for signs of *Pseudomonas* burn wound sepsis. Surface swabs are not adequate to determine the amount of *Pseudomonas* within a burn and wound biopsies utilizing quantitative culture and histological techniques are necessary to follow burn wounds. The presence of focal dark brown or black discoloration of the wound is an important sign that wound sepsis may soon occur. Another useful sign is the conversion of an area of partial-thickness injury to full-thickness necrosis. Other clinical signs that wound sepsis is about to occur include hypothermia, hyperglycemia, or hypoglycemia. Altered mental status is also a useful sign. If on histological examination of a specimen there is extensive vasculitis around vessels with characteristic perivascular cuffing, this should be highly suggestive that burn wound sepsis will occur. Pruitt et al. (1983) have suggested that focal or multifocal invasive *Pseudomonas* burn wound infections may be treated by local excision or subeschar injection of a diffusible semisynthetic penicillin such as ticarcillin. Most other workers, however, would institute therapy with an anti-*Pseudomonas* penicillin, ticarcillin, azlocillin, or piperacillin, and an aminoglycoside by the parenteral route, administering large doses of both drugs. Selection of the aminoglycoside will depend upon local resistance factors, but tobramycin would be the initial choice in most institutions in the United States. Recently cefsulodin and ceftazidime have proved useful in the treatment of some patients with burn wound infection, either singularly or combined with aminoglycosides.

Bacteremia

The first case of *P. aeruginosa* sepsis was reported by Finkelstein in 1896. In the period between 1906 and 1925 Fraenkel established the histopathology of *Pseudomonas* bacteremia (Fraenkel, 1912). He noted there was a characteristic colonization of bacilli in the walls of blood vessels in the disease focus. Most reports of *Pseudomonas* septicemia in the 1940s and 1950s were noted in children. The origin of the sepsis in these children was the gastrointestinal tract, the skin, and ear sites. During the same period most *Pseudomonas* sepsis in adults originated after urological surgical procedures. Curtin et al. (1961) reviewed the cases of bacteremia due to *Pseudomonas* at the Johns Hopkins Hospital going back to 1940. They noted that the majority of the patients succumbed to their infection because of severe underlying illness. Flick and Cluff (1976) reported a 67% mortality in bacteremia due to *Pseudomonas* in the newborn and a similar mortality in patients with hematological disorders. These investigators noted a mortality of 92% when bacteremia due to *Pseudomonas* occurred in individuals with solid tumors. Overall the reported experience with *Pseudomonas* bacteremia from 1951 to 1975 has shown a mortality of 37—77% with a mean of 57% (Trapper and Armstrong, 1974; Singer et al., 1977; Bishop et al., 1981; Baltch and Griffin, 1977; Spengler et al., 1978; Setia and Gross, 1977). During this same period *Pseudomonas* has accounted for 7—18% with a mean of 13%, of reported bacteremias. More recent studies by several groups (Reynolds et al., 1975; Schimpff et al., 1973; Bodey, 1982) have indicated much higher survival in *Pseudomonas* bacteremia when appropriate therapy has been used.

Pseudomonas bacteremia may develop from many different sites, such as the urinary or respiratory tract, or from cutaneous lesions. However, the origin of the *Pseudomonas* bacteremia is not always apparent. Several cutaneous manifestations of *Pseudomonas* may suggest that bacteremia is due to this organism. *Pseudomonas* septicemia has been associated with skin lesions referred to as ecthyma. These lesions vary from pustular lesions with hemorrhagic centers to areas of flat, demarcated cellulitis which rapidly spreads out from a central area. Biopsy of such lesions will show massive numbers of bacilli. It is important, however, to realize that similar skin lesions have been seen in neutropenic patients infected with other microorganisms. Furthermore, ecthyma gangrenosum has been uncommon in recent years, since antimicrobial therapy is instituted much earlier in patients prone to *Pseudomonas* infection.

Patients with *Pseudomonas* bacteremia characteristically become tremulous and sweaty, disoriented, and delirious. The skin is often warm before the patient develops hypotension, collapse, and shock. A recent study by Grief et al. (1983) of *Pseudomonas* septicemia in childhood showed that the majority of patients were debilitated by underlying disease or had recent major invasive procedures. Furthermore, most of the patients had received broad-spectrum antibiotic therapy. In this study ecthyma gangrenosum was infrequent, and overall mortality was 53%. Bryan et al. (1983) analyzed 1186 episodes of gram-negative bacteremia and again confirmed that mortality due to *Pseudomonas* was frequently higher than that due to other organisms. However, severity of underlying disease was frequently greater in patients with *Pseudomonas* infection. Jackson et al. (1983), reviewing *P. aeruginosa* septicemia in childhood cancer patients, noted that severe myelosuppression was almost always present and that previous antibiotic therapy had been used in most patients. Disruption of the skin and mucosa in the anogenital region was a source of entry in a number of the patients, as was the gastrointestinal tract. Jackson et al. (1983) did not find that synergistic antibiotic combinations improved survival.

There are now many drugs available for the treatment of *Pseudomonas* bacteremia. These include carbenicillin, ticarcillin, azlocillin, and piperacillin as anti-*Pseudomonas* penicillins. Also available would be ceftazidime and aztreonam when these latter two drugs become commercially available. Cefoperazone, which is commercially available, can be used, but it must be used in doses larger than 2 g twice daily. A minimum dose would be 8 g/day. The aminoglycosides which could be used would be gentamicin or tobramycin at a dose of 2 mg/kg, every 8 hr, or at a modified dosage in the patient with reduced renal function. Amikacin would be used at a dose of 5–8 mg/kg, every 8 hr, if one suspected *Pseudomonas* sepsis. *Pseudomonas* bacteremia must be treated with two drugs if one wishes to achieve the greatest reduction in mortality and morbidity (Anderson et al., 1978). It is obvious, also, that the site of the *Pseudomonas* bacteremia must be determined to prevent persistent bacteremia.

Endocarditis

Endocarditis due to *P. aeruginosa* has been known throughout this century. In 1960 Forkner reported 20 cases in the world's literature. *Pseudomonas* endocarditis remains an uncommon entity, except in the narcotic addict and infrequently following open heart surgery. *Pseudomonas* endocarditis in addicts appears to be a disease in which there is a geographical localization. The disease has been most common in Detroit and New York, and less frequently reported in other areas of the United States or abroad (Reyes et al., 1973). Mixed streptococcal–*Pseudomonas* or staphylococcal–*Pseudomonas* endocarditis has been recognized.

The clinical presentation of *Pseudomonas* endocarditis has not differed from that seen with endocarditis due to *S. aureus* in the narcotic addict. Frequently the patient is a young addict in his or her 20s or 30s who seeks medical attention because of fever or chest pain (Reyes et al., 1973, 1978). The most common valve involved is the tricuspid valve. Thus there are frequently embolic lesions in the lung with multiple infiltrates noted on the chest x-ray which are cavitary in type. Such a disease cannot be differentiated from tricuspid endocarditis due to *S. aureus*. The pathogenesis of endocarditis due to *Pseudomonas* is unknown. The association of *Pseudomonas* with other organisms suggests that prior damage to the valve may allow *Pseudomonas* endocarditis to develop. There are reported cases of *Pseudomonas* endocarditis in burn patients that have followed prolonged staphylococcal septicemia which may have been a source of initial damage to the valve.

Mortality from *Pseudomonas* endocarditis remains high despite optimal use of antimicrobial agents (Reyes and Lerner, 1983). Infection of the tricuspid valve tends to be more subacute, whereas involvement of the mitral or aortic valve often precipitates fulminant valvular cardiac failure. *Pseudomonas* endocarditis involving the right side of the heart can occasionally be cured medically with an anti-*Pseudomonas* penicillin combined with an aminoglycoside. Both agents must be administered at maximum dosage. However, many of these patients will require subsequent surgery. Refractory right-sided endocarditis can be treated with removal of the valve without replacement of a new valve. Left-sided infection does not respond to medical therapy and therefore immediate valve replacement accompanied by a 6-week course of high-dose antimicrobial therapy is necessary in order to achieve cure.

Pulmonary Infections

There are three major forms of pulmonary infections: tracheal bronchitis, pneumonia, and infections in patients who have underlying cystic fibrosis. Tracheal bronchitis is a common occurrence in patients who have undergone endotracheal intubation and who have received antimicrobial drugs which do not inhibit *Pseudomonas*, such as ampicillin or cephalosporin antibiotics such as cephalothin, cefoxitin, or cefamandole. In some situations *Pseudomonas* is a cause of exacerbation of chronic bronchiectasis or of chronic bronchitis, but this is a fairly infrequent occurrence. Tracheal bronchitis can progress to pneumonia or resolve with therapy directed at improving pulmonary clearance secretions, even without the use of specific anti-*Pseudomonas* antimicrobial agents.

Pseudomonas is a distinctly uncommon cause of bacterial pneumonia in the community; however, it, along with other gram-negative bacilli, is a frequent cause of hospital-acquired pneumonia. Pneumonia due to *Pseudomonas* occurs primarily in individuals with compromised host defenses or structural lung disease (Pennington et al., 1973; Stevens et al., 1974). *Pseudomonas* pneumonia is particularly frequent in intensive care units and in cancer patients. There are two major pathogenic mechanisms of *Pseudomonas* pneumonia. In one, aspiration of oral pharyngeal contents results in bilateral or unilateral bronchopneumonia in the lower lobe. There may be extensive necrosis of lung tissue and cavitation in the lung. The other mechanism occurs in patients with underlying hematological malignancy. There is a hematogenous spread of bacilli, probably from the gastrointestinal tract to the lungs, or in the individual with *Pseudomonas* endocarditis the organism directly enters the lungs.

There are no specific clinical or radiological findings that will sufficiently characterize *Pseudomonas* pneumonia to differentiate it from other pneumonias. In the past various reports suggested that patients with *Pseudomonas* pneumonia had nodular infiltrates throughout the lung. This, however, can be seen with many other microorganisms. Establishing a diagnosis of *Pseudomonas* pneumonia is much more difficult than one would think. The reason for this is that many cultures of sputum from critically ill patients will yield *Pseudomonas*. It is difficult to differentiate *Pseudomonas* colonization in tracheotomized patients or critically ill patients from *Pseudomonas* producing true peneumonia. Although it may be useful to obtain blood cultures, bacteremia is not seen in more than 25% of patients with proven *Pseudomonas* pneumonia.

Current therapy of *Pseudomonas* pneumonia, similar to therapy of *Pseudomonas* bacteremia, requires the use of two antimicrobial agents. This would be an anti-*Pseudomonas* penicillin combined with an aminoglycoside. Ideally, the most synergistic combination should be used. Therapy should be with the largest doses of drugs tolerated and therapy probably must be continued for a minimum of 3 weeks. Our group has had excellent results of therapy of *Pseudomonas* pneumonia with abcess formation with ceftazidine or aztreonam, 2 g administered every 8 hr. We used these drugs as single therapy in the investigational studies of these compounds. It is probably better to combine them with an aminoglycoside, preferably tobramycin.

Respiratory Infections in Cystic Fibrosis Patients

Pseudomonas aeruginosa has replaced *S. aureus* as the major pathogen in patients with cystic fibrosis (Kulczycki et al., 1978). Bacterial infection in cystic fibrosis is limited to the lungs in which colonization leads to chronic bronchitis followed by mucous plugging, atelectasis, bronchiectasis, and extensive destructive pneumonitis. Many studies have been performed in an attempt to understand the colonization of cystic fibrosis patients with *Pseudomonas*, particularly mucoid strains (Reynolds et al., 1976).

The clinical symptoms and signs of *Pseudomonas* pulmonary infection in patients with cystic fibrosis are extremely variable. Some patients have upper respiratory infections with cough, shortness of breath, weight loss, reduced appetite, and decreased activity. Pulmonary infections in cystic fibrosis patients tend to be chronic, with progressive infection leading to increasing destruction of lung tissue. The location of the bacteria in the airways has made microbial therapy less than satisfactory. Recurrent episodes of bronchopneumonia do not result in dissemination of the organism throughout the body, but rather the lungs are slowly destroyed and children or adults die not of disseminated infection but as the result of pulmonary insufficiency, cor pulmonale, and anoxia. Since *Pseudomonas* is present in the sputum of these patients at all times, it is frequently difficult to decide when there has been an exacerbation of the illness. However, increasing production of sputum and change in general sense of wellbeing frequently are useful parameters to follow. In our experience mere use of pulmonary toilet will not improve these patients and antimicrobial therapy is necessary, although the optimal program is not known.

We and others have achieved excellent results with a variety of different programs which include ticarcillin, azlocillin, piperacillin, and, more recently, drugs such as cefsulodin, ceftazadine, aztreonam, and imipenem (Parry et al., 1977; Parry and Neu, 1978; Pancoast et al., 1981; Moeller et al., 1982; Hoogkamp et al., 1983; Permin et al., 1983; Scully and Neu, 1984). Under investigation is the use of ciprofloxacin in the treatment of theses patients. Preliminary results by our group are encouraging, although even with this agent there is an increase in the MIC values of the *Pseudomonas*. It is not possible to eradicate *Pseudomonas* from secretions of patients with cystic fibrosis.

It is extremely important to realize that the pharmacokinetics of both aminoglycosides and cephalosporins are altered in cystic fibrosis patients, since they have a larger extracellular volume due to frequent malnutrition and they also appear to have increased clearance of these drugs; hence the doses, particularly of aminoglycosides, must be larger than those that are normally used in other patients (Neu, 1982b). Although nebulization programs of aminoglycosides or even anti-*Pseudomonas* penicillins have been used to treat these patients, this is not an established form of therapy.

Central Nervous System Infections

Meningitis due to *P. aeruginosa* is distinctly uncommon, but it may follow direct introduction of the organism into the nervous system at the time of a neurosurgical procedure or lumbar puncture, or as the result of infection in a contiguous focus, such as occurs in ottitis media or mastoiditis (Wise et al., 1969; Bray and Calcaterra, 1976). Meningitis due to *Pseudomonas* has followed spinal anesthesia as a result of contaminated solutions, wounds of the head, infected spina bifida, myelography, laminectomy, and placement of ventriculoperitoneal shunts for relief of increased intercranial pressure.

In general, *Pseudomonas* meningitis has been a much more indolent type of disease than that due to the common pathogens *Streptococcus pneumoniae* or *Neisseria meningitidis*. As with other types of meningitis, there is nuchal rigidity, fever, and a marked polymorphonuclear response in the spinal fluid with an increase in spinal fluid protein and decrease in spinal fluid sugar content. Unfortunately, microorganisms usually are not seen on the initial Gram stain, but they will grow in culture within 24 hr. Many cases of *Pseudomonas* meningitis have been associated with ventriculitis, and it is probable that the major reason this infection is so difficult to eradicate is because of the ventriculitis.

Subdural empyema, epidural empyema, and brain absess all have been caused by *Pseudomonas*. Diagnosis of these conditions is difficult, as is the treatment in most situations. The etiology of these infections has been a contiguous focus of infection and thus therapy is a combination of antibiotics and surgical drainage.

There is no known optimal therapy for *Pseudomonas* infections of the central nervous system. In the majority of situations it is necessary to use intrathecal aminoglycoside therapy in order to achieve a cure. An anti-*Pseudomonas* penicillin and aminoglycosides should also be given intravenously. There have been reports of successful treatment of *Pseudomonas* meningitis with azlocillin (Davidson et al., 1982) and ceftazadine, and it may be possible to treat these infections with new quinolones, administered either as parenteral therapy or orally. If lumbar injection of aminoglycosides is used, it is necessary to give a dose of 8 mg of gentamicin or tobramycin daily.

Otolaryngological Infections

Pseudomonas is the major cause of external otitis, the so-called swimmer's ear. *Pseudomonas* is rarely cultured from dry skin, but it flourishes in the moist environment of the ear, particularly in tropical climates and following minor trauma to the ear. It has occurred after hot tube use (Harelaar et al., 1983). Examination of the patient shows that the skin of the external auditory canal is devoid of wax and is covered by exudates of various hues, usually greenish. The underlying skin is reddened and denuded and there is often a crust and a weeping, edematous nature to the skin which bleeds easily.

The external canal is usually extremely sensitive, and manipulation of the external ear produces marked pain. Although the drum may appear to be inflamed, it moves well with a pneumatic otoscope, indicating that disease is limited to the external ear.

Topical treatment usually will be sufficient unless there is extensive accompanying cellulitis. The external canal should be cleaned and topical antibiotic drops containing polymyxin, neomycin, and hydrocortisone may be used locally.

Malignant otitis externa is an extremely serious disease that was first described in the late 1950s and whose natural history was described by Chandler (1977). This disease is seen most frequently in elderly patients, most of whom have underlying diseases such as diabetes (Dawson, 1978; Doroghazi et al., 1981). However, in recent years malignant otitis external has been seen in younger patients, including small children. The infection begins as an external otitis that does not respond to topical therapy. Infection progresses to gain access to the deep tissues of the external auditory canal through defects in the cartilage which form the floor of the external auditory canal. As a result, there is development of osteomyelitis of the temporal bone with involvement of areas where the cranial nerves exit. Further extension of the osteomyelitis may advance to the petrous ridge, with the development of cerebral or parietal lobe brain abcess.

The diagnosis of *Pseudomonas* malignant otitis should be suggested by complaints of severe pain and the appearance of persistent granulation tissue in the floor of external auditory meatus near the junction of the osseous and cartilaginous portions. The patients will have a purulent discharge, edema, and swelling of the tissue of the external ear. The patient usually is afebrile and general laboratory tests are not helpful, since the white count is often normal. Most patients have markedly elevated sedimentation rates. Initially computer-tomographic scans of the mastoid are not helpful, but as the disease progresses the mastoid will be shown to have undergone extensive damage (Curtin et al., 1982; Strashun et al., 1984). Radionuclide scans, however, with technestium-99 may be extremely helpful, as may be gallium scans on occasion (Parisier et al., 1982; Lucente et al., 1983).

Before the adequate recognition of this disease and the application of appropriate antimicrobial therapy, mortality exceeded 50%. Recently, with the prompt administration of appropriate antimicrobial therapy, mortality is between 20 and 25%. One must realize, however, that the patient who presents with facial nerve paralysis has a much smaller chance of surviving than the patient who has no nerves involved at the time the diagnosis is made. If more than one nerve is involved at the time of presentation, mortality approaches more than 75%.

Once the diagnosis of this infection is made, treatment is a combination of high-dose anti-*Pseudomonas* penicillin combined with aminoglycoside or one of the new anti-*Pseudomonas* cephalosporins, cefsulodin, ceftazidime, or aztreonam. Therapy must be for a minimum of 6 weeks, with attention given to a decline in the erythrocyte sedimentation rate. Although it is stated that surgery may not be necessary in most patients now that adequate antimicrobial agents are available, this has not been our experience and we believe that debridement of granulation tissue in abscessed cavities is necessary to achieve cure. Since many of the patients are diabetics, it is particularly important to pay close attention to renal function and to the auditory status, since toxicity due to aminoglycoside may develop in these patients.

Pseudomonas aeruginosa is infrequently cultured from the mouth of normal individuals. Patients with leukemia, however, do develop periodontal disease which can progress to osteomyelitis of the mandible. Ulceration of the pharynx, buccal mucosa, and tonsils with the formation of extensive necrotic membranes that resemble diphtheria are seen in such patients. *Pseudomonas* nosocomial sinusitis may develop in patients who have long-term nasopharyngeal tubes following resuscitation in severe trauma (Caplan and Hoyt, 1982). This infection usually will respond to removal of nasal tubes and brief treatment with anti-*Pseudomonas* antibiotics and decongestants. Other serious sinus infections due to *Pseudomonas* are infrequent (Fried et al., 1984).

Osteomyelitis and Septic Arthritis

There appear to be several different populations of patients who develop *Pseudomonas* osteomyelitis. There are those who develop hematogenous disease. These patients are often heroin addicts or individuals receiving long-term intravenous therapy, particularly those with underlying hematological malignancy (Holzman and Bishko, 1971; Lewis et al., 1973; Miskew et al., 1983; Klein et al., 1983). Some patients develop *Pseudomonas* osteomyelitis in association with previous surgery or instrumentation and the use of prolonged antimicrobial therapy, particularly with drugs lacking *Pseudomonas* activity, such as the first- or second-generation cephalosporins. There are also individuals who develop osteomyelitis of the calcaneus as a result of puncture wounds.

Overall, chronic osteomyelitis due to *Pseudomonas* is the most common form of *Pseudomonas* bone disease. It can follow any operation on previous osteomyelitis. The illness characteristically is chronic and usually is the result of previous antimicrobial therapy and inability to close a wound completely, resulting in colonization of sinus tracts with *Pseudomonas* which eventually penetrate deeply into the bone tissue. It is important to realize that in chronic osteomyelitis culture of *Pseudomonas* from a sinus tract does not establish *Pseudomonas* as the etiological agent; rather, it is necessary to perform a biopsy of bone to establish that *Pseudomonas* is truly the organism involved in the bone disease and not merely a sinus tract pathogen. In most situations the presence of a foreign body such as a rod or other device, or poor blood supply, predisposes the patient to infection with *Pseudomonas*.

It is questionable whether patients are ever truly cured of *Pseudomonas* ostemyelitis, since the organism often can be cultured from a sinus tract many years after an apparent cure. Current therapy for this disease is a combination of anti-*Pseudomonas* antibiotics (Olive et al., 1983), usually an anti-*Pseudomonas* penicillin and an aminoglycoside. Excellent results have recently been obtained with ceftazidime and cefsulodin and most recently with aztreonam. Therapy must be for long periods, that is, up to 6 weeks. Investigation of the use of anti-*Pseudomonas* quinolones such as ciprofloxacin is underway. Preliminary results are encouraging and it may be possible to achieve higher cures in this problematic disease.

Vertebral osteomyelitis has long been associated with urinary tract infection. It has been thought to be the result of spread via lumbrar paravertebral veins that communicate with the veins of the pelvis. Since the end of the 1960s there have been numerous reports of vertebral osteomyelitis in heroin addicts. Infection of vertebrae due to *Pseudomonas*, whether in normal individuals or heroin addicts, usually begins with low back pain of several weeks duration. Fever often is low grade or even absent. The patient rarely has a leukocytosis, although there characteristically is an elevation of the erythrocyte sedimentation rate. Blood cultures are infrequently positive. Although the spine is the area most frequently involved, sacroiliac joints and sternal clavicular joints have also been the sites of involvement. The initial radiological signs show narrowing of the intervertebral disk space and irregular cortical erosion of the vertebral end plates and subchondral areas of erosion. The vertebral end plates of both sides of a disk space may be involved. The use of scintillation bone scans have markedly increased our ability to diagnose this illness. Nonetheless, it is essential to establish the precise etiology by biopsy of bone or aspiration of the effected disk space, since organisms other than *Pseudomonas* may be involved.

Therapy has up to the present been the use of an anti-*Pseudomonas* penicillin, either singly or in combination with aminoglycosides. The new anti-*Pseudomonas* β-lactam drugs have been used infrequently. However, these drugs should prove equally effective. The precise role of bedrest in the treatment of this illness has not been established, although it has been our recommendation that the patient be placed in a bivalve spica cast to encourage healing.

Puncture wounds of the heel as a cause of osteomyelitis are seen most frequently in children (Green and Bruno, 1980; Jacobs, et al., 1982). The clinical picture differs from that seen with hematogenous osteomyelitis in that following puncture of the foot there is a temporary decrease in pain with the development of tenderness and swelling of the foot after several weeks. Usually the patient does not have a leukocytosis or fever. By the time the patient is seen by a physician, the x-ray usually will show loss of osseous structure in the calcaneus (Sieber et al., 1982). Whether the initial infection was caused by *Pseudomonas* is unclear. Indeed, it is possible that *Pseudomonas* was introduced into the wound through the common practice of soaking puncture wounds. This disease is one in which surgical intervention, as well as antimicrobial therapy, is needed, since the devitalized tissue must be removed in order to allow for proper drainage. An anti-*Pseudomonas* penicillin, azlocillin, piperacillin, or ticarcillin, combined with an aminoglycoside would be appropriate therapy, but ceftazidime or aztreonam would also be effective.

Septic arthritis due to *Pseudomonas* is uncommon and accounts for less than 3% of septic arthritis (Greico, 1972; Goldenberg and Cohen, 1976). It has been seen most often in patients with diabetes, in individuals with hematological malignancy, and as a cause of sternal clavicular arthritis after cardiac surgery and hyperalimentation. It had been a particular problem in heroin addicts in the 1960s and 1970s. *Pseudomonas* arthritis has also occurred as a nosocomial infection when joint irrigation fluid has beome contaminated. This has occurred in spite of the use of polymyxin B in irrigating fluid.

Diagnosis can only be established by the growth of the organism, since there is no clinical way to differentiate *Pseudomonas* septic arthritis from infectious arthritis due to staphylococci or to other gram-negative bacilli. In the majority of cases of *Pseudomonas* arthritis the initial Gram stain is negative, although organisms will grow after 24 hr of incubation. There may not be a polymorphonuclear response in patients with hematological malignancy. Treatment in the past has been with aminoglycosides. It should be noted that in infected joint fluid at pH 6 aminoglycosides are 30-fold less active than they are at pH 7.5. Furthermore, the aminoglycoside activity is reduced through complexing with the cellular debris in the joint and with the degenerating white blood cells. Joint fluid cultures often remain positive for more than 72 hr. Thus it would be necessary to instill aminoglycosides into the joint if these agents are used. Combined systemic therapy with anti-*Pseudomonas* penicillins and with aminoglycosides selected on the basis of susceptibility has been the usual form of therapy, with a 3-week minimum period. Recently we have used aztreonam or ceftazidime, 2 g, every 8 hr, or imipenem, 1 g, every 6 hours, with much earlier sterilization of joint fluid and a better outcome than in the past when we relied only on aminoglycosides.

Urinary Tract Infections

Pseudomonas is a major cause of noscomial urinary tract infections (Turck and Stamm, 1981). *Pseudomonas* is the second most common nosocomial organism, at 11.5%, after *E. coli* at 31.7%. Most *Pseudomonas* urinary tract infections have been associated with manipulations of the bladder, either because of urethral catheter, cystoscopy, or prostate surgery (Strand et al., 1982; Sherertz and Sarubbi, 1983). The clinical course of *Pseudomonas* urinary infection, although similar to that due to organisms such as *E. coli*, frequently is much more complicated and cure is less frequent. *Pseudomonas* urinary infections may be associated with the presence of bladder or kidney stones, although this occurs less frequently than with *Proteus mirabilis* infections. Pylonephritis due to *Pseudomonas* usually develops as a result of ascending infection, although hematogenous pyelonephritis and perinephric abscesses have occurred in burn patients, narcotic addicts, and patients with hematological malignancies. In general, the urinary tract has been a much more important source of infection than any organ system involved in systemic infections.

Many of the antimicrobial agents discussed in this article have been used to treat urinary tract infections. Cure rates have been generally lower than those seen with other urinary pathogens, such as *E. coli* or *Klebsiella*. The reason for this is that infections due to *Pseudomonas* characteristically occur in individuals in whom it is impossible to remove indwelling urethral catheters or where there are abnormalities of the urinary tract that are not amenable to surgical reconstruction. Indeed, many studies have shown that *P. aeruginosa* urinary tract infections more commonly occur on neurological and neurosurgical services (Sherertz and Sarubbi, 1983). At the present time, no one particular agent would be recommended as the preferred therapy of *Pseudomonas* urinary tract infections. The availability of anti-*Pseudomonas* oral compounds, quinolones, may make it possible to make further progress in the therapy of urinary infections due to this organism.

Ophthalmological Infections

Pseudomonas infections of the eye have been known since the end of the 1800s. The most common form of infection has been corneal ulceration which may progress to panophthalmitis (Ayliffe et al., 1966). Superfacial eye infections and blepharoconjunctivitis may follow the use of lenses or mascara (Wilson and Ahearn, 1977). Recently *Pseudomonas* infection has complicated extended-wear contact lenses (Adams et al., 1983). Patients usually have undergone recent contact lense manipulation or they have failed to properly disinfect their contact lenses. Rosner et al.(1983) reported 14 cases of corneal abscesses in contact lense wearers. Wearers of silicone contact lenses accounted for 21% of the corneal abscesses. There have been other reports of *Pseudomonas* corneal ulcers with extended-wear soft contact lenses used for myopia (Hassman and Sugar, 1983).

Neonates also have been reported to acquire *Pseudomonas* ophthalmia neonatorum following trauma to the eyes as the result of wearing a positive pressure mask, from inadvertant lesions following the installation of antibiotics in the eye, or from pads used to protect the eye from phototherapy used in the treatment of hyperbilirubinemia (Cole et al., 1980). Hilton et al. (1983) reported the development of nosocomial *Pseudomonas* eye infections due to spillage of tracheal secretions into the eyes of tracheotomized patients.

Therapy of *Pseudomonas* eye infections depends upon the degree of infection. Corneal infections may be treated with the topical use of gentamicin or tobramycin; however, more serious infections must be treated with subconjunctival installation of aminoglycosides as well as topical administration. In endophthalmitis aminoglycosides have been injected intravitreously and systemic therapy with aminoglycosides and anti-*Pseudomonas* penicillins has been used (Leveille et al.,

1983). There is some evidence that ceftazidime may achieve therapeutic concentrations within eye tissue and it may be useful in the therapy of *Pseudomonas* eye infections.

Skin Infections

Green nail syndrome, toe web infections, hot tub folliculitis, superinfections in chronic antibiotics-treated acne, and infectious eczematoid dermatitis are examples of mild cutaneous infection due to *P. aeruginosa*. These infections occur in otherwise healthy individuals. Acute dermatological infection may follow the use of a hot tub or spa. The lesions may range from macular to pustular eruption or folliculitis. Infection has been associated with infrequent or inadequate cleaning of hot tub water and the presence of carbon-rich debris that collects after the use of many people (Gustafson et al., 1983; Smith, 1982). The *Pseudomonas* strains that are involved often are type 011, and are thought to enter the skin as a result of dilation of pores caused by the heat (Khabbaz et al, 1983). Therapy of *Pseudomonas* folliculitis as the result of hot pool contamination can be primarily with topical measures. There is no evidence that the use of topical iodine preparations will increase or speed healing. Prevention of this illness can be achieved by correct attention to the chlorination of water and frequent removal of water from these hot tub devices.

Decubitus ulcers in chronically ill patients frequently are caused by multiple bacteria, including *Pseudomonas*. Use of drugs such as azlocillin or piperacillin may prove effective, since these agents also inhibit enterococci which are frequently found in these infections (Sabbaj et al., 1983).

Peritonitis

Pseudomonas as a cause of peritonitis may follow surgery in which other antimicrobial agents have been used to treat infection. This is particularly true when first- or second-generation cephalosporins have been the antibiotics rather than a combination with an aminoglycoside. *Pseudomonas aeruginosa* also can produce peritonitis at the catheter site in patients on chronic peritoneal dialysis (Perrott et al., 1982). These episodes of peritonitis were associated with contaminated polxamer—iodine solution. Peritonitis can be treated by several different mechanisms. Peritonitis associated with chronic peritoneal dialysis can be effectively treated by installation of an aminoglycoside into the peritoneal cavity during dialysis. Use of a 10-μg/ml solution will effectively eradicate the organisms. Peritonitis due to *Pseudomonas* which follows other forms of surgery in which installation of antibiotic into the peritoneum is not feasible should be treated with a combination of an anti-*Pseudomonas* penicillin and an aminoglycoside. Alternatively, one of the new anti-*Pseudomonas* cephalosporins such as ceftazidime, cefsulodin, and cefoperazone may be used, and aztreonam is a possible agent when it becomes available. Krothapalli et al.(1982)

noted that *Pseudomonas* accounted for 38% of gram-negative peritonitis in continuous ambulatory peritoneal dialysis. Tenchkhoff catheters had to be removed in all situations in order to achieve cure of the peritoneal infection.

Other Infections

Pseudomonas aeruginosa, like other gram-negative bacilli, has been a cause of other forms of infection such as pericarditis, abscesses, and vascular infections. In the majority of these situations the infection has been acquired nosocomially and has followed the extensive use of the first- or second-generation cephalosporin antibiotics which lack activity against *Pseudomonas*. There are no particular clinical or laboratory clues which would unequivocally allow one to suspect *Pseudomonas* as the sole infecting agent in these various situations. *Pseudomonas aeruginosa* has been associated with outbreaks of diarrhea in neonatal units in a number of countries (Jellard and Churcher, 1967; Falcao et al., 1972). Overall, many conditions will continue to predispose to *Pseudomonas* infections (Table 10).

Table 10 Conditions Known to Predispose to *Pseudomonas* Infection

Condition	Major infection
Burn	Cellulitis, septicemia
Corneal trauma	Corneal ulcers, corneal abscess, panophthalmitis
Cystic fibrosis	Pneumonitis
Drug addiction	Osteomyelitis, endocarditis
Diabetes	Urinary infection, malignant otitis
Leukemia with neutropenia	Septicemia, deep tissue infection, septic pneumonitis
Neonatal antibiotic therapy	Diarrhea, sepsis
Orthopedic surgery with cephalosporin therapy	Urinary infection, osteomyelitis
Surgery of central nervous system	Meningitis, shunt infection
Tracheotomy	Pneumonia
Urinary catheterization and antibiotic therapy	Urinary tract infection
Vascular catheters	Suppurative phlebitis

GENERAL CONSIDERATIONS IN THE TREATMENT OF INFECTIONS

The treatment of P. aeruginosa infections was revolutionized with the introduction in 1963 of gentamicin. Further improvements were made with the introduction of carbenicillin in 1967. At the present time it is the feeling of most infectious disease experts that septicemica should be treated with either an anti-Pseudomonas penicillin and an aminoglycoside or with an anti-Pseudomonas cephalosporin and an aminoglycoside, especially if the patient has underlying host defects (Love et al., 1980; Wade and Schimpff, 1982). It has not been possible to demonstrate the clear superiority of one anti-Pseudomonas penicillin over another, or of one anti-Pseudomonas aminoglycoside over another in terms of clinical efficacy, even though marked differences are seen in the in vitro activities of these agents. It is also exceedingly difficult to interpret the results of therapy of respiratory tract infection due to Pseudomonas. In the presence of underlying structural disease of the lung due to chronic bronchiectasis or bronchitis, or in the patient with cystic fibrosis, Pseudomonas is rarely eradicated, even though there is clinical and radiographic improvement. In urinary tract infections, comparative studies have not shown the superiority of any individual agent, since the patients who develop Pseudomonas urinary infection are bound to develop recurrences. Nonetheless, new anti-Pseudomonas penicillins and anti-Pseudomonas cephalosporins or monobactams offer an alternative to aminoglycosides, which may not achieve adequate levels in the renal tissue of patients with depressed renal function.

Resistance of Pseudomonas to the different antimicrobial agents has developed in a number of situations; for example, we have noted the development of resistance in Pseudomonas from patients with cystic fibrosis treated even with ceftazidine (Scully and Neu, 1984). Similarly, in a study by Gentry et al (1983) of Pseudomonas infections, cure rates with ceftazidime were lowest for Pseudomonas compared to other gram-negative organisms. Thus, even with drugs which have excellent in vitro activity, the type of patient who develops a Pseudomonas infection is such that failure and development of resistance will continue to be common. We have noted for all classes of antimicrobial agents including penicillin, cephalosporin, monobactam, thienamycin, aminoglycoside, and quinolone that Pseudomonas infection in the presence of a major structural disease such as chronic lung infection, complicated urinary infection, or osteomyelitis with foreign bodies will develop resistance. True cure of a number of Pseudomonas infections probably never is truly achieved. This is particularly true for osteomyelitis, urinary tract infection, and pulmonary infections. In each of these situations anti-Pseudomonas therapy should be directed at reducing the number of organisms so that the individual is able to continue a productive life.

VACCINES

Immunization with a variety of *Pseudomonas* vaccines have been shown to enhance the resistance of experimental animals to *P. aeruginosa* infection (Pier et al., 1978; Sensakovic and Bartell, 1977; Tegtmeir and Anderson, 1983; Pennington, 1979; Homma et al., 1983). There has been a reduction in deaths due to *P. aeruginosa* with the use of lipopolysaccharide vaccines in patients with burns, but no evidence of protection has been noted in patients with cancer (Young et al., 1973). Tegtmeir and Anderson (1983) have shown that active immunization of mice with type-specific *Pseudomonas* lipopolysaccharide vaccine will produce protection against *Pseudomonas* sepsis in granulocytopenic mice. In their study mice immunized nonspecifically with a heterologous lipopolysaccharide in given specific antisera survived significantly longer than controls, supporting the hypothesis that active immunization with lopopolysaccharide induces a nonspecific cellular response that is antibody dependent. This study seems to imply that passive immunization or active immunization with vaccines that only induce antibody may not be effective in neutropenic patients. Nonetheless, further testing of vaccines and hyperimmune sera and the development of monoclonal antibodies may be extremely important in specific situations.

CONCLUSION

Pseudomonas aeruginosa remains an extremely important organism as a cause of numerous forms of clinical infection in man. Although many new antimicrobial agents have been developed to deal with *Pseudomonas*, the underlying host defects of patients who develop *Pseudomonas* infections are such that therapy tends to be much less successful than that achieved against other bacteria. Furthermore, *Pseudomonas* has the unique ability to become resistant to many different antimicrobial agents simultaneously, either through chromosomal or plasmid-mediated mechanisms of resistance. Clearly, much further research is needed into our understanding of this intriguing organism.

REFERENCES

Adams, Jr., C. P., Cohen, E. J., Laibson, F. R., Galentin, P., and Arenstsen, J. J. (1983). *Am. J. Ophthalmol.* 96:705.
Alexander, J. W., and Fisher, M. W. (1970). *J. Trauma* 10:565.
Alexander, J. W., Fisher, M. N., and MacMillan, G. B. (1971). *Arch. Surg.* 102:31.

Allen, J. R., Heightower, A. W., Martin, S. M., and Dixon, R. E. (1981). *Am J. Med.* 70:389.
Anderson, E. T., Young, L. S., and Hewitt, W. L. (1978). *Chemotherapy* 24:45.
Andriole, V. T. (1971). *J. Infect. Dis. Suppl.* 124:46.
Ayliffe, G. A. J., Barry, R. D., Lowbury, E. J. L., Ropert-Hall, M. J., and Walker, W. M. (1966). *Lancet* 1:1113.
Baltch, A. L., and Griffin, P. E. (1977). *Am. J. Med. Sci.* 274:119.
Baltch, A. L., Griffin, P. E., and Hammer, M. (1979). *J. Lab. Clin. Med.* 93:600.
Bishop, J. F., Schmipff, S. C., Diggs, C. H., and Wiernik, P. H. (1981). *Ann. Intern. Med.* 95:549.
Bodey, G. P. (1970). *Am. J. Med. Sci.* 260:82.
Bodey, G. P. (1982). In *Combination Antibiotic Therapy in the Comprimised Host* (J. Klastersky and M. J. Stuquet, eds.), Raven Press, New York, p. 147.
Bray, P. A., and Calcaterra, T. C. (1976). *Laryngoscope* 86:1386.
Bryan, L. E., and Van Den Elzen, H. M. (1977). *Antimicrob. Agents Chemother.* 12:163.
Bryan, L. E., Haraphonge, R., and Van Den Elzen, H. M. (1976) *J. Antibiot.* 29:743.
Bryan, C. S., Reynolds, K. L., and Brenner, E. R. (1983). *Rev. Infect. Dis.* 5:629.
Buck, A. C., and Cooke, E. M. (1969). *J. Med. Microbiol.* 2:521.
Caplan, E. S., and Hoyt, N. J. (1982). *J. Am. Med. Assoc.* 247:639.
Chandler, J. R. (1977). *Ann. Otol.* 86:417.
Chin, N. X., and Neu, H. C. (1983). *Antimicrob. Angents Chemother.* 25:319.
Chin, N. X., and Neu, H. C. (1984). *Antimicrob. Agents Chemother.* 25:319.
Cole, G. J., Davies, D. P., and Austin, D. J. (1980). *Br. Med. J.* 281:440.
Costerton, J. W., Lam, J., Lam, K., and Chan, R. (1983). *Rev. Infect. Dis. Suppl.* 5:867.
Cross, A. S., Sadoff, J. C., Iglenski, B. H., and Sokol, P. A. (1980). *J. Infect. Dis.* 142:538.
Cross, A., Allen, J. R., Burke, J., et al. (1983). *Rev. Infect. Dis. Suppl.* 5:837.
Crowe, K. E., Bass, J. A., Young, V. M., and Strauss, D. C. (1982). *J. Clin. Microbiol.* 15:115.
Curtin, J. A., Petersdorf, R. C., and Bennett, Jr., I. L. (1961). *Ann. Intern. Med.* 54:1077.
Curtin, J. D., Wolfe, P., and May, M. (1982). *Radiology* 145:383.
Davidson, S., Yellin, E. D., Shaked, I., and Rubenstein, E. (1982). *Infection Supp.* 3:168.

Dawson, D. A. (1978). *J. Laryngol. Otol.* 92:803.
Doroghazi, R. M., Nadol, Jr., J. B., Hyslop, Jr., N. E., Baker, A. S., and Axelrod, L. (1981). *Am. J. Med.* 71:603.
Falcao, D. P., Mendonca, C. P.. Scrassalo, A., deAlmeida, B. B., Hart, L., Farmer, L. H., and Farmer, J. J. (1972). *Lancet* 2: 38.
Favero, M. S., Carson, L. A., Bond, W. W., and Peterson, N. J. (1971). *Science* 173:836.
Feeman, R., and McPeake, P. K. (1983). *Thorax* 37:732.
Flick, M. R., and Cluff, L. E. (1976). *Am. J. Med.* 60:501.
Forkner, Jr., C. E. (1960). *Pseudomonas aeruginosa Infections*, Grune and Stratton, New York.
Fraenkel, E. (1912). *J. Hyg.* 72:486.
Fried, M. P., Kelly, J. H., and Strome, M. (1984). *Laryngoscope* 94:192.
Fu, K. P., and Neu, H. C. (1978). *J. Antibiot.* 31:135.
Furth, A. (1979). *Beta-Lactamases* (J. M. T. Hamilton-Miller and J. T. Smith, eds.), Academic, London, p. 403.
Gentry, L. D., Douthit, M. D., Childs, S. J., and Madsen, P. O. (1983). *J. Antimicrob. Chemother. Supp. A* 12:53.
Goldenberg, D. L., and Cohen, A. S. (1976). *Am. J. Med.* 60:369.
Golfarb, W. B., and Margraf, H. (1967). *Ann. Surg.* 165:104.
Green, N. E., and Bruno, J. M. (1980). *South. Med. J.* 73:146.
Greico, M. H. (1972). *J. Bone Joint Surg.* 54:1693.
Grief, Z., Merzbach, D., and Freundlich, E. (1983). *Isr. J. Med. Sci.* 19:977.
Grogan, J. B. (1966). *J. Trauma* 6:639.
Gross, P. A., Neu, H. C., Aswapokee, P., VanAntwerpen, C., and Aswapokee, P. (1980). *Am. J. Med.* 68:219.
Gustafson, T. L., Band, J. D., Hutcheson, Jr., R. H., and Schaffner, W. (1983). *Rev. Infect. Dis.* 5:1.
Hague, H., and Russell, D. (1974). *Antimicrob. Agents Chemother.* 6:200.
Hassman, G., and Sugar, J. (1983). *Arch. Ophthalmol.* 101:1549.
Harelaar, A. H., Bosman, M., and Borst, J. (1983). *J. Hyg.* 90: 489.
Hilton, E., Adams, A. A., Uliss, A., Lesser, M. L., Samuels, S., and Lowy, F. D. (1983) *Lancet* 1:1318.
Holby, N. (1977). *Acta Pathol. Scand. Suppl. C* 266:3.
Holzman, R. S., and Bishko, F. (197). *Ann. Intern. Med.* 75:693.
Homma, J. Y., Abe, C., Yangawa, R., and Noda, H. (1983). *Rev. Infect. Dis. Suppl.* 5:858.
Hoogkamp, Korstanje, J. A., and van der Lang, J. (1983). *J. Antimicrob. Chemother.* 12:175.
Iglewski, B. H., and Kabat, D. (1975). *Proc. Nat. Acad. Sci. U.S.A.* 72:2284.

Jackson, M. A., Wong, K. Y., and Lampkin, B. (1983). *Pediatr. Infect. Dis.* 1:239.
Jacobs, R. F., Adelman, L., Sacks, C. M., and Wilson, C. B. (1982). *Pediatrics* 69:432.
Jellard, C. H., and Churcher, G. M. (1967). *J. Hyg.* 65:219.
Jouvenot, M., Bonn, P., and Michel-Briand, Y. (1983). *J. Antimicrob. Chemother.* 12:451.
Khabbaz, R. F., McKinley, T. W., Goodman, R. A., Hightower, A. W., Highsmith, A. K., Tait, D. A., and Band, J. D. (1983) *Am. J. Med.* 74:73.
Klein, B., Mittelman, M., Katz, R. and Djaldetti, M. (1983). *Chest* 83:143.
Kominos, S. D., Copeland, G. E., Grosiak, B., and Postic, B. (1972). *Appl. Microbiol.* 24:567.
Kominos, S. D., Copeland, G. E., and Delento, C. A. (1977). In *Pseudomonas aeruginosa* (V. M. Young, ed.), Raven Press, New York, p. 59.
Krothapalli, R., Duffy, W. B., Lacke, C., Payne, W., Patel, H., Perez, V., and Senekjian, H. O. (1982). *Arch. Intern. Med.* 142:1862.
Kulczycki, L. L., Murphy, T. M., and Bellanti, J. A. (1978). *Am. Med. Assoc.* 240:30.
Kumada, T., and Neu, H. C. (1983). In *Ofoxacin, DL8280 Broad Spectrum Agent,* Excerpta Medica Amsterdam, p. 1.
Leake, E. S., Wright, M. J., and Kerger, A. S. (1978). *Exp. Mol. Pathol.* 29:241.
Leveille, A. S., McMullen, F. D., and Cavanagh, H. D. (1983). *Ophthalmology* 90:38.
Lewis, R., Gorbach, S., and Attner, S. (1973). *N. Engl. J. Med.* 286:1303.
Liu, P. V. (1966). *J. Infect. Dis.* 116:112.
Livermore, D. M. (1983). *J. Antimicrob. Chemother.* 168.
Love, L. J., Schimpff, S. C., Schiffer, C. A., and Wiernik, P. H. (1980). *Am. J. Med.* 68:643.
Lowbury, E. J. L., Thom, B. T., Lilly, J. A., Babb, J. R., and Whithall, K. (1970). *J. Med. Microbiol.* 3:39.
Lucente, F. E., Parisler, S. C., and Som, P. M. (1983). *Laryngoscope* 93:279.
Marrie, T. J., Major, Gurwith, M., Ronald, A. R., Harding, G. K., Forrest, G., and Forsythe, W. (1978). *Can. Med. Assoc. J.* 119:563.
Miller, G. H., Hare, R. S., Sabatelli, F. J., and Waitz, J. A. (1980). *Ind. Microbiol.* 21:91.
Mirelman, D., Nuchamowitz, Y., and Rubenstein, E. (1981). *Antimicrob. Agents Chemother.* 19:687.
Miskew, D. B., Lorenz, M. A., Pearson, R. L., and Pankovich, A. M. (1983). *J. Bone Joint Surg.* 65:829.

Mitsuhashi, S., Kawabe, H., and Fuse, A. (1975) In *Microbial Drug Resistance* (S. Mitsuhashi and H. Hosimote, eds.), University Tokyo, Tokyo, p. 515.
Mitsuhashi, S., and Kawabe, H. (1982). In *Aminoglycoside Antibiotics* (A. Welton and H. C. Neu, eds.), Marcel Dekker, New York, p. 97.
Moelhring, Jr., R. C. (1983). *Rev. Infect. Dis. Supp.* 2:212.
Moller, N. E., Koch, C., Vesterhauge, S., and Jensen, K. (1982). *Scand. J. Infect. Dis.* 14:207.
Neu, H. C. (1982a). *Annu. Rev. Pharmacol. Toxicol.* 22:599.
Neu, H. C. (1982b). In *The Aminoglycosides* (A. Welton and H. C. Neu, eds.), Marcel Dekker, New York.
Neu, H. C. (1982c). *Med. Clin. North Am.* 66:61.
Neu, H. C. (1982d). *Rev. Infect. Dis. Supp.* 4:288.
Neu, H. C. (1984). *Am. J. Med.* 76:11.
Neu, H. C. (1983). *Rev. Infect. Dis. Supp.* 2:319.
Neu, H. C., and Fu, K. P. (1979). *Antimicrob. Agents Chemother.* 15:646.
Neu, H. C., and Kamimura, T. (1981). *Antimicrob. Agents Chemother.* 19:1013.
Neu, H. C., and Labthavikul, P. (1981). *J. Antimicrob. Chemother. Suppl. E* 8:111.
Neu, H. C., and Labthavikul, P. (1982a). *Antimicrob. Agents Chemother.* 21:11.
Neu, H. C., and Labthavikul, P. (1982b). *Antimicrob. Agents Chemother.* 21:180.
Neu, H. C., and Labthavikul, P. (1982c). *Antimicrob. Agents Chemother.* 21:906.
Neu, H. C., and Labthavikul, P. (1982d). *Antimicrob. Agents Chemother.* 22:23.
Neu, H. C., Fu, K. P., Aswapokee, N., Aswapokee, P., and Kung, K. (1979). *Antimicrob. Agents Chemother.* 16:150.
Noone, M. R., Pitt, T. L., Bedder, M., Hewlett, A. M., and Rogers, K. B. (1983). *Br. Med. J.* 1:341.
Olive, G., Mogabgab, W. J., Holmes, B., Pollock, B., Pauling, B., and Beville, R. (1983). *J. Antimicrob. Chemother Suppl. B.* 11:153.
Pancoast, S., Prince, A. S., Francke, E. L., and Neu, H. C. (1981). *Arch. Intern. Med.* 141:1447.
Parisier, S. C., Lucente, F. E., Som, P. M. Hirschman, S. Z., Arnold, L. M., and Roffman, J. D. (1982). *Laryngoscope* 92:1016.
Parry, M. F., and Neu, H. C. (1978). *Am. J. Med.* 64:961.
Parry, M. F., Neu, H. C., Merlino, M., Gaerlan, P. F., Ores, C. N., and Denning, C. R. (1977). *J. Pediatr.* 90:144.
Pennington, J. E. (1979). *J. Infect. Dis.* 140:73.
Pennington, J. E., Reynolds, H. Y., and Carbone, P. P. (1973).

Am. J. Med. 55:155.
Permin, H., Koch, C., Holby, N., Christensen, H. O., Moller, A. F., and Moller, S. (1983). J. Antimicrob. Chemother Suppl. A 12:313.
Perrott, P. L., Terry, P. M., Whitworth, E. N., Frawley, L. W., Coble, R. S., Wachsmuth, I. K., and McGowan, Jr., J. E. (1982). Lancet 2:603.
Peterson, P. K. (1980). Pseudomonas aeruginosa (L. D. Sabath, ed.), Huber, Vienna, p. 103.
Petras, G., and Bognar, S. (1981). Acta. Microbiol. Acad. Sci. Hung. 28:367.
Pier, G. B., Sidberry, H. F., and Sadoff, J. C. (1978). Infect. Immun. 22:919.
Pollack, M. (1983). Rev. Infect. Dis. Suppl. 5:979.
Pollack, M., and Young, L. S. (1979). J. Clin. Invest. 63:276.
Pollack, M., Callahan, L. T., and Taylor, N. S. (1976). Infect. Immun. 14:942.
Pollack, M., Huang, A. I., Prescott, R. K., Young, L. S. Hunter, K. W., Greuss, D. F., and Tsai, C. M. (1983). J. Clin. Invest. 72:1874.
Prince, A. S., and Neu, H. C. (1981). Antimicrob. Agents Chemother. 20:545.
Pruitt, Jr., B. A. (1974). J. Infect. Dis. 130:58.
Pruitt, Jr., B. A., Lindberg, R. B., McManus, W. F., and Mason, Jr., A. (1983). Rev. Infect Dis. Suppl. 5:889.
Reyes, M. P., Brown, W. J., and Lerner, A. M. (1978). Medicine 57:57.
Reyes, M. P., and Lerner, A. M. (1983). Rev. Infect. Dis. 5:314.
Reyes, M. P., Paliutke, W. A., Wylin, R. F., and Lerner, A. M. (1973). Medicine 52:173.
Reynolds, H. Y. (1974). J. Infect. Dis. Suppl. 130134.
Reynolds, H. Y. Levine, A. S., Wood, R. E., Zierdt, C. H., Dale, D. C., and Pennington, J. E. (1975). Ann. Intern. Med. 82:819.
Reynolds, H. Y., Disant'Agnese, P. A., and Zierdt, C. H. (1976). J. Am. Med. Assoc. 233:2190.
Rosner, M. Treistern, C., and Blumenthal, M. (1983). Ann. Ophthalmol. 15:949.
Sabath, L. D., Jago, M., and Abraham, E. P. (1965). Biochem. J. 96:739.
Sabbaj, J., Torres, M., and Loza, L. (1983). J. Antimicrob. Chemother. Suppl. B 11:175.
Schimpff, S. C., Moody, M., and Young, V. M. (1970). Antimicrob. Agents Chemother. 1969:240.
Schimpff, S. C., Greene, W. H., Young, V. M., and Wiernik, P. (1973). Eur. J. Cancer 9:449.
Schroth, M. N., Cho, J. J., Green, S. K., and Kominos, S. D. (1977). In Pseudomonas aeruginosa (V. M. Young, ed), p. 1.

Schultz, D. R., and Miller, K. D. (1974). *Infect. Immun.* 10:128.
Scott, R. E., and Robson, H. C. (1976). *Antimicrob. Agents Chemother.* 10:646.
Scully, B. E., and Neu, H. C. (1984). *Arch. Intern. Med.* 1:57.
Sensakovic, J. W., and Bartell, P. F. (1977). *Infect. Immun.* 18:304.
Setia, V., and Gross, P. S. (1977). *Arch. Intern Med.* 137:1698.
Shannon, K., and Phillips, I. (1982). *J. Antimicrob. Chemother.* 9:91.
Sherertz, R. J., and Sarubbi, F. A. (1983). *J. Clin. Microbiol.* 18:160.
Sieber, W. T., Dewan, S., and Williams, Jr., T. M. (1982). *Am. J. Med. Sci.* 283:83.
Singer, C., Kaplan, M. H., and Armstrong, D. (1977). *Am. J. Med.* 62:731.
Smith, G. L. (1982). *Cutis* 29:378.
Southern, Jr., P. M., Mays, B. B., Pierce, A. K., and Sanford, J. P. (1970). *J. Lab. Clin. Med.* 76:548.
Spengler, P. F., Greenough III, W. B., and Stolley, P. D. A. (1978). *Johns Hopkins Med. J.* 142:77.
Stamm, W. E., Weinstein, R. A., and Dixon, R. E. (1981). *Am. J. Med.* 70:393.
Stevens, R. M., Teres, D., Skillman, J. J., and Finegold, D. S. (1974). *Arch. Intern. Med.* 134:106.
Stoodley, B. J., and Thom, B. T. (1970). *J. Med. Microbiol.* 3:367.
Strand, C. L., Bryant, J. D., Morgan, J. W., Foster, Jr., J. G., McDonald, Jr., H. P., and Morgenstern, S. L. (1982). *J. Am. Med. Assoc.* 248:1615.
Strashun, A. M., Nejatheim, M., And Goldsmith, S. J. (1984). *Radiology* 150:541.
Sutter, V. L., and Hurst, V. (1966). *Ann. Surg.* 163:597.
Sutter, V. L., Hurst, V., and Lane, C. W. (1967). *Health Lab. Sci.* 4:245.
Sykes, R. B., and Bush, K. (1982). In *Chemistry and Biology of β-Lactam Antiobiotics* (R. B. Morin, and M. Gorman, eds.), Academic, New York, p. 155.
Tapper, M. L., and Armstrong, D. (1974). *J. Infect. Dis. Suppl.* 130:14.
Tegtmeir, B. R., and Anderson, B. R. (1983). *Rev. Infect. Dis. Suppl.* 5:963.
Turck, M., and Stamm, W. (1981). *Am. J. Med.* 70:651.
Vasil, M. L., Kabat, D., and Iglewski, B. H. (1977). *Infect. Immun.* 16:353.
Verstrate, W., Voets, J. P., and Vanstaen, H. (1975). *Environ. Pollut.* 8:275.
Wade, J. C., and Schimpff, S. C. (1982). In *Combination Antibiotic Therapy in the Compromised Host* (J. Klastersky and M. J. Slaquet,

eds.), Raven Press, New York, p. 125.
Wilson, L. A., and Ahearn, P. G. (1977). *Am. J. Ophthal. 84*:12.
Winston, D. J., Gale, R. P., Meyer, D. V., and Young, L. S. (1979). *Medicine 58*:1.
Wise, B. L., Mathis, J. L., and Jawetz, E. (1969). *J. Neurol. 32*: 432.
Woods, D. E., Bass, J. A., Johanson, Jr., W. G., and Straus, D. C. (1980). *Infect. Immun. 30*:694.
Woods, D. E., Straus, D. C., Johanson, Jr., W. G., and Bass, J. A. (1981). *J. Clin. Invest. 68*:1435.
Woods, D. E., Straus, D. C., Johanson, Jr., W. G., and Bass, J. G. (1983). *Rev. Infect. Dis. Supp. 5*:846.
Wretlind, B., and Pavlorskis, O. R. (1981). *Scand. J. Infect. Dis. Suppl. 29*:13.
Wretlind, B., and Pavlorskis, O. R. (1983). *Rev. Infect. Dis Suppl. 5*:998.
Young, L. S. (1981). *Am. J. Med. 70*:398.
Young, L. S., Meyer, R. D., and Armstrong, D. (1973). *Ann. Intern. Med. 79*:518.
Zimakoff, J., Holby, N., Rosendal, K., and Guibert, J. P. (1983). *J. Hosp. Infect. 4*:31.
Zimmerman, W. (1980). *Antimicrob. Agents Chemother. 18*:94.

5
Ecology, Clinical Significance, and Antimicrobial Susceptibility of *Acinetobacter* and *Moraxella*

ROBERT W. LYONS *St. Francis Hospital and Medical Center, Hartford; University of Connecticut School of Medicine, Farmington; and Yale University School of Medicine, New Haven, Connecticut*

INTRODUCTION

Neither *Acinetobacter* nor *Moraxella* stand in the first rank of the many pathogenic organisms with which most clinicians deal. Indeed, the changes in the taxonomy and nomenclature which have befallen both genera have served to make them even less familiar to the average physician than they would otherwise be. There are few clinicians who associate *Herellea vaginicola* and *Mima polymorpha* that they learned in medical school with *Acinetobacter calcoaceticus* var. *anitratus* and *Acinetobacter calcoaceticus* var. *lwoffi*, respectively. The organisms belonging to *Moraxella* have never been particularly familiar to physicians. A recent edition of a standard two-volume textbook of medicine mentions neither genus (Beeson et al., 1979).

Nevertheless, organisms belonging to the genus *Acinetobacter* have been acquiring more clinical importance in recent years, especially as nosocomial pathogens (Centers for Disease Control, 1979). Significant *Moraxella* infections remain uncommon, and the clinical literature on that organism continues to consist mostly of single-case reports, but the clinician who encounters it must be wary of dismissing it as a contaminant.

ACINETOBACTER

Ecology

Acinetobacter calcoaceticus can be found living free in nature and has been isolated from soil, water, and sewage (Henriksen, 1973). It is usually found in moist areas, but it will survive drying. It has been isolated from air samples in an iron foundry in which the air was heavily contaminated with metal dusts (Cordes et al., 1981).

In hospitals *A. calcoaceticus* has been isolated from respirators and mist tents (Cunha et al., 1980; Snydman et al., 1977). On one occasion it was found to have contaminated the penicillinase being added to blood cultures, and this gave rise to a pseudoepidemic that persisted for 11 months before being discovered (Faris and Sparling, 1972). The organism has also been isolated from the hands of hospital personnel, and hand carriage seems to be an important way in which the organism is spread in nosocomial epidemics (Buxton et al., 1978; French et al., 1980).

Both *A. calcoaceticus* varieties are human commensal organisms. They inhabit moist skin areas, including toe webs, groin, axilla, and antecubital fossa (Taplin et al., 1963). They have also been occasionally recovered from saliva, conjunctival fluid, and the pharynx. DeBord (1939), in his original description of what is now variety *lwoffi*, reported isolation from the vagina and from the conjunctiva and he expressed fear that the organism might be confused with *Neisseria gonorrhoeae*.

The colon is one area of the body that does not seem to readily harbor the organism, even in patients receiving antibiotics. Despite the use of selective media, Grehn and von Graevenitz (1978) were able to find the organism in the stools of only 2 of 50 hospitalized patients and in none of the stools of 50 healthy volunteers.

With the exception *Acinetobacter* organisms have been isolated from most of the usual clinical specimens, although, as with other commensal organsims, the patient's clinical condition must be known before deciding if the isolate represents colonization or infection. Even with knowledge of the patient's condition, that determination is not always easily made.

Acinetobacter var. anitratus

Those strains of *A. calcoaceticus* which oxidize carbohydrates are classified as variety *anitratus* and are more common human pathogens than are the nonoxidizing strains, classified as variety *lwoffi*. They are also more resistant to antibiotics than are the *lwoffi* strains.

Seasonal Incidence

Nosocomial infections with *A. calcoaceticus* var. *anitratus* show a curious seasonal variation. During a 3-year period in the mid-1970s the National Nosocomial Infection Study (NNIS) conducted by the Centers for Disease Control reported 1372 cases of nosocomial infection with *A. calcoaceticus* in 81 hospitals, and a marked preponderance of these infections occurred in late summer (Retailliau et al., 1979).

This late summer peak did not correspond to any factors that the authors of the study could discover. Many of the hospitals did not distinguish between *anitratus* and *lwoffi* strains, but in those that did, the seasonal incidence was associated only with the *anitratus* strains.

The same seasonal incidence of *A. calcoaceticus* var. *anitratus* from nonblood clinical specimens was also reported from Shands Teaching Hospital in Florida, where the mean number of nonblood isolates of that organism in July and August was five times that seen in January (Ramphal and Kluge, 1979). The investigators could offer no explanation for this phenomenon, beyond the fact that it is hotter in Florida in the summer than it is in the winter.

Pneumonia

Pneumonia has been the most common serious clinical problem caused by *A. calcoaceticus* var. *anitratus*. While *Acinetobacter* pneumonia in nonhospitalized patients continues to be something of a medical curiosity, nosocomial pneumonia with this organism seems to be becoming more common.

Nosocomial Pneumonia: In the NNIS study 28.9% of the *A. calcoaceticus* strains isolated were from the lower respiratory tract (Retailliau et al., 1979). This was more than from any other site.

Nosocomial pneumonia with *A. calcoaceticus* var. *anitratus* is chiefly a problem for patients in intensive care units who are receiving ventilatory assistance. The disease may be endemic in a unit, or it may occur as a brief epidemic due to contaminated equipment.

The dramatic nature of epidemics encourages their publication, while endemics probably often go unrecorded, save in the minds of those dealing with them daily. A good account, however, of endemic *Acinetobacter* disease was that published from the Massachusetts General Hospital, recording their experience in the early 1970s (Glew et al., 1977).

At the Massachusetts General Hospital from January 1972 to December 1974 there were 25 cases of nosocomial pneumonia and 11 cases of nosocomial tracheobronchitis among the 53 patients infected with *A. calcoaceticus* var. *anitratus*. All of the pneumonia patients and 9 of the 11 tracheobronchitis patients had either endotracheal tubes in place or had tracheostomies. Most of them (32 of the 36) were in intensive care units and all but one had received prior antibiotic therapy. The chest x-rays of the patients with *Acinetobacter* pneumonia usually showed bronchopneumonia. Cavitation of the lung was seen in only two of the patients.

In another study of endemic infection from a hospital in Buffalo, New York, half of the 80 *Acinetobacter* isolates came from the respiratory tract, and that was the most common site from which *A. calcoaceticus* var. *anitratus* was isolated (O'Connell and Hamilton, 1981).

While the authors do not clearly separate nosocomial from community-acquired infections, they do note that *Acinetobacter* was commonly found in seriously ill patients requiring respiratory assistance.

An outbreak of epidemic *Acinetobacter* respiratory disease at Hartford Hospital in Hartford, Connecticut, was associated with the introduction of a new type of mechanical ventilator, a Wright respirator, into the intensive care units (Cunha et al., 1980). Ten patients were infected with *A. calcoaceticus* var. *anitratus,* all of the 8J serotype. The authors showed that respirators in use were contaminated with *Acinetobacter*, and they showed that artificially contaminated Wright respirators could disseminate an *Acinetobacter*-contaminated aerosol. They did not show how the respirators were becoming contaminated.

A total of 38 patients in an intensive care unit in Salt Lake City, Utah, were infected with *A. calcoaceticus* var. *anitratus* over a 6-month period from late 1978 to early 1979 (Carlquist et al., 1982); 22 of the patients had *Acinetobacter* pneumonia. No environmental sources were found. Respirators were only infected if they were being used by patients with previous *Acinetobacter*-positive sputum. Personnel cultures were not done. The epidemic was only halted when strict isolation, including handwashing with an antibacterial cleanser, was enforced.

At Norwalk Hospital in Norwalk, Connecticut, there was an epidemic of *A. calcoaceticus* respiratory tract infections in 1975—1976 (the authors did not specify the variety of *A. calcoaceticus*) (Buxton et al., 1978). This epidemic occurred in the intensive care unit and involved 16 patients. Five of the patients had pneumonia, seven had tracheobronchitis, and four were colonized without being infected. All the patients were intubated.

At Norwalk the investigators implicated skin carriage by hospital personnel as the source of the nosocomial outbreak. Transient carriage on the hands was noted in 12 of 34 nurses and in 10 of 32 respiratory therapists. One respiratory therapist, who had chronic dermatitis of his hands, was found to be a persistent carrier of *A. calcoaceticus*. Since this therapist assembled the ventilators prior to their use in the intensive care unit, it was thought that he contributed to the outbreak.

Community-Acquired Adult Pneumonia: There have been only 22 cases of community-acquired *A. calcoaceticus* var. *anitratus* pneumonia occurring in adults reported in the medical literature. Most of these were bacteremic pneumonias (Cordes et al., 1981; Rudin et al., 1979; Glick et al., 1959; Stockwell et al., 1964; Hammett, 1968; Wands et al., 1973; Wallace et al., 1976; Goodhart et al., 1977; Altman and Sacks, 1979; Rosen et al., 1979; Holton and Shorvon, 1982; Guerrero et al., 1980). In those patients who were not bacteremic the diagnosis was made on sputum culture (Goodhart et al., 1977; Gardner et al., 1960; Buscaglia, 1978).

As can be seen in Table 1, the vast majority of these pneumonias occurred in middle-aged males. Ten of the patients were alcoholic, and some of these had other serious illnesses as well. It is noteworthy, however, that about one-third of the patients did not have any serious underlying disease (Table 2).

The disease in outpatients seemed to be much more severe than that in hospitalized patients. Of these 22 patients, 19 were bacteremic. In contrast, in the Massachusetts General study of endemic *A. calcoaceticus* disease, only 3 of the 25 patients with pneumonia and 2 of the 11 with tracheobronchitis were bacteremic (Glew et al., 1977). This may represent a selection bias in that more severe cases tend to be reported, particularly in single-case reports, or it may be that strains of bacteria that are capable of infecting patients not on respirators are more virulent.

Leukopenia is a common finding in almost all the reported cases. In only 2 of the 22 cases was the initial white blood cell count over 15,000 cells per cubic millimeter, while in 12 it was below 5000 cells per cubic millimeter. Two cases had initial white blood cell counts of less than 100 cells per cubic millimeter. This leukopenia is probably a reflection of the overwhelming nature of the infection in these patients. Leukopenia is often seen in infection in alcoholics, but it is seen in *Acinetobacter* pneumonia in patients who are not alcoholic.

The diplococcal appearance of this plemorphic organism can be confusing, and the *Acinetobacter* in the Gram stains of sputum from patients with community-acquired pneumonia are sometimes thought to be *Streptococcus pneumoniae* (Cordes et al., 1981; Goodhart et al., 1977; Rosen et al., 1979). This mistake, due to either failure to

Table 1 Community-Acquired Pneumonia *A. Calcoaceticus* var. *Anitratus*

Number of cases	22
Number of male cases	20
Average age	49.2 ± 2.5[a]
Average white blood cell count	$(7.0 \pm 1.8) \times 10^3$ [a]
Number bacteremic	19
Number of deaths	10

[a] Mean ± SEM.

Table 2 Community-Acquired Pneumonia *A. Calcoaceticus* var. *Anitratus* Underlying Diseases

Alcoholism	10
Chronic obstructive pulmonary disease	2
Pneumoconiosis	2
Pancreatitis	1
Lymphoma	1
Chronic renal disease	1
Pulmonary alveolar proteinosis	1
None	6

decolorize the stain enough or to a belief that the stain has been decolorized too much, can result in catastrophic treatment errors.

The Hartford Foundry Epidemic: Three cases of community-acquired *A. calcoaceticus* var. *anitratus* pneumonia (Table 3) occurred in an iron foundry in Hartford, Connecticut, under circumstances peculiar enough to warrant special mention (Cordes et al., 1981). This epidemic is the only apparent common source outbreak of *Acinetobacter* pneumonia outside of a hospital.

In the spring of 1979 three middle-aged men who worked in the same foundry in Hartford, Connecticut, became ill with bacteremic pneumonia due to *A. calcoaceticus* var. *anitratus*. Each man's illness was separated from the next man's by a period of 4—5 weeks. Two of the men died. None of the three were alcoholics. Although two of the men had evidence of pneumoconiosis, none of them suffered from incapacitating lung disease or any other serious illness.

All three patients presented to the hospital in a similar fashion. All were severely ill with high fever and hypoxia. Patients 2 and 3 were leukopenic on admission, or with a few hours of admission, and patient 1 had a low normal white blood count.

The first two patients were misdiagnosed as having pneumococcal pneumonia and were treated with penicillin. Patient 1 died in 24 hr while still receiving penicillin. Patient 2 was changed to gentamicin and carbenicillin within a day of admission because of his clinical deterioration and recovered. Patient 3 died despite treatment with tobramycin and ticarcillin, to which his organism was sensitive. All three patients grew *A. calcoaceticus* var. *anitratus* from blood cultures and sputum cultures obtained on admission. These isolates were all determined to be serotype 7J by the Centers for Disease Control (CDC).

After the second case had been hospitalized an epidemiological investigation of the foundry was begun by the microbiology and infectious desease sections of the two hospitals to which the patients had been admitted and the Connecticut State Health Department. After the third case occurred teams for the CDC and from the National Institute for Occupational Safety and Health (NIOSH) became involved.

The three men not only worked in the same foundry, but they worked within 10 yards of one another. Patients 1 and 2 were grinders who ground the rough edges off castings using high-speed abrasive wheels. Patient 3 was a welder. There was no central air conditioning in the factory. Ventilation was provided through large windows throughout the building. The environment was dry, but once or twice a day the building would be filled with steam when large iron pieces, heated whitehot, were plunged into a water-filled "quenching pit." The air of the foundry was thick with dust. Test of the air breathed by 18 chippers and grinders showed it to exceed national safety standards for chromium and nickel in 17, for total particulate matter in 6, and for iron oxide in 5. Of 46 samples, 8 exceeded the standards for free silica. Tests of the air breathed by welders showed similar findings.

Of 24 air samples cultured, 3 grew *A. calcoaceticus* var. *anitratus*. One of these was a 7J serotype, the same as that isolated from all three patients. The other two air sample isolates were type 1F. Of 27 environmental surface cultures, 1 grew serotype 3F and 1 water sample grew serotype 10F.

Table 3 Hartford Foundry Epidemic *Acinetobacter* Pneumonia (1979)

	Patient 1	Patient 2	Patient 3
Age	54	63	56
White blood cell count	5300	1100	1900
pO_2	28	69	44
Admission diagnosis	S. pneumonia	S. pneumonia	A. calcoaceticus
Ill	4/4/79	4/30/79	6/11/79
Hospitalized	4/9/79	5/3/79	6/13/79
Died	4/10/79	—	6/20/79

Antecubital and throat cultures were done on employees. Of 134 antecubital cultures, 3 grew *Acinetobacter*, but none of the 3 were the 7J serotype. None of 133 throat cultures were positive.

Serological studies were done on blood from workers in the Hartford foundry, from an unrelated foundry in Bridgeport, Connecticut, and on blood from Hartford residents submitted to the state health department for premarital serology. Using an indirect fluorescent antibody test on absorbed serum, it was found that 28 of 120 workers at the Hartford foundry and 27 of 47 workers at the Bridgeport foundry had antibody at greater than 1:64 dilution, compared to only 2 of 93 Hartford residents. This would indicate that workers, at least at these two foundries if not at foundries in general, have a higher than usual experience with *Acinetobacter*.

It is intriguing to suppose that the high level of metallic dust in the air of foundries, and in the lungs of those who work there, might contribute to some diminution in host defenses and to some increase in the virulence of the invading organisms.

Occupation histories are given in only 5 of the other 19 cases of community-acquired *A. calcoaceticus* var. *anitratus* pneumonia. Two of these patients had jobs where exposure to silica and metallic dusts might be expected; one was a construction worker (Wands et al., 1973) and one was a molder (Stockwell et al., 1964). There has recently been another case of an unusual outpatient gram-negative pneumonia in a foundry worker in the Hartford area, this one due to *Serratia marcescens*. This patient had no connection with the foundry in which the *Acinetobacter* pneumonia had occurred (H. Sherzer, personal communication, 1983).

The source of the Hartford epidemic was not found. Extensive improvements were made to protect the workers against dust, and no further cases have appeared at that foundry in the past 5 years.

Pediatric Respiratory Infections: In 1963 Reynolds and Cluff reported the case of an 18-month-old boy with fever and x-ray evidence of pneumonia from whom a nasopharyngeal culture yielded "only *Herellea* species." The child improved on tetracycline.

In 1981 there was a report of four cases of upper respiratory infection and brochiolitis in children due to *A. calcoaceticus* var. *anitratus* in which the diagnosis was made by recovery of "nearly pure" cultures of the organism from throat or nasal swabs (O'Connell and Hamilton, 1981). All the children were 2 years old or younger, but the authors do not give any other clinical details.

It is difficult to draw any conclusions about the pathogenic role played by the organism in the diseases described in these reports. The organism may have been simply colonizing the sites cultured, rather than infecting the patients. It is not a commonly reported organism in pediatric respiratory infection.

Endocarditis

Acinetobacter calcoaceticus var. *anitratus* endocarditis is very uncommon. In a review of the literature from 1945 to 1977, Cohen et al., (1980) found nine cases of *Acinetobacter* endocarditis. Four of these were casued by variety *anitratus*, four by variety *lwoffi*, and one undetermined. Infection with either variety of organism occurred in five patients with rheumatic or congenital heart disease, as well as in patients with prosthetic valves and with normal valves. Of the nine patients reported, five died. Six of the nine had evidence of arterial embolization. Infection with variety *anitratus* was seen as a fatal acute endocarditis in one patient who was addicted to heroin (Thompson, 1971).

Polymicrobial endocarditis has been reported with *Acinetobacter* in association with *Klebsiella pneumoniae* in one instance (Noble et al., 1981), and with *Bacteroides fragilis* in another (Juffe et al., 1977). The latter case occurred in a patient 1 week after replacement of his aortic and mitral valves and may have represented bacteremia without endocarditis. In neither case was the variety of *Acinetobacter* stated.

Five cases of *Acinetobacter* endocarditis have been reported from the Indian subcontinent in the last few years (Rao et al., 1980; Pal et al., 1981). Three were in children and two were in young adults; all had congenital or rheumatic heart disease. Unfortunately, the clinical data provided with these cases are sparse. None of the four patients in one report had continuous bacteremia, usually considered a hallmark of endocarditis (Rao et al., 1980). All patients survived.

Meningitis

Meningitis with *A. calcoaceticus* var. *anitratus* is a problem on neonatal care units and on neurosurgical wards. In 1962 Daly reported three cases of meningitis following neurosurgical procedures at the Boston City Hospital. In a survey of all meningitides seen at hospitals affiliated with Boston University from 1968 to 1978 it was found that there were five cases caused by *A. calcoaceticus* var. *anitratus*; all cases were in neurosurgical patients (Berk and McCabe, 1981). Two other cases of *A. calcoaceticus* var. *anitratus* following neurosurgical procedures were reported from South Africa (Berkowitz, 1982).

Five of these neurosurgical patients had had craniotomies. One had had a transnasal resection of a craniopharyngioma. One had had surgical decompression of a cervical fracture. One had had a ventriculoperitoneal shunt inserted. In one case the infection followed a myelogram. The cases were widely separated in time, and there was no suspicion of an epidemic at any of the hospitals.

The meningitis developed within 2 weeks and usually within a few days of the surgical procedure and was marked by high fever and purulent spinal fluid with hypoglycorrhachia. Four of the nine patients in these three reports died.

Successful therapies were intrathecal polymyxin B, intravenous kanamycin and intrathecal streptomycin, intravenous and intrathecal gentamicin, intravenous penicillin and gentamicin, oral chloramphenicol and intravenous gentamicin, and intrathcal gentamicin alone.

In an outbreak of *A. calcoaceticus* var. *anitratus* meningitis in a neonatal intensive care unit in England four infants became infected over a 5-day period (Morgan and Hart, 1982). No source was found for the outbreak. All infants were treated with intravenous carbenicillin (800 mg/kg per day), and all survived.

A 4-day-old premature girl (gestational age 28 weeks) died in South Africa of *Acinetobacter* meningitis. She was treated with co-trimoxazole because it was thought that the organism seen on Gram-stain was *Neisseria*. Her strain of *A. calcoaceticus* var. *anitratus* was resistant to co-trimoxazole (Berkowitz, 1982).

This tendency to misread the spinal fluid Gram stain in patients infected with *Acinetobacter* is a recurrent theme in case reports of meningitis, just as it is in pneumonia. One sees what one expects to see; in sputums pneumococci, and in spinal fluid mningococci or *Haemophilus influenza*. One can easily mistake *Acinetobacter* for any of these. The therapeutic errors that can follow these mistakes may be irretrievable.

That the disease can occur in normal hosts was seen in a recent report of a fatal case from India in a 10-year-old girl. This child died despite chloramphenicol therapy to which the organism was sensitive (Sachdev and Deb, 1980). *Acinetobacter calcoaceticus* var. *anitratus* was grown from both her spinal fluid and her blood.

Skin and Wound Infections

Since *A. calcoaceticus* var. *anitratus* forms part of the normal skin flora (Taplin et al., 1963), it is surprising that it does not cause wound infections more frequently than it does.

Six cases of postoperative wound infections were reported from the Massachusetts General Hospital (Glew et al., 1977). Five of the six patients had foreign bodies in place, and one of the six became bacteremic.

Two cases of mixed bacterial cellulitis from which *Acinetobacter* and *Staphylococcus aureus* were recovered were also reported from the Massachusetts General Hospital (Glew et al., 1977). One was in an elderly man with pancytopenia who had the organisms recovered by needle aspiration of the involved skin, and the second was a cellulitis surrounding a stasis ulcer in a woman. The former patient died of his infection. The latter, after failing to improve on antistaphylococcal therapy, cleared her infection when the therapy was changed to tetracycline to which the *Acinetobacter* was sensitive.

At St. Francis Hospital and Medical Center in Hartford, Connecticut, we recently saw an elderly black woman with bilateral, painful ankle ulcers from which *A. calcoaceticus* var. *anitratus* was grown repeatedly. Biopsy of the ulcer edge showed only acute inflammatory changes. The lesions healed with intravenous gentamicin and ticarcillin therapy after outpatient treatment with oral drugs failed.

Bacteria classified as *"Mimeae-Herellea-Bacterium-Alcaligenes"* were recovered in 29 cultures of 63 war wounds of the extremity (Tong, 1972). Two patients were bacteremic with *"Bacterium anitratum."*

Daly (1962) reported a case of infection of the elbow in a 52-year-old man following blunt trauma and also several cases of septic phlebitis occurring in hospitalized patients with intravenous catheters in place.

Urinary Tract Infection

Most urinary tract infections with *A. calcoaceticus* var. *anitratus* occur in hospitalized patients with indwelling urinary catheters; although the organism has also been recovered from the urine of outpatients, its significance in these circumstances is not always clear.

In the NNIS study *Acinetobacter* was isolated from the urine in 370 patients. Secondary bacteremia occurred in only two patients for a rate of 0.5%, considerably lower than the 2.7% rate of secondary bacteremia seen with other gram-negative rod urinary tract infections (Retailliau et al., 1979).

Glew et al. (1977) found 5 cases of urinary tract infection among 53 patients infected with the organism at the Massachusetts General Hospital. All of these infections were hospital acquired, and three of the patients had indwelling urinary catheters.

In Kansas City, Missouri, at the V.A. Hospital, *A. calcoaceticus* var. *anitratus* was recovered from the urine of 24 men, but only 6 patients had urinary tract symptoms, and only 4 of those had pyuria. Eight other patients with positive cultures had asymptomatic pyuria (Robinson et al., 1964).

An epidemic of *Acinetobacter* urinary tract infection on an English urological ward was traced to contamination of the patients' urinary catheter bags (Lowes et al., 1980). The investigators believed that the patients' urine jugs were contaminated by a faulty bed-pan washer, and that the organisms were transferred to the patients' urinary catheter bags when the bags were being emptied into the jugs. Eight patients were infected, but only two patients required treatment for symptomatic infection.

In an extensive study of outpatient urinary tract infection, Danish investigators isolated *A. calcoaceticus* var. *anitratus* from 13 of 1822 patients with significant bacteriuria ($\geq 10^5$ organisms per milliliter) (Hoffmann et al., 1982). Seven of these people were women, four men, and in two the sex was not determined. It is not clear how many of these patients were symptomatic.

It appears that in general *A. calcoaceticus* var. *anitratus* is a low-virulence pathogen in the urinary tract and may often be there as a commensal organism.

Peritonitis

Acinetobacter has been implicated as the cause of peritonitis in chronic renal failure patients undergoing peritoneal dialysis in a number of reports.

In an extensive study Andersen and Kolmos (1981) looked at 159 cases of peritonitis occurring over 5 years among 164 patients undergoing chronic peritoneal dialysis at a center in Denmark. *Acinetobacter calcoaceticus* was isolated from 13 cases; 12 times in pure culture and once in association with *S. aureus*. Seven of these cases were part of an epidemic caused by a contaminated water bath used to preheat the dialysis fluid.

A contaminated water bath was also the source of an outbreak of *A. calcoaceticus* var. *anitratus* peritoneal dialysis contamination described by Abrutyn et al. (1978). In this 4-month outbreak cultures of the peritoneal drainage fluid of 14 patients grew the organism, although the patients remained asymptomatic and did not have signs of peritonitis. The investigators showed that bacteria in the warming bath could be transferred to the peritoneal dialysis fluid. Since the fluid was placed in the warming bath in unopened bottles, the most likely mode of contamination was from fluid draining down the sides of the bottles onto the spike used to puncture the rubber ring.

The same source was implicated in one of two cases reported by Said et al. (1980). This case was due to *A. calcoaceticus* var. *lwoffi*, while the other, in which a soruce was not found, was due to variety *anitratus*. Both of these patients were symptomatic and both were treated successfully with intraperitoneal gentamicin.

Fifteen dialysis patients were infected with *Acinetobacter* in the mid-1970s at a hospital in Seattle due to contamination of the dialysate from an unstated source (Larson, 1984).

Roxe and Santhanam (1983) report one case of *Acinetobacter calcoaceticus* var. *anitratus* peritonitis which they attribute to a break in the peritoneal dialysis catheter.

The frequency with which this organism is involved in this rather specialized infection probably reflects its dual nature as both a free-living water organism and a human commensal organism. It would grow readily in the warm water baths that are part of the peritoneal dialysis routine; and it would also grow on the skin of patients, where it could infect the catheter.

Acinetobacter var. lwoffi

Some studies do not clearly distinguish between the two varieties of *Acinetobacter*, but I think it is safe to say that variety *lwoffi* can cause anything variety *anitratus* can, although, with the exception of meningitis, it does so less often.

Acinetobacter calcoaceticus var. *lwoffi* has been a cause of endocarditis, indistinguishable from that caused by variety *anitratus* (Cohen et al., 1980). It has been implicated as the cause of skin disease in two patients, both of whom had a pustular eruption on a raised, sharply demarcated erythematous base (Dexter et al., 1958). It has rarely been proven to be the cause of pneumonia.

Meningitis

Acinetobacter calcoaceticus var. *lwoffi* more frequently causes meningitis in normal hosts than does variety *anitratus*. The disease resembles meningococcal meningitis both in its age distribution and in the frequency with which a petechial rash accompanies it (DeBord, 1948; Schuldberg, 1953; Townsend et al., 1954; Olaffson et al., 1958; Fred et al., 1958; Peyla and Burke, 1965; Burrows and King, 1966).

A 19-year-old air force enlisted man died of the Waterhouse-Friderichsen syndrome caused by this organism (Townsend et al., 1954). The patient died within a few hours of becoming ill. His body was covered with a petechial rash, and at autopsy he had extensive adrenal hemorrhage.

A petechial rash was also noted in a case of *A. calcoaceticus* var. *anitratus* bacteremia unaccompanied by meningitis in a 7-month-old girl reported by Reynolds and Cluff (1963).

Postneurosurgical infection with variety *lwoffi* has been reported following spinal cord surgery in a 76-year-old woman (Burrows and King, 1966) and in a 49-year-old man who had had a ventricular drain placed because of cerebral hemorrhage (Kobayashi et al., 1983). In general, however, this variety of *Acinetobacter* seems to cause less nosocomial meningitis than does variety *anitratus*.

This variety of *Acinetobacter* was originally given the genus name *Mima* and assigned to the new tribe *Mimeae* by DeBord because he though it "mimicked" *Neisseria* (Henriksen, 1973). The tendency to misidentify these organisms as *Neisseria meningitidis* on Gram stain is a common problem, just as it is with variety *anitratus*. While the occasional strain may be sensitive to penicillin (Fred et al., 1958), most strains are not, and misidentification can lead to serious therapeutic errors.

Urinary Tract Infection

In their large study from Denmark Hoffmann and his colleagues (1982) found 93 strains of *A. calcoaceticus* var. *lwoffi* in the urine of 85 of the 1822 people sampled. It is not at all clear how many of these people were actually infected. Of the 85 people, 80 were women and *A. calcoaceticus* var. *lwoffi* has been found in the vaginas of normal women (Henriksen, 1973). It is interesting, however, that the number of people with variety *lwoffi* in their urine was more than six times the number with variety *anitratus*. Urinary tract infection in outpatients with either variety of *Acinetobacter* is uncommon in the United States.

Nosocomial urinary tract infection with variety *lwoffi* is less common than that with variety *anitratus*. In a report from Buffalo, New York, variety *lwoffi* was isolated only twice from the urine of hospitalized patients over a 21-month period when variety *anitratus* was isolated 11 times (O'Connell and Hamilton, 1981). Reynolds and Cluff (1963) reported a variety *lwoffi* urinary tract infection in a 69-year-old man with obstructive uropathy due to uric acid stones.

Antibiotic Sensitivities

Both varieties of *A. calcoaceticus* are resistant to many of the commonly used antibiotics. Although variety *lwoffi* may be more sensitive than variety *anitratus*, there is a great deal of variation from country to country and even from hospital to hospital. Hoffmann et al. (1982) demonstrate striking differences between the antibiotic sensitivities of their Danish strains and strains studied elsewhere.

The antibiotic sensitivities of *A. calcoaceticus* can be quite variable, even within one hospital. Murray and Moellering (1980) have reported that some strains of this organism are the most resistant gram-negative rods isolated at the Massachusetts General Hospital. They have identified a variety of plasmid-mediated enzymes that enable the organism to inactivate aminoglycosides. In a 6-month hospital epidemic in Salt Lake City, investigators found that the *Acinetobacter* strains isolated became progressively more resistant to aminoglycosides as the epidemic went on. This resistance, however, was not due to aminoglycoside-inactivating enzymes, but may have been due to changes in the bacteria's cell wall permeability (Carlquist et al., 1982).

Among strains of *A. calcoaceticus* var. *anitratus* resistance to penicillin, ampicillin, and cephalothin are almost universal, and many strains are resistant to chloramphenicol as well. Tetracycline resistance has been variable (Glew et al., 1977; Garcia et al., 1983). *Acinetobacter calcoaceticus* var. *lwoffi* is more likely than variety *anitratus* to be sensitive to ampicillin and chloramphenicol, but cephalothin resistance is still the rule (O'Connell and Hamilton, 1981).

Minocycline has been shown to be effective in vitro against *Acinetobacter* in two studies of 21 and 65 strains, respectively (Maderazo et al., 1975; Kuck, 1976).

The aminoglycosides—gentamicin, tobramycin, amikacin, netilmicin, sissomicin—usually have a minimum inhibitory concentration (MIC) for *Acinetobacter* well below their achievable serum levels (Garcia et al., 1983; Daschner and Nopper, 1980). In the early 1970s, however, at a time when gentamicin was new, investigators at the Massachusetts General Hospital reported that 30% of 100 strains of *A. calcoaceticus* var. *anitratus* were resistant to gentamicin (Glew et al., 1977).

In a study of new β-lactam antibiotics against 51 strains of variety *anitratus* and 23 strains of variety *lwoffi*, Garcia et al. (1983) found that N-formimidoyl thienamycin was highly effective, inhibiting the growth of 90% (MIC_{90}) of both varieties at a concentration of 0.39 μg/ml. Piperacillin had an $MIC_{90\%}$ of 50 μg/ml for variety *anitratus* and of 25 μg/ml for variety *lwoffi*, making it marginally better than ticarcillin ($MIC_{90\%}$ of 50 μg/ml for both) and mezlocillin ($MIC_{90\%}$ of 100 μg/ml for both).

Among the new cephalosporins Garcia found ceftazidime (MIC_{90} of 12.5 μg/ml for both varieties), ceftriazone ($MIC_{90\%}$ of 25 μg/ml for variety *anitratus* and of 12.5 μg/ml for variety *lwoffi*), and ceftizoxime ($MIC_{90\%}$ of 25 μg/ml for variety *anitratus* and of 12.5 μg/ml for variety *lwoffi*) to be the most effective. In contrast to this, however, Neu (1982) found the $MIC_{90\%}$ for both ceftizoxime and ceftraxone to be in excess of 128 μg/ml and the $MIC_{90\%}$ of ceftriaxone to be 32 μg/ml.

Trimethoprim-sulfamethoxazole (TMP-SMX) was used successfully as oral therapy in a case of endocarditis (Noble et al., 1981), but sensitivity to that antibiotic is variable. A total of 89% of variety *anitratus* strains were sensitive to TMP-SMX by disk testing in one study (O'Connell and Hamilton, 1981). In another study the $MIC_{50\%}$ for both varieties was a clinically achievable level but the $MIC_{90\%}$ of 8 mg of trimethoprim and 152 mg of sulfamethoxazole was more than can be achieved by the usual oral therapy (Garcia et al., 1983).

For treatment of serious *Acinetobacter* infections an aminoglycoside and ticarcillin or piperacillin should be used until the strain's antibiotic sensitivities are known. Combinations of this sort have been shown to be synergistic against some strains which showed moderate resistance to aminoglycoside alone (Glew et al., 1977).

Cefuroxime has recently been used successfully to treat two cases of nosocomial *Acinetobacter* pneumonia in Belgium (Guerisse, 1981), but clinical experience in treating these infections with third-generation cephalosporins is extremely limited. They should not be used unless one has evidence that the strain causing the infection is sensitive to the drug in question.

MORAXELLA AND KINGELLA

Organisms placed in the genus *Moraxella* were previously called *Mima polymorpha* var. *oxidans* and considered to be oxidase-positive variants of *Mima*. The new genus *Kingella* contains the organism once called *Moraxella kingae*, now called *Kingella kingae*.

If that were not confusing enough, there was a period of 6 years during which *M. kingae* was called *Moraxella kingii* (Henriksen and Bøvre, 1968; Bøvre et al., 1974). The organism had been named for Dr. Elizabeth O. King, and the change in the ending of the species name was made when someone realized it was inappropriate for Dr. King's gender. Neither genus is of major clinical importance.

Ecology

Unlike *Acinetobacter*, organisms in these two genera are not soil or water organisms. Most species have been isolated from the nose, throat, or conjunctiva of humans. *Moraxella osloensis* has been isolated from the human genitourinary tract. *Moraxella bovis* has been isolated from the eyes of cattle with conjuctivitis (Henriksen 1973).

Clinical Importance

Moraxella was first isolated by Mora (1896) from the eye of a patient with conjunctivitis. For decades after that these organisms were considered important causes of conjunctivitis, but a recent investigation casts doubt on that association (Sadoff, 1979). In one study *Moraxella* was isolated from only 0.4% of 236 patients with blepharoconjunctivitis and from 3% of 64 cases of angular conjunctivitis. The author noted a much higher recovery rate from nasal swabs from the same patients (van Bijsterveld, 1972).

In his review of gram-negative endocarditis from 1945 to 1977, Cohen et al. (1980) found three cases of endocarditis due to *Moraxella*. One patient had infection of a normal valve.

There have been seven cases of endocarditis due to *Kingella* (Christensen and Emmanouilides, 1967; Miridjanian and Berrett, 1978; Geraci and Wilson, 1982; Sage et al., 1983; Rabin et al., 1983; Le, 1983). Three of the seven cases had infection of artificial heart valves. Three patients were treated with penicillin, one with penicillin and an aminoglycoside, two with ampicillin and an aminoglycoside, and one with cephalothin.

There has been a recent case report of endocarditis caused by an organism called M6 which is a *Moraxella*-like organism (Simor and Salit, 1983). The patient was a 31-year-old woman with mitral valve prolapse. Echocardiography showed a large vegetation on the mitral valve, but she did not have systemic embolization. Her infection was controlled with ampicillin, but she required cardiac surgery because of increasing congestive heart failure.

Appelbaum et al. (1974) reported a case of a purulent pericarditis due to *Moraxella* in a 44-year-old woman. The patient had had a left radical mastectomy for cancer. The pericarditis was thought to have developed by direct extension from the ulcerated chest wall which had been severely damaged by radiation therapy. The patient was successfully treated with cephaloridine and surgical drainage. The organism was not speciated.

Meningitis due to *Moraxella* has been reported. One report documented the disease in a 16-year-old boy and a 4-year-old girl. The girl had a petechial rash. Neither child was shown to be bacteremic, and both survived (Herman and Melnick, 1965).

Moraxella osloensis was isolated from the blood and in pure culture from the mouth, lips, and gums of a Belgian child with a severe stomatitis (Butzler et al., 1974). The organism was resistant to penicillin, and the child was treated successfully with ampicillin.

A *Moraxella nonliquefaciens* bacteremia in association with an acute arthritis occurred in a woman with multiple myeloma (Brorson et al., 1983). The patient responded rapidly to broad-spectrum antibiotics. Her joints were not cultured.

Two cases of *Moraxella* infection have been reported that mimicked disseminated gonococcal disease (Lasser and Goldman, 1978; Rosebaum et al., 1980).

Bosworth (1983) has reported three *Kingella* infections occurring in three Iowa City children over a 2-month period. One was a child with a septic hip, one with a septic knee, and one with endocarditis. She could find only five other cases of *Kingella* infections in children in the literature. There was no explanation for the curious cluster of cases in Iowa City.

Two more cases of septic arthritis in children caused by *K. kingae* were added in a subsequent report by Powell and Bass (1983), bringing the total number of reported bone and joint infections with this organism to 10.

Antibiotic Sensitivity

Organisms in these two genera are usually very sensitive to penicillin and are also sensitive to cephalothin, erythromycin, ampicillin, and tetracycline. *Moraxella osloensis* may sometimes be penicillin resistant (Henriksen, 1973; Butzler et al., 1974).

Aminoglycoside sensitivity has been variable. Amikacin has been the most effective aminoglycoside, although some strains can inactivate that agent with an N-acetylating enzyme (Sadoff 1979).

SUMMARY

Of the two varieties of *A. calcoaceticus*, variety *anitratus* is more important as a cause of nosocomial infection, especially pneumonia. It also causes rare but fulminant pneumonia outside the hospital, mostly in middle-aged men. A curious association of *A. calcoaceti-*

cus var. *anitratus* pneumonia with foundry workers has been observed on one occasion.

Acinetobacter calcoaceticus var. *lwoffi* is most important for its capacity for causing a meningitis that closely resembles meningococcal meningitis.

The diplococcal appearance of *Acinetobacter* strains often causes them to be mistaken for pneumococci or meningococci on Gram stain. Because *Acinetobacter* is resistant to penicillin and to many other antibiotics, such a mistake can lead to serious therapeutic errors.

Moraxella organisms may cause a disease that mimicks disseminated gonococcal disease. They also are rare causes of endocarditis, meningitis, or bacteremia.

Kingella kingae is emerging as a cause of bone and joint infections, especially in children.

Most strains of *Moraxella* and *Kingella* are sensitive to penicillin as well as to most other antibiotics.

ACKNOWLEGMENTS

I would like to thank Mrs. Nancy A. Bianchi, clinical librarian, of the St. Francis Hospital and Medical Center library for her help in compiling the bibliography for this chapter. I would like to thank Miss Susan M. Brock, MT(ASCP), for her help in investigating the Hartford foundry epidemic.

REFERENCES

Abrutyn, E., Goodhart, G. L., Roos, K., Anderson, R., and Buxton, A. (1978). *Am. J. Epidemiol.* 107:328.
Altman, K. A., and Sacks, F. (1979). *N.Y. State J. Med.* 79:1434.
Andersen, K. E. H., and Kolmos, H. J. (1981). *J. Artif. Organs* 4:281.
Appelbaum, A., Giladi, A., and Borman, J. B. (1974). *J. Cardiovasc. Surg.* 15:479.
Beeson, P. B., McDermott, W., and Wyngaarden, J. B. (1979). *Cecil Testbook of Medicine,* 15th ed., W. B. Saunders, Philadelphia.
Berk, S. L., and McCabe, W. R. (1981). *Arch. Neurol.* 38:95.
Berkowitz, F. E. (1982). *S. Afr. Med. J.* 61:448.
Bøvre, K., Henriksen, S. D., and Jonsson, V. (1974). *Int. J. Syst. Bacteriol.* 24:307.
Bosworth, D. E. (1983). *Am. J. Dis. Child.* 137:650.
Brorson, J. E., Falsen, E., Nilsson-Ehle, H., Rodjer, S., and Westin, J. (1983). *Scand. J. Infect. Dis.* 15:221.

Burrows, S., and King, M. J. (1966). *Am. J. Clin. Pathol.* 46:234.
Buscaglia, A. J. (1978). *Ann. Intern. Med.* 89:1010.
Butzler, J. B., Hansen, W., Cadranel, S., and Henriksen, S. D. (1974). *J. Pediatr.* 84:721.
Buxton, A. E., Anderson, R. L., Werdger, D., and Atlas, E. (1978). *Am. J. Med.* 65:507.
Carlquist, J. F., Conti, M., and Burke, J. P. (1982). *Am. J. Infect. Cont.* 10:43.
Centers for Disease Control (1979). *Morbid. Mortal. Week. Rep.* 28:177.
Christensen, C. E., and Emmanouilides, G. C. (1967). *N. Engl. J. Med.* 277:803.
Cohen, P. S., Maguire, J. H., and Weinstein, L. (1980). *Prog. Cardiovasc. Dis.* 22:205.
Cordes, L. G., Brink, E. W., Checko, P. J., Lentnek, A., Lyons, R. W., Hayes, P. S., Wu, T. C., Tharr, D. W., and Fraser, D. W. (1981). *Ann. Intern. Med.* 95:688.
Cunha, B. A., Klimek, J. J., Gracewski, J., McLaughlin, J. C., and Quintiliani, R. (1980). *Postgrad. Med. J.* 56:170.
Daly, A. K., Postic, B., and Kass, E. H. (1962). *Arch. Intern. Med.* 110:86.
Daschner, F., and Nopper, S. (1980). *J. Antimicrob. Chemother.* 6:415.
DeBord, G. G. (1939). *J. Bacteriol.* 38:119.
DeBord, G. G. (1948). *J. Bacteriol.* 55:764.
Dexter, H. L., Glacy, J., Leonard, J., Dexter, M. W., and Lawton, A. (1958). *Arch. Dermatol.* 77:109.
Faris, Jr., H. M., and Sparling, F. F. (1972). *J. Am. Med. Assoc.* 219:76.
Fred, H. L., Allen, T. D., Hessel, H. L., and Holtzman, C. F. (1958). *Arch. Intern. Med.* 102:204.
French, G. L., Casewell, M. W., Roncoroni, A. J., Knight, S., and Phillips, I. (1980). *J. Hosp. Infect.* 1:125.
Garcia, I., Fainstein, V., LeBlanc, B., and Bodey, G. (1983). *Antimicrob. Agents Chemother.* 24:297.
Gardner, D. L., Pines, A., and Stewart, S. M. (1960). *Br. J. Med.* 1:1108.
Geraci, J. E., and Wilson, W. R. (1982). *Mayo Clin. Proc.* 57:145.
Glew, R. H., Moellering, Jr., R. C., and Kunz, L. J. (1977). *Medicine* 56:79.
Glick, L. M., Moran, G. P., Coleman, J. M., and O'Brien, G. F. (1959). *Am. J. Med.* 27:183.
Goodhart, G. L., Abrutyn, E., Watson, R., Root, R. K., and Egert, J. (1977). *J. Am. Med. Assoc.* 238:1516.
Grehn, M., and von Graevenitz, A. (1978). *J. Clin. Microbiol.* 8:342.

Guerisse, P. (1981). *Lancet* 2:96.
Guerrero, M. L. F., Fernandez, J. L. D., Preito, J. deM., and Garces, J. L. G. (1980). *Chest* 78:670.
Hammett, J. B. (1968). *J. Am. Med. Assoc.* 206:641.
Henriksen, S. D. (1973). *Bacteriol. Rev.* 37:522. (1973).
Henriksen, S. D., and Bøvre, K. (1968). *J. Gen. Microbiol.* 51: 377.
Herman, III, G., and Melnick, T. (1965). *Am. J. Dis. Child.* 315.
Hoffmann, S., Mabeck, C. E., and Vejlsgaard, R. (1982). *J. Clin. Microbiol.* 16:443.
Holton, J., and Shorvon, P. J. (1982). *J. Infect.* 4:263.
Juffe, A., Miranda, A. L., Rufilanchas, J. J., Maronas, J. M., and Figuero, D. (1977). *Arch. Surg.* 112:151.
Kobayashi, T. K., Yamaki, T., Toshino, E., Terawaki, S., Tara, K., Nishida, K., and Sawargi, I. (1983). *Acta Cytol.* 27:281.
Kuck, N. A. (1976). *Antimicrob. Agents Chemother.* 9:493.
Larson, E. (1984). *Am. J. Infect. Cont.* 12:14.
Lasser, A. E., and Goldman, E. J. (1978). *Cutis* 21:657.
Le, C. T. (1983). *Am. J. Dis. Child.* 137:1212.
Lowes, J. A., Smith, J., Tabaqchali, S., and Shaw, E. J. (1980). *Br. Med. J.* 1:722.
Maderazo, E. G., Quintiliani, R., Tilton, R., Bartlett, R. C., Joyce, N. C., and Andriole, V. T. (1975). *Antimicrob. Agents Chemother.* 8:54.
Miridjanian, A., and Berrett, D. (1978). *West. J. Med.* 129:344.
Morax, V. (1896). *Ann. Inst. Pasteur* 10:337.
Morgan, M. E. I., and Hart, C. A. (1982). *Arch. Dis. Child.* 57: 557.
Murray, B. E., and Moellering, Jr., R. C. (1980). *Antimicrob. Agents Chemother.* 17:30.
Neu, H. C. (1982). *Ann. Intern. Med.* 97:408.
Noble, R. C., Cooper, R. M., Jarvis, A. L., Caples, P. L., and Todd, E. P. (1981). *South. Med. J.* 74:1299.
O'Connell, C. J., and Hamilton, R. (1981). *N.Y. State J. Med.*, 750.
Olaffson, M., Lee, Y. C., and Abernethy, T. J. (1958). *N. Engl. J. Med.* 258:465.
Pal, R. B., Sujatha, V., and Kale, V. V. (1981). *Lancet* 2:313.
Peyla, T. L., and Burke, E. C. (1965). *Mayo Clin. Proc.* 40:236.
Powell, J. M., and Bass, J. W. (1983). *Am. J. Dis. Child.* 137:974.
Rabin, R. L., Wong, P., Noonan, J. A., and Plumly, D. D. (1983). *Am. J. Dis. Child.* 137:403.
Ramphal, R., and Kluge, R. M. (1979). *Am. J. Med. Sci.* 277:57.
Rao, K. N. A., Kotian, M., and Prabhu, S. G. S. (1980). *J. Postgrad. Med.* 26:186.
Retailliau, H. F., Hightower, A. W., Dixon, R. F., and Allen, J. R. (1979). *J. Infect. Dis.* 139:371.

Reynolds, R. C., and Cluff, L. E. (1963). *Ann. Intern. Med.* 58: 759.
Robinson, R. G., Garrison, R. G., and Brown, R. W. (1964). *Ann. Intern. Med.* 60:19.
Rosen, J. H., Muren, O., Gander, G. W., Irby, S. K., and Cashion, C. F. (1979). *Virginia Medical.* 106:660.
Rosenbaum, J., Liberman, D., and Katz, W. A. (1980). *Ann. Rheum. Dis.* 39:184.
Roxe, D. M., and Santhanam, S. (1983). *Nephron* 34:267.
Rudin, M. L., Michael, J. R., and Huxley, E. J. (1979). *Am. J. Med.* 67:39.
Sachdev, H. S., and Deb, M. (1980). *Indian Pediatr.* 17:551.
Sadoff, J. C. (1979). In *Principles and Practice of Infectious Diseases* (G. L. Mandell, R. G. Douglas, Jr., and J. E. Bennett, eds.), Wiley, New York, p. 1670.
Sage, M. J., Maslowski, A. H., and MacCulloch, D. (1983). *N. Z. Med. J.* 96:795.
Said, R., Krumlovsky, F. A., and del Greco, F. (1980). *J. Dialysis* 4:101.
Schuldberg, I. I. (1953). *Am. J. Clin. Pathol.* 23:1024.
Simor, A. E., and Salit, I. E. (1983). *J. Clin. Microbiol.* 17:931.
Snydman, D. R., Maloy, M. F., Brock, S. M., Lyons, R. W., and Rubin, S. J. (1977). *Am. J. Epidemiol.* 106:154.
Stockwell, B. A., Whitaker, A. N., and Cheong, M. (1964). *Med. J. Aust.* 2:370
Taplin, D., Rebell, G., and Zaias, N. (1963). *J. Am. Med. Assoc.* 186:952.
Thompson, W. R. (1971). *J. Am. Med. Assoc.* 215:982.
Tong, M. J. (1972). *J. Am. Med. Assoc.* 219:1045.
Townsend, F. M., Hersey, D. F., and Wilson, F. W. (1954). *U.S. Armed Forces Med. J.* 5:673.
Van Bijsterveld, O. P. (1972). *Am. J. Ophthalmol.* 74:72.
Wallace, Jr., R. J., Awe, R. J., and Martin, R. R. (1976). *Am. Rev. Respir. Dis.* 113:695.
Wands, J. R., Mann, R. B., Jackson, D., and Butler, T. (1973). *Am. Rev. Respir. Dis.* 108:964.

6
Ecology, Clinical Significance, and Antimicrobial Susceptibility of Infrequently Encountered Glucose-Nonfermenting Gram-Negative Rods

ALEXANDER VON GRAEVENITZ *University of Zürich, Zürich, Switzerland*

GENERAL FEATURES OF INFREQUENTLY ENCOUNTERED NONFERMENTING GRAM-NEGATIVE RODS

Ecology and Transmission

The natural habitat of "nonfermenters" is soil, water, and plants (391). With the exception of the obligately parasitic mammalian species *Pseudomonas mallei*, they are free-living, ubiquitous bacteria. Only *Pseudomonas pseudomallei* seems to occur in a limited geographical area (see below). Since they are not fastidious, they are able to survive in moist environments, particularly food (118, 221, 230) and hospital water sources such as infusion fluids, distilled water, tap water, sink drains, puddles, wet surfaces, lubricants, cosmetics, solutions (including those for contact lenses), incubators, inhalation therapy equipment (nebulizers), humidifiers, catheters, and even disinfectants (40, 49, 195, 219, 280, 296, 297, 340, 342, 368, 426). Antibiotic-resistant strains may contaminate tissue cultures and selective media (213, 483). Psychrophilic strains may be found in refrigerated food (239) and contribute to its spoilage (118). They may also survive in bank blood (41, 352, 435) and platelet pools (54).

Survival and growth in disinfectants will be mentioned in the individual species sections. Sources for the organisms include unsterile distilled water (65, 354), saline solutions (344), and caps of plastic containers (79). In some instances the otherwise effective disinfectant had been inactivated by contact with gauze, absorbent cotton, or cork (tannin effect?) from bottle stoppers (9, 293); in other instances such inactivators could not be incriminated (25, 56).

Man acquires nonfermenters outside the hospital from contaminated soil and water sources. Transmission inside hospitals is significantly related to the sources listed above and may also involve carriers. Direct man-to-man transmission in hospitals is also possible, for example, from hand to wound or catheter. These organisms are also encountered in veterinary specimens (281).

Pathogenicity

Pseudomonas aeruginosa is, of course, the nonfermenter species most often found in clinical samples. It is followed in frequency by *Acinetobacter calcoaceticus* and *Pseudomonas maltophilia* (30, 300, 341).

A systematic search for unusual nonfermenters in the normal flora of the human body has never been undertaken. On occasion, the organisms have been recovered from the gastrointestinal or genital tracts of healthy individual (212, 448, 449). The clinical significance of individual strains is often difficult to determine. Many of them are obvious epiphytes, often occurring in mixed cultures (154, 300, 340). An evaluation of their pathogenicity has to take into account association with symptoms, repeated isolation, clinical response to antimicrobial therapy, serological reactions, and postmortem isolation. Other strains are opportunistic, such as those reactivating melioidosis after trauma or intercurrent infection (392). It also seems that highly virulent strains or very large inocula can overwhelm even normal host defenses.

In contrast to *P. aeruginosa* and *P. pseudomallei*, unusual nonfermenters are not pathogenic for mice and guinea pigs if injected intraperitoneally (176). Boivin extracts of *P. maltophilia*, however, had about the same LD_{60} as those of *P. aeruginosa* (176).

Although unusual nonfermenters may cause any type of disease, they figure prominently in urinary and wound infections, septicemia with and without endocarditis, and necrotizing pneumonitis. Except in melioidosis, the prognosis depends largely on the restitution of local and/or general defense mechanisms or on the removal of contaminated material from the patient.

Antimicrobial Susceptibility

Antimicrobial susceptibilities of unusual nonfermenters show inter- and intraspecies variations which call for the testing of every significant human strain. However, a few general statements can be made, using data obtained in tube dilution and disk sensitivity tests. Application of the Kirby—Bauer method (26) to the testing of nonfermentative gram-negative rods showed that good correlations between zone diameters and minimal inhibitory concentrations are found for polymyxin B (88%), gentamicin (82%), tetracycline (91%), cephalothin (95%), and carbenicillin (92%) if only susceptible—resistant disagreements are considered (386).

In this review only commonly used drugs will be covered. The categories "sensitive" and "resistant" correspond to breakpoints issued by the National Committee for Clinical Laboratory Standards (313). "Intermediate" strains were put into the "resistant" category. The "sensitive" data presented are, of course, subject to revision, as more resistant strains may emerge.

Generally, ususual nonfermenters are (with few exceptions, such as groups II-E, II-F, and II-J and occasional strains of *Pseudomonas alcaligenes*, *Pseudomonas testosteroni*, and *Pseudomonas paucimobilis*) resistant to penicillin; likewise, the penicillinase-resistant penicillins and vancomycin are ineffective. Occasional strains of flavobacteria are susceptible to clindamycin; more species of nonfermenters are susceptible to erythromycin. Aminoglycoside-resistant strains are mostly non-enzyme producers unable to accumulate these drugs intracellularly (358). There may or may not be cross-resistance between aminoglycosides (295). Synergisms between aminoglycosides and β-lactam antibiotics have also been investigated (90, 91, 223, 224).

UNUSUAL PSEUDOMONAS SPECIES

In this section all human *Pseudomonas* species with the exception of *P. aeruginosa* will be reviewed.

Pseudomonas fluorescens and Pseudomonas putida

Ecology

Pseudomonas fluorescens and *Pseudomonas putida* were first described in 1886 by Fluegge, who isolated them from rotting material (142). Soil, water, plants, and contaminated foodstuffs, including milk, are the main habitats of the two species (128, 219, 222, 239, 391).

Hospital water sources have been found to be contaminated with *P. fluorescens* (212, 355, 378, 384). Survival of a *Pseudomonas* species likely to be *P. fluorescens* in benzalkonium chloride has been noted (355). Sutter et al. (449) isolated *P. putida* as part of the oropharyngeal flora in 1.7%, *P. fluorescens* in 0.6%, and *P. aeruginosa* in 6.6% of 350 healthy individuals. With the help of a mineral—acetate—methionine agar, Rosenthal (383) isolated *P. putida* at 37°C from inanimate hospital sources (sinks, floors) with a frequency third only to those of *P. aeruginosa* and *P. maltophilia* among nonfermenters. He did not isolate *P. putida* from any patient sources.

Important habitats of psychrophilic *Pseudomonas* species in hospitals can be bank blood contaminated through the skin of the donor, unsterile bleeding equipment, or hairline fissures in the bottle. The blood may look normal on macroscopic inspection, as it did in the first documented case of Pseudomonas ("Pseudomonas geniculata") contamination (435). Pittman (352) described eight strains of Pseudomonas species and four strains of unidentified gram-negative rods which caused severe and in some instances fatal transfusion reactions; four strains may have been P. aeruginosa, two P. fluorescens, and two P. putida. Bourgain et al. (41) isolated 40 strains of Pseudomonas species from 2310 units of stored blood, 33 of which formed fluorescein and failed to grow at 37°C but grew between 4 and 25°C. A mixed invasion of transfusion blood with Erwinia species (Enterobacter agglomerans) and P. fluorescens was observed by Felsby et al. (133). The authors assumed that the acid—citrate—dextrose stabilizer in the pilot tube had been contaminated prior to the addition of blood. However, fluorescent pseudomonads are not known to fix nitrogen; thus survival in the nitrogen-free stabilizer is difficult to explain.

Clinical Significance

Most strains of P. fluorescens and P. putida have been of indeterminate clinical significance (36, 219, 277, 340, 380, 448, 489). Isolation has been infrequent if compared to P. aeruginosa, sources being urine, stool, pus, and the environment (36, 260, 277). A few apparently significant strains have been recorded—from a pleural exudate which was secondarily infected (378), urinary tract infections (64, 162, 388, 489), septic arthritis (267), osteomyelitis (162), wound infections (162, 388), pelvic inflammatory disease (144), and septicemia (351, 380, 448, 489). Most of these rare cases occurred in an opportunistic setting and/or showed a few symptoms of infection. Hypothermia was found to be an important risk factor for septicemia in cancer patients (351). In the sputum of children with cystic fibrosis, P. fluorescens and P. putida have been found mixed with pathogens commonly isolated in this condition (21).

On the basis of our numerically small observations (21, 489) and those of other authors quoted above, it can be inferred that the virulence of these bacteria for normal hosts is low unless they enter the circulation directly and in large amounts. Liu (259) has shown that the reason for the lack of systemic reactions to P. fluorescens infection in warm-blooded animals (mice) is the inability of the bacteria to grow at their internal body temperature. Dermal necrosis, however, can be elicited. The LD_{50} for mice is high (at least 2×10^8 to 5×10^8 bacteria intraperitoneally) (260). Rapid elimination of P. fluorescens from the internal organs follows infection of burns in mice (259), and a similar process may occur in man. *Pseudomonas fluorescens* and P. putida may be able to survive but not to multiply significantly and may eventually be eliminated. Septicemic symptoms following transfuons or infusions contaminated with these bacteria can be explained by endotoxin release. Endotonix from psychrophilic pseudomonads has the same toxic effect on rabbits as an equivalent dose of *Escherichia coli* endotoxin (43).

Antimicrobial Susceptibility

Most strains are susceptible to amnioglycosides, norfloxacin, piperacillin, and polymyxin B; over half also the tetracyclines; and some *P. fluorescens* to trimethoprim—sulfamethoxazol. Resistance to carbenicillin is notable. Susceptibility to third-generation cephalosporins is variable (36, 126, 128, 163, 164, 221, 227, 304, 319, 321, 483). Other antibiotics usually have no effect. Cefotaxime or azthreonam plus aminoglycosides gave synergistic effects (91, 224).

Pseudomonas pseudomallei

Ecology and Transmission

In the past 30 years it was established that *P. pseudomallei*, like other pseudomonads, occurs in soil and water, albeit in a more limited geographical area. Chambon (67) found it in 5 of 150 mud and water samples in Vietnam. Using direct hamster inoculations, Strauss et al. (441, 443, 444) found *P. pseudomallei* in surface water and soil samples from East and West Malaysia: 12 of 375 samples from forests (3.2%), 112 of 1269 samples from cleared fields (8.8%), and 110 of 753 samples from wet rice fields (14.7%) in West Malaysia contained *P. pseudomallei*. Soil from forests was positive in 5 of 43 samples and soil from cleared fields was positive in 5 of 18 samples. The authors explain the relatively high isolation rates from wet rice fields by virtue of the high moisture and water temperatures (40—43°C), which provide optimal growth conditions. In contrast, in the forest areas, the soil temperature was only 22—25°C, and in the cleared areas it was 26—37°C. The optimal terrain for *P. pseudomallei* seems to be flooded low-lying plains in a climate with high moisture (103). *Pseudomonas pseudomallei* was usually isolated from stagnant water following periods of rainfall, and rarely during dry periods (67). An association with soil or water pH could not be established. Similar findings have been recorded in Queensland, Australia (460). The bacteria were still viable in water specimens 20 months after collection (441) and have been shown to multiply in tap water for 4 weeks (290).

In 1966 Redfearn et al. (366) observed that human and animal infections with *P. pseudomallei* occurred only in a zone between 20° North and 20° South, for example, in Southeast Asia (Burma, Thailand, Laos, Cambodia, Vietnam, Malaysia, and Indonesia); Ceylon and the Philippines; Guam and Northern Australia (Queensland); Madagascar,

Upper Volta, Niger, and Chad; and Ecuador, the Netherlands Antilles, and Panama. More recently, however, several exceptions to this geographical rule have been found. One is Iran, where P. pseudomallei was found in rice fields (103). The other one is the Oklahoma panhandle where the first indigenous isolations from the United States of a bacterium at least resembling P. pseudomallei were made (284). In the latter instance culturally and biochemically identical strains with susceptibility patterns typical for P. pseudomallei were recovered from a soil-infested laceration of a patient and from the surrounding soil. They showed, however, less virulence for guinea pigs than usual laboratory strains, their fatty acids were different from those of P. pseudomallei, and the patient's serum showed no titer rise against control strains of P. pseudomallei or against his own isolate (284). Thus any conclusion as to the final character of the organism should be tentative. A similar organism (which reacted with specific antiserum) was recently found in the noninfected anophthalmic orbit of a patient who had previously sustained multiple facial fractures following a car accident, with clay contamination of the orbit; he was seen in Augusta, Georgia (322). Further exceptions are cases of melioidosis from Turkey (119) and from a newborn in the United States (330). In 1975 and 1976 P. pseudomallei was isolated from a few diseased animals (wild horse, zebra) in zoos in Paris (103) and Vincennes (298). Soil, air, and some free-living animals (rats, pigeons, and cats) in the areas surrounding the Paris (Jardin de Plantes) zoo were also found to harbor P. pseudomallei (298). Melioidosis was later seen in an employee of the zoo (153). Other confirmed cases of melioidosis either originated in endemic areas (35, 283) or represented laboratory infections (175, 399). The only area with a high incidence of P. pseudomallei isolation and infection is Southeast Asia, and only there has soil been systematically examined for P. pseudomallei (490, 491, 493, 494).

Pseudomonas pseudomallei is usually transmitted to man from infected soil or water. Vaucel (476) was able to infect guinea pigs by sterile scarification and subsequent exposure to water from a suspicious pool. Most cases in the French army occurred in soldiers who had waded through flooded terrain, and many U.S. soldiers infected with P. pseudomallei in Southeast Asia were infantry men (385). Histories of soil contamination of wounds or abrasions are common in patients with melioidosis. Airborne transmission from soil is suggested by cases of melioidosis pneumonitis observed in helicopter crews (151, 211). In laboratory infections contact and airborne transmission is possible (175, 399). The only documented case of direct man-to-man transmission (by sexual route from a patient with P. pseudomallei prostatistis) was published in 1975 (283). Animal-to-man transmission or epidemics have not been reported so far. Hospital-acquired infection in two patients with urethral catheters (via dust?) in an endemic area was recently reported (14).

Infection in Animals, Toxin Production, and Immunological Studies

Pseudomonas pseudomallei has a wide range of natural animal hosts, including horses, cows, pigs, sheep, goats, cats, dogs, hamsters, and animals in zoos (macaques, orangutans, tree-climbing kangaroos) (103, 145, 442). Melioidosis in animals is as sporadic as it is in man and has been found in the same geographical areas, albeit rarely. A survey of 68 heads of cattle and goats for hemagglutinating antibodies in the endemic area of Carey Island, Selangor (Malaysia), did not reveal any positive titers (441, 443). Wild rats, once suspected to be the reservoir for *P. pseudomallei* (432), have only rarely been found infected (4, 74). The three wild horses with fatal melioidosis reported from the Paris Jardin des Plantes were probably infected by fodder mixed with contaminated soil (103).

The most susceptible animals for experimental infections are hamsters and ferrets, which can be infected, in order of effectiveness, by intraperitoneal, subcutaneous, or peroral routes (289). The LD_{50} values of virulent strains for hamsters have been 6 (intraperitoneal), 10 (subcutaneous), and 70 (respiratory) bacteria (289); figures for peroral infections vary. Somewhat less susceptible are guinea pigs, wild and white mice, and rabbits (145, 289). Guinea pigs develop periorchitis and orchitis upon intraperitoneal injection of *P. pseudomallei* within 48—72 hr (Straus reaction), unless they die earlier (439). In white mice high doses (1.5×10^6 bacteria) of the virulent strain 103-67, applied by the respiratory route, produced acute fatal disease of 1—3 days' duration, while low doses (100 bacteria) produced a chronic disease of 2—8 weeks with some degree of immunity against reinfection. In hamsters the low dose caused acute disease as well (86). Such experimental respiratory infection begins as pneumonitis and ends in speticemia with excretion of *P. pseudomallei* in stool and urine. Organ lesions resemble those observed in man. Inapparent infections, with persistence of *P. pseudomallei* and nonprogressive lesions, resulted when avirulent strains were aerosolized. Starvation or administration of cortisone (2.5 mg for 3 days before and after the infection) led to focal necrotic lesions resembling those following infections with virulent strains (87). Rats, birds, monkeys, cattle, and dogs are relatively more resistant to experimental infections (145).

While *P. pseudomallei* has endotoxic properties (363), two thermolabile exotoxins have also been demonstrated in broth culture filtrates (193, 316): one is lethal and dermonecrotic, and the other lethal and immunogenic for mice and hamsters. Both are probably low molecular weight proteins. Local tissue toxicity was found associated with proteolytic activity, and lethal toxicity with anticoagulant activity (192). Nigg et al. found toxin formation (316) and virulence for mice (318) unrelated to colonial morphology (S and R), while Chambon and Fournier (68) claimed virulence to be dependent on the presence of the K antigen which determines the colonial type. Passage through mice enhances virulence (318).

Low doses of *P. pseudomallei*, causing chronic melioidosis in mice (see above), provided a 40-fold increase in immunity against respiratory reinfection (86). Immunization by the subcutaneous or respiratory route with avirulent *P. pseudomallei* strains increased resistance to the establishment of melioidosis 4—17 times, but had only a small effect on the resistance to progression of already established disease (87, 88). Evaluation of immunity to *P. pseudomallei* infection is complicated by the fact that immunization of mice with an auxotrophic mutant protects against parenteral but not against respiratory infection (256). Previous bacille Calmette-Guérin vaccination increases the survival time of mice after intravenous challenge with *P. pseudomallei* two- to fourfold (199).

In humans impairment of T-cell function in melioidosis has been documented, but whether this is a primary or secondary event is as yet unclear (339).

Human Melioidosis: Clinical Features

In 1912 Whitmore and Krishnaswami (500) described a rapidly fatal granulomatous disease occurring in Rangoon, Burma. In the following year Whitmore gave an extensive account of the disease called melioidosis and its bacteriology in a now classic article (501).

Clinical aspects are covered in textbooks of internal and tropical medicine and in monographs (e.g., see Ref. 145), as well as in individual articles (16, 50, 75, 123, 140, 167, 211, 338, 339, 392, 429, 493). Melioidosis is a protean disease which presents most frequently as pneumonitis (two-thirds to four-fifths of all cases) (140, 429), septicemia, localized or spreading suppurative lesions, or a combination of these. Less frequent localizations have been summarized recently (339). In view of serological findings, inapparent disease probably occurs more often than overt illness. The incubation period varies widely, from a few days to months. The disease occurs mainly in male adults; only a few cases in children have been recorded (338, 339). A useful classification was developed by Alain et al. (4). These authors recognized acute, subacute, and chronic forms. The acute form presents, as a rule, as septicemia with metastatic lesions; pneumonia may or may not be present. The fatality rate without treatment is over 95% (493), with death occurring within a few days of the onset of symptoms. Even antimicrobial treatment is often ineffective. The main differential diagnostic consideration is plague.

The subacute form presents most frequently as a bacteremic pneumonitis with upper lobe infiltrates and/or cavitation (123). Less frequent are cellulitis and lymphangitis originating from an abrasion. Both types may develop into generalized infection. The subacute form has been observed in the majority of U.S. military personnel with symptomatic melioidosis in Southeast Asia (429). It usually runs a course of several weeks, responds well to antimicrobial treatment, and has a case fatality rate of about 15% (140). The main differential diagnostic considerations are chronic lung infections, for example, tuberculosis or pulmonary mycoses.

The chronic form presents either as a localized suppurative process or as a chronic pneumonitis. It is preceded either by a silent infection or by the subacute form of the disease, with a possible silent interval of 3 weeks to 9 years (392). This form was characteristically observed in service men who had returned from Southeast Asia to the United States (429). The case fatality rate is very low (55, 348).

The overall case fatality rate from melioidosis decreased strikingly from over 90% to about 20% in French soldiers serving in Indochina when chloramphenicol was introduced in the treatment of the disease (140). Of 187 cases reported in American soldiers in Vietnam between April 1965 and December 1969, 13 cases (7%) were fatal (392).

The opportunistic potential of pseudomonads in general and the relative frequency of asymptomatic infections as revealed by serotests have led some observers to believe that exposure to *P. pseudomallei* is followed by overt disease only in individuals with low resistance and that *P. pseudomallei* has little invasive power, that is, it is an opportunistic pathogen (366). This concept is supported by many, but by no means by all, recorded case histories. Most of Whitmore's patients were morphine addicts (501). Three of six patients reported by Rimington (376) from North Queensland were diabetics, and one had chronic nephritis, and another cystic lung disease. Diabetes mellitus was present in other patients with melioidosis as well (14, 16, 404). Further preexisting conditions were cancer, alcoholism (16), and lupus erythematosus (419). Activation of unrecognized infection has also been observed. Flemma et al. (140) recorded 15 patients in whom burns had preceded the outbreak of melioidosis; the interval between the burn and the outbreak of melioidosis lasted from a few days to 5 months. In recrudescent melioidosis, recrudescence was precipitated by trauma associated with surgical treatment, diabetic ketoacidosis, burns, pneumococcal pneumonia, or influenza (264, 392, 404). Association between surgical trauma or infectious disease and the outbreak of melioidosis has also been noted by Alain et al. (4). In one case development of a bronchogenic carcinoma precipitated pulmonary melioidosis 26 years after presumed exposure (282).

Nigg (315) found that 28 (8.3%) of 337 healthy young males and 1 of 78 healthy young females from Thailand had positive complement fixation tests for *P. pseudomallei*, whereas none of 138 sera from a control group in the United States was positive. Of 372 unselected healthy U.S. soldiers in Vietnam in 1966, 4 (0.9%) had significant hemagglutination titers (above 1:40) for *P. pseudomallei* (429). In another series studied at least 5 years later, significant titers were observed in 8.9% of 412 exposed soldiers, versus 2.9% of 606 control persons (73). Strauss et al. (440) found significant hemagglutination titers in 1.9—15.8% of Malaysian army recruits; the highest percentage of positive titers was in individuals hailing from rice-growing areas. Of Commonwealth soldiers serving in Malaysia, 2% had significant hemagglutination titers; in their histories exposure to surface waters as well as fever of unknown origin were more frequently noted than in the histories of the other 98% (459). On the other hand, most of the U.S. soldiers who contracted melioidosis in Vietnam had no preexisting disease (140, 151). Thus *P. pseudomallei* cannot be an exclusively opportunistic pathogen. Highly virulent strains and large inocula may well overcome normal defense mechanisms. The high virulence of most *P. pseudomallei* strains for certain animals (see above) may also argue against an exclusively opportunistic role of the organism. The much higher frequency of melioidosis among males (122) has been explained by greater exposure of soldiers to contaminated soil and water (145), but the predilection for males in civilian populations cannot be explained on that basis.

Human Melioidosis: Serodiagnosis

Antisera against *P. pseudomallei* cross-agglutinate with *P. mallei* and some also in low dilutions with *Yersinia pestis* (102). Agglutinins against *P. pseudomallei* have been found in normal sera up to a titer of 1:320 (82, 83). They may cross-react with *Salmonella* species and with *Legionella pneumophila* (51, 233a). Experience with indirect fluorescent antibody tests is limited (271). As of this writing, the most relevant serotests are the complement fixation (CF) and indirect hemagglutination (IHA) tests. Recently, IgM indirect fluorescent antibody titers have shown correlation with clinical disease, being absent in subclinical cases (15).

An optimal antigen for the IHA test seems to be the supernatant of mated microtitration tests have been employed using either pyruvic aldehyde-stabilized or freshly sensitized erythrocytes (183). In one series titers of 1:10 to 1:20 were found in 3.5% of 200 persons never exposed to *P. pseudomallei* (440); in another series titers of 1:40 were observed in only 1.4% of 145 healthy individuals and rarely in individuals with other infectious diseases, for example, due to *Pseudomonas stutzeri* (6). Titers of 1:40 or higher were seen in 445 patients with proven overt melioidosis 2 weeks after the onset of the disease (6). These titers persisted for at least 9 months and sometimes for as long as 3½ years and were not related to the type of the patient's infection, except that patients with localized pneumonitis or wound infection caused by *P. pseudomallei* gave variable responses. In a further series of 114 culturally proven cases, 97.5% were eventually IHA positive, with titers of 1:40 or above (6).

A good antigen for the CF test is prepared according to Nigg and Johnston (317) as an aqueous extract of disintegrated *P. pseudomallei* cells. The authors obtained high (1:8000) titers in experimental animals 9—11 days after intravenous infection. In human subjects titers

1:4 or higher are considered significant. Titers of 1:8 are more specific and only somewhat less sensitive; higher ones are only occasionally observed in disease caused by *P. aeruginosa* and *P. stutzeri* (6). Of 401 sera from patients with proven melioidosis, 71% showed 1:8 titers by the end of the first week of disease (6). Persistence of the CF antibody parallels that of IHA antibody (6); titers may reach 1:1024. Variable responses occur more rearely than in the IHA test. Of 114 culturally proven cases of melioidosis, 99% were eventually CF positive (6). A combination of the IHA and CF tests is extremely valuable since transient drops in titers, seen in about 20% of patients with melioidosis, usually occur in one test at one time. The serological response has no prognostic significance (6).

Skin tests are no longer used because of unreliability (339).

Human Melioidosis: Pathology

Lesions are observed in decreasing order of frequency in the following organs: lung, liver, spleen, lymph nodes, kidney, and skin (50, 175, 348, 385). In acute cases with a very short course, lesions are often microscopic. When grossly visible lesions appear, they present as multiple nodules with raised yellow centers and sharply defined hemorrhagic (lung) or nonhemorrhagic borders (other organs), measuring from a few millimeters to a few centimeters in diameter. They may coalesce later. Microscopically, formation of an inflammatory exudate with histiocytes and neutrophils is seen first. Soon a central coagulation necrosis develops which becomes surrounded by degenerating histiocytes; peripherally neutrophils and sometimes a few epithelioid cells are seen. In the alveoli the peripheral capillaries show congestion and hemorrhage. In older lesions multinucleated or multilobated giant cells appear. The granuloma then resembles that of lymphogranuloma venereum, sporotrichosis, tularemia, or cat scratch disease. Organisms are found most frequently in acute lesions and least frequently in chronic lesions.

Antimicrobial Susceptibility of Pseudomonas pseudomallei

Based on the generally accepted breakpoints for drugs used at present (NCCLS), most *P. pseudomallei* strains are susceptible to tetracyclines, chloramphenicol, and kanamycin (but not to other aminoglycosides). Data on ampicillin and carbenicillin are contradictory. Some strains are susceptible to nalidixic acid. All other single drugs are generally ineffective, except for novobiocin (6, 115, 122, 138, 145, 147, 148, 178, 182, 198, 237, 271, 517). Of the third-generation cephalosporins, ceftazidime has shown effectiveness against four strains (482).

Eickhoff et al. (115) tested 12 combinations involving chloramphenicol, kanamycin, tetracycline, sulfadiazine, novobiocin, ampicillin, and dicloxacillin. They observed antagonism between kanamycin and chloramphenicol, kanamycin and tetracycline, and chloramphenicol and sulfadiazine. Synergism was found between ampicillin and large doses of dicloxacillin, and between sulfadiazine and kanamycin. All other combinations were additive. The authors assumed that the synergistic effect of the two penicillins was due to competitive inhibition of a *P. pseudomallei* penicillinase through binding to dicloxacillin, which allows ampicillin to act. The combination of trimethoprim and sufamethoxazole was found effective against 75% of 33 strains at therapeutically achievable levels (2 + 40 g/liter), while none of those strains was inhibited by trimethoprim alone and only about one-third were inhibited by sulfamethoxazole alone (122). Other reports have listed most strains as sensitive to the combination of 1 + 20 g/liter (23, 151).

Antimicrobial Agents in Experimental Melioidosis

The response to these drugs depends on the experimental animal used, the inoculum and virulence of the strain, the route of administration, and the dosage and schedule of the drugs. Thus strains may show different in vivo responses to drug(s) to which they are sensitive in vitro (178). Evaluations should take into account ED_{50}, extension of survival time, and eradication of the organism from internal organs. Rifampin, sulfadiazine, nalidixic acid, chlortetracycline, tetracycline, minocycline, doxycycline, and novobiocin extended the survival time in mice (138, 178, 197, 198, 237). In spite of its in vitro ineffectiveness, rifampin was the only drug which led to an eradication rate of *P. pseudomallei* from internal organs within 30 days that was significantly higher than that observed in controls (138, 237). Unsatisfactory in every respect were penicillin and all of its derivatives, streptomycin, oxytetracycline, the polymyxins, gentamicin, nitrofurantoin, lincomycin, and erythromycin, as well as some drugs that are effective in vitro, namely, peroral tetracycline, methacycline, sulfisoxazole, and sulfisomidine (138, 178, 197, 237, 291).

Antimicrobial Agents in Human Melioidosis

The most frequently used drugs have been found effective only if given over a period of 4 weeks in high doses, for example, tetracycline alone or in combination with novobiocin, chloramphenicol, or both (16, 55, 271). Novobiocin alone does not appear to be effective. For severely ill patients, a combination of chloramphenicol, kanamycin, and novobiocin for at least 2 weeks had been proposed (75); however, one may question the effectiveness of such a regimen on grounds of drug toxicity, drug antagonism (see above) and possible resistance of the strain to kanamycin. More adequate and successful, seems to be the combination regimen of tetracycline and chloramphenicol (55). Relapses after all these types of treatment are not uncommon (429).

Recently several cases of melioidosis have been successfully treated with sulfamethoxaxole—trimethoprim (16, 151, 220, 339). One case was successfully treated with ceftazidime (419), while another demonstrated failure of ceftriaxone in spite of a minimal inhibitory concentration of 3 mg/liter (404).

Pseudomonas mallei

The malleus bacterium was first described by Loeffler in 1882. It is an obligate parasite of animals (horses, mules, donkeys; more rarely, goats, sheep, dogs, and cats) and is occasionally transmitted to man. Man-to-man transmission is very rare. The portal of entry seems to be an abrasion of the skin, of the mucosa of the respiratory tract, and, rarely, of the mucosa of the gastrointestinal tract. Intratracheal or intranasal injection of P. mallei results in acute disease in animals. Pulmonary lesions may follow oral infection.

In animals, "glanders" refers to a primary infection of the respiratory tract with subsequent spread, while "farcy" refers to a primary skin lesion with subsequent lymphangitis and formation of subcutaneous abscesses. For details of the human infection, textbooks of internal medicine should be consulted. The rapidly spreading nodular form is most often fatal. Prognostically better are the chronic respiratory and ulcerating/abscess-forming types. The diagnosis is confirmed by cultures of pus, sputum, or (terminally) blood and by the Straus reaction (analogous to the reaction with P. pseudomallei) (439), the mallein skin test, a complement fixation reaction (cross-reactive with P. pseudomallei) (83), and by tissue slides (tuberculosis-like granulomata without caseation). There is no immunity.

Susceptibility to sulfonamides, streptomycin, kanamycin, tetracycline, chloramphenicol, novobiocin, and erythromycin and resistance to penicillin, ampicillin and the polymyxins have been observed in a few strains (275, 291). In experimental infections sulfadiazine was effective, while streptomycin and penicillin were ineffective (210, 291). In man sulfonamide therapy has been reported successful, but little information is available on the effect of other drugs.

Pseudomonas cepacia

Ecology

Pseudomonas cepacia was first described by Burkholder in 1950 (57) as a yellow-pigmented pseudomonad that caused sour skin, an onion bulb rot. Subsequently, similar bacteria were isolated from other sources but were not recognized as P. cepacia (e.g., strains from soil and river water in Trinidad) (309). Milk may occasionally be contaminated with P. cepacia (222). An important source of P. cepacia is the moist hospital environment (surfaces and instruments, water sources, flower vases, solutions, etc.), as detailed below. Strains isolated from water reservoirs of unheated nebulizers were able to multiply in doubly ionized or doubly distilled water, in 5% glucose, and in 0.9% saline, but not in doubly deionized and then doubly distilled water or in commercial hypertonic intravenous nutrition solutions (156). Organisms kept in distilled water were less temperature sensitive and smaller than those subcultured to tryptic(ase) soy media and showed no flagella (65). They did not cause turbidity even when grown in distilled water up to a count of 10^7 per milliliter. Survival in distilled water was 48 hr and 21 days at 50 and 10°C, respectively (65).

Experimentally, the disinfectant Savlon (British Pharmacopoeia; 0.05% chlorhexidine + 0.5% centrimide) inhibited *P. cepacia* only at a 1:320 dilution, large inocula even surviving at a 1:30 dilution (25). In a 1% peptone solution in distilled water, multiplication from an initial concentration of 10^1 organisms per milliliter to a final concentration of 2.5×10^4 organisms per milliliter occurred at room temperature in 4 weeks (25). The adaptation was pH dependent, however, taking place at pH 6.0 (distilled water) but not at pH 7.2 (tap water) (22). In 0.05% aqueous chlorhexidine alone, concentrations of 10^5-10^7 cells per milliliter have been found (428). Many strains grow on cetyl trimethylammonium bromide (0.08%) in agar (414). In one instance, 0.15% dimethyl benzylammonium chloride in water and phenoxypolyethoxy–ethanol (Detergicide, British Pharmacopoeia), used as a preservative/disinfectant in a commercial catheter kit, contained 10^3 *P. cepacia* cells per milliliter (186) which could grow to approximately 10^6 per milliliter (280). Detergicide was also unable to rid equipment of *P. cepacia*. Resistance to benzalkonium chloride has been observed as well (25, 146, 155, 161, 227) and can be created by graded exposure to that disinfectant (280). Survival in 0.15% wt/wt dibromopropamidine isoenthonate (Brulidine, British Parmacopoeia) cream has also been reported (436). Later studies have shown survival in p-hydroxybenzoate (410), lidocaine (278, 307), water baths (306, 334, 420), aqueous cocaine (278), 10% povidone–iodine (32, 81), pHisohex (278), and possibly 2% formaldehyde for 2 hr in a dialysis machine (32).

Clinical Significance

Pseudomonas cepacia infections are significantly hospital associated. The ability of the organism to survive under conditions of minimal nutrition or in the presence of certain disinfectants was responsible for colonization or infections caused by contaminated solutions applied directly or via equipment (e.g., bronchoscopes, intravenous medications). In most noncompromised patients, such application results either in mere colonization or in a short-lived infection, for example, bacteremia. In a few patients with no immune defects and in many with compromised defense mechanisms, severe infections may result which require antibiotic treatment. Such infections may occur in the urinary tract, in the respiratory tract (pneumonitis), or as wound infections, peritonitis, or septicemia (24, 32, 63, 81, 109, 113, 146, 161, 186, 227, 241, 278, 288, 306, 312, 334, 362, 364, 394, 410, 412, 414, 420, 428, 433, 436, 512). Pseudobacteremia due to contaminated benzalkonium chloride or povidone—iodine has also been observed (32, 81, 227). The use of serogrouping (O and H) has been helpful to elucidate such epidemics (194, 301). There are, however, also infections that occur outside of epidemics with largely unknown sources: urinary tract infections in the presence of renal calculi (379) or necrotizing pneumonia/lung abscess in patients with chronic granulomatous disease (39, 84, 96, 411), or other preexisting conditions (131, 356, 494). In one such patient, a diabetic, lung abscess followed ultrasonic nebulization treatment for a pneumonia of unspecified origin; the nebulizer reservoir was found contaminated with P. cepacia (356). Administration of a polymyxin B aerosol to the upper airways of 292 patients in a respiratory—surgical intensive care unit (with the aim of preventing P. aeruginosa pneumonitis) led to colonization with P. capacia in 8 patients and to subsequent pneumonitis in 2 others (131). Occurrence of P. cepacia in patients with cystic fibrosis has been documented several times (21, 37, 139, 244, 263, 461). In some patients pneumonia and septicemia developed (170, 382); and in spite of in vitro effectiveness, ceftazidime was unable to control acute exacerbations of pulmonary symptoms (170).

Nine cases of postoperative wound infection caused by local application of a contaminated solution of Savlon (see above) have been reported (25); local antiseptic treatment led to a satisfactory response. Likewise, septic arthritis followed intra-articular injection of a contaminated multidose preparation of methylprednisolone (238). Taplin et al. (452) isolated P. cepacia from the toewebs of 43 of 51 (85%) army ranger trainees after completion of swamp training. Previous toe cultures of the soldiers had been negative for P. cepacia. The soldiers' feet had been frequently immersed in swamps and river water. Water from one swamp contained P. cepacia in a concentration of less than 10 bacteria per milliliter. The authors maintain that within the spectrum of jungle rot or foot rot disease, there is a clinical entity involving the toewebs and occasionally the plantar surface of the feet, giving rise to hyperkeratosis, maceration, and sometimes induration and fissuring. This entity is said to be associated with P. cepacia. The authors, however, used a fungal isolation medium inhibitory for most other bacteria.

A few cases of P. cepacia septicemia are reported in which no external source for the organism could be found. One patient had pancreatic carcinoma (113), the other one a burn infection (506). Both had previously received antibiotics.

The first paper on P. cepacia endocarditis described the organism as a "variant of the genus Herellea" (423). A similar case was subsequently reported as Flavobacterium endocarditis (398). Both patients died. The identity of the causative agents was determined later (414). Several cases of *P. cepacia* endocarditis have been reported since (184, 273, 314, 360, 401, 427). One preexisting condition was heroin addiction. The aortic, mitral, or tricuspid valves may be affected. Ecthyma gangrenosum appeared in one patient (273). Treatment is difficult. Trimethoprim—sulfamethoxazole (184, 320, 401) or this combination plus kanaymcin (427) or polymyxin B (273, 314, 360) eliminated the organism only in some cases, in spite of an in vitro bactericidal effect of the combinations. Previous therapy with chloramphenicol had not been successful in these patients.

Two cases of skin or lymph node abscesses due to *P. cepacia* in patients with chronic granulomatous disease have been described (39), as has been abscess formation due to the same bacterium in a heroin addict (162). Conjunctivitis was caused after the use of humidifiers contaminated with *P. cepacia* (364). Finally, colonization of a Holter valve in a hydrocephalic child led, 9 months after insertion, to septicemia originating from the shunt (24).

Even more than 10^7 *P. cepacia* are rarely pathogenic for mice or guinea pigs intraperitoneally (368, 423, 437a). In the skin of burned mice, *P. cepacia* persisted but was not invasive (437a).

Antimicrobial Susceptibility

Multiple studies have shown that *P. cepacia* is, at this time, consistently susceptible to a limited number of antimicrobials, among them thrimethoprim—sulfamethoxazole, piperacillin, mezlocillin, azlocillin, minocyclin (but not tetracycline!), cefoperazone, moxalactam, ceftazidime, ceftizoxime, and thienamycin. A large percentage of strains is also still susceptible to carbenicillin, cefotaxime, azthreonam, ceftriazone, cefsulodin, chloramphenicol, and nalidixic acid. Other antimicrobials—including polymyxin B and the aminoglycosides—are ineffective (12, 13, 100, 126, 128, 302—304, 319—321, 340, 397, 414, 438, 464, 468—470, 482, 505). Synergism between aminoglycosides and cefotaxim, piperacillin, or ceftriaxone could not be shown; aminoglycosides plus azthreonam, however, occasionally produced synergistic effects (90, 91, 223, 224).

Pseudomonas stutzeri

Pseudomonas stutzeri was first described as *Bacillus denitrificans* II by Burri and Stutzer in 1895 (58). The authors found it in soil, manure, canal water, and straw. Survival in cosmetics (49) and evacuated blood collection tubes (490) has been observed. Human sources, first identified in the 1960s, include the respiratory tract, wounds, blood, the urogenital tract, and spinal and joint fluid (44, 162, 166, 242, 340, 454, 478).

Most human strains were then found in mixed culture and were not considered etiologically significant. Septicemia in debilitated patients

Infrequently Encountered Gram-Negative Rods

has been reported a few times (134, 169, 242, 300, 478); contaminated infusion fluids (134) or deionized water used to prepare the dialysate for hemodialysis (169) were among the sources. Pseudobacteremia due to contamination of liquid soap was observed recently (229). An ulcer leading to an abscess on a previously scarred cornea (45), cervical lymphadenopathy (etiology?) (243), endocarditis (157a), and septic arthritis (266) have each been reported once. Gilardi has isolated *P. stutzeri* in pure culture from otitis media and from postoperative/post-traumatic infections of the extremities; improvement occurred with ampicillin or tetracycline, to which the strains had been susceptible (166). In these three cases, soil-borne *P. stutzeri* was strongly implicated as the causative agent. Bret and Durieux (44) found *P. stutzeri* and *Acinetobacter lwoffi* in the lochia of a primipara with jaundice and oliguria who delivered a stillborn child. Blood cultures were not reported from either mother or offspring. On injection into the peritoneal cavity of pregnant guinea pigs, the *P. stutzeri* strain led to miscarriage, but is causal relationship to the fetal death in the case reported remains unclear in view of the unclear nature of the maternal disease. In another patient the authors isolated *P. stutzeri* from the vagina (44). Finally, Tan et al. (451) have reported 14 patients (i.e., 3% of all cases in 1974—75) with urinary tract infection due to *P. stutzeri* in Singapore. Similar reports from elsewhere are lacking.

Most strains of *P. stutzeri* are resistant to cephalothin, cefamandole, and cefoxitin. Susceptibility to ampicillin, chloramphenicol, and trimethoprim—sulfamethoxazole varies. The other presently used antimicrobials are usually effective in vitro (12, 13, 126, 128, 221, 387, 483).

Pseudomonas maltophilia

Ecology

The species *P. maltophilia* was first outlined by Hugh and Ryschenkow in 1961 (213). Despite its unusual nutritional requirements, *P. maltophilia* is a free-living, ubiquitous bacterium, having been isolated from vegetable and water sources (92, 431), contaminated milk (222), soil in petroleum areas (215), and in the hospital environment from distilled water, incubator reservoirs, nebulizers (296, 297), intralipid emulsion (287), and evacuated blood collection tubes (403, 480), as well as from tissue cultures (213, 431). It survives for a few days on moist inanimate vectors (296) and in Savlon (see above) (503).

Clinical Significance

Pseudomonas maltophilia is the most frequently isolated unusual *Pseudomonas* species in clinical laboratories (30, 149, 154, 160, 340, 350, 383, 477, 516). Common sources are the respiratory tract, urine, pus, vaginal secretions, and blood (30, 149, 154, 279, 340, 350, 425,

431, 448, 477, 516, 518). The majority of those strains have been recovered in mixed culture and are etiologically insignificant or indeterminate; most of them have disappeared without appropriate drug treatment. A few cases of infections in *non*compromised hosts are known from which pure cultures of P. *maltophilia* were isolated: meningitis (94, 337), external eye infection (28), mastoiditis (187), traumatic wound or postoperative hip infection (162), and pneumonitis (394). In patients with serious underlying disease, P. *maltophilia* may cause pneumonia, meningitis, urinary tract infection, and suppurative lesions (95, 154, 160, 162, 337, 340, 350, 448, 471, 503, 518). Transient bacteremia (477) and true septicemia (36, 270), particularly after hemodialysis, in association with P. *maltophilia* pneumonitis, and after infections of the biliary, genital, or urinary tract (141, 160, 422, 518) have been reported. Sepsis also evolved after contamination of intralipid emulsion (287) and blood pressure monitoring devices (137), most patients showing few if any symptoms. Pseudobacteremia was traced to contamination of blood collection tubes (403). One hospital outbreak of urinary tract infections, bacteremia, and omphalitis was traced to the use of Savlon (see above) prepared with contaminated deionized water; a minority of the patients showed symptoms which reacted promptly to antimicrobial treatment (503).

Pseudomonas maltophilia bacteremia was also observed in association with a prolapse of the mitral valve without endocarditis in a previously healthy adult (311). Endocarditis with P. *maltophilia* followed insertion of prosthetic valves and/or intravenous drug abuse; the left ventricle has thus far been the only one affected, and six of the eight patients recovered under combination therapy (101, 136, 403, 511, 514). Colonization and pulmonary infection with P. *maltophilia* occurred in 7 and 2, respectively, of 292 intensive care patients who were treated prophylactically with a polymyxin B aerosol (131).

The significance of P. *maltophilia* strains in patients with cystic fibrosis is unclear (21, 244). A search for P. *maltophilia* antigens in Crohn's disease tissue proved fruitless (502), which may disprove the etiological role of L forms of the organism in this disease (335, 336). Finally, P. *maltophilia* was found in pure culture in an infected wound of an abattoir worker who has sustained a comminuted fracture and lacerations of the foot (112).

While many strains of P. *maltophilia* are hospital acquired (154), there is a good number that have been isolated from outpatients or from patients at the time of admission (425). An association with previous antimicrobial treatment was evident in one series of hospitalized patients (154), but much less clear-cut in another one (425). Respiratory isolates have been associated with previous tracheostomy (154). Recently one new source of the organism has been discovered: feces. The authors isolated P. *maltophilia* from 14 of 218 outpatients stools (483).

Large inocula (5 × 10⁸ organisms, injected intraperitoneally, have not been lethal to mice (176); a contradictory report has also appeared (141).

Antimicrobial Susceptibility

Most strains of *P. maltophilia* show multiple resistance. Consistent sensitivity in vitro has so far only been shown to moxalactam, chloramphenicol, minocycline and doxycycline (but not to tetracycline!), and trimethoprim—sulfamethoxazole. Sensitivity to polymyxin B and nalidixic acid is still seen more often than resistance. The other antimicrobials are either consistently ineffective or, like the aminoglycosides and cephalosporins, rarely effective (12, 13, 100, 126, 128, 132, 302—304, 319—321, 340, 397, 414, 438, 464, 468—470, 482, 505). Combinations between sulfamethoxazole or trimethoprim with colistin (320), and ceftriaxon, azthreonam, cefotaxim, or piperacillin with animoglycosides (90, 91, 223, 224) have shown—albeit unpredictably—synergistic or additive effects. The combinations of gentamicin or trimethoprim—sulfamethoxazole with carbenicillin plus rifampicin have shown consistent synergism (132, 513).

Pseudomonas putrefaciens

This organism was first described by Derby and Hammer in 1931 (97). The authors isolated it from dairy products, especially from tainted butter. Later studies by Long and Hammer (261) were done on strains from milk, natural water sources, and soil. Petroleum (353) and foodstuffs like milk, meat, and haddock, in which *Pseudomonas putrefaciens* acts as a spoiler (97, 239, 254, 261, 353), are other sources.

Pseudomonas putrefaciens has been isolated from various human sources, for example, from the respiratory tract, pus, urine, feces, and pleural fluid (10, 93, 94, 162, 168, 202, 230, 236, 255, 276, 372, 384, 462, 480, 484, 488). Most strains were found in mixed culture and had an indeterminate clinical significance. In several cases of purulent ulcerations of the legs (93, 94, 162, 400, 474) and chronic otitis media (162, 202, 488), *P. putrefaciens* was the only potential pathogen isolated. The otitis strains belonged to group II (255). Leg ulcers were the source of septicemia with group II strains (400, 474). Other cases of septicemia were also associated with lesions on the legs (120, 481) or had no clear source (481). A case of meningitis after head injury has been reported recently (246).

Pseudomonas putrefaciens is generally susceptible to most antimicrobials (including erythromycin) *except* cephalothin and, sometimes, ampicillin, carbenicillin, and cefamandole (126, 128, 163; see the literature cited above).

Pseudomonas alcaligenes and Pseudomonas pseudoalcaligenes

In 1928 Monias first described *Bacterium alcaligenes* (299), which was later named *P. alcaligenes* by Ikari and Hugh (216). The name *Pseudomonas pseudoalcaligenes* was proposed by Stanier et al. (431) in 1966 for a closely related species. Both have been isolated from water sources (216, 453), iodinated swimming pools (129), and contaminated milk (222), as well as from clinical specimens (212, 340). Very few cases, however, of true infections due to *P. alcaligenes* have been established; empyema (340), eye infection (28), and endocarditis on a rheumatic aortic valve (472) are thus far the only ones reported. *Pseudomonas pseudoalcaligenes* has been isolated in pure culture from a postoperative wound infection (162), pneumonitis in a drug addict (162), septicemia (300), and meningitis in a child with a respiratory disorder (77). Furthermore, it was isolated from the uterine tissue of a patient with septic abortion (248), although it was thought that clostridia were the major local pathogens in this case. Likewise, it occurred in a mixed respiratory culture in a child with cystic fibrosis (21). Strains of both species show variable susceptibilities. Cephalothin is rarely effective; susceptibility to the second-generation cephalosporins, ampicillin, carbenicillin, chloramphenicol, trimethoprim—sulfamethoxazole, and the aminoglycosides varies. The third-generation cephalosporins, moxalactam, the ureidopenicillins, tetracycline, and polymyxin B have generally been effective (126, 128).

Pseudomonas diminuta and Pseudomonas vesicularis

In 1954 the species *Pseudomonas diminuta* was proposed by Leifson and Hugh (253), who isolated three strains from stream water. Later, strains were isolated from human sources such as urine, sputum, wounds, and blood (212, 340), as well as from respiratory equipment (296). There is only one report, thus far, in which etiological significance is ascribed to *P. diminuta* (a case of septicemia) (300). Most strains are susceptible to chloramphenicol, erythromycin, tetracycline, the aminoglycosides, carbenicillin, mezlocillin, piperacillin, cefotaxime, ceftizoxime, and trimethoprim—sulfamethoxazole, variable in their susceptibility to polymyxin B and cefoperazone, and resistant to other antimicrobials used at present (126, 128).

Pseudomonas vesicularis has been isolated from a medicinal leech (59) and from human cervical specimens (331). No strain has been clinically significant so far. Susceptibility resembles that of *P. diminuta*, except that more strains are sensitive to the cephalosporins except cefamandole (126, 128, 164).

Pseudomonas acidovorans and Pseudomonas testosteroni (Comamonas terrigena)

The first description of a strain belonging to this group goes back to Guenther (179), who, in 1894, isolated it from soil and named it

Vibrio terrigenus. Both bacteria are widely distributed in nature (soil, water) and have been isolated from animal (212) and human sources [urine, respiratory tract (340), and eye (28)] as well. In one reported case of septicemia in an apparently immunocompetent host, the etiological role of *P. testosteroni* was established (17). Another case of septicemia and purulent hip joint arthritis also affected a "normal" host (418). Sonnenwirth (422) cultured *Comamonas terrigena* from two blood cultures of a patient with suspected endocarditis, but was unable to confirm its etiological role.

Bacteremia from a contaminated pressure transducer due to *Pseudomonas acidovorans* and *Enterobacter cloacae* has also been reported (497), as well as *P. acidovorans* as the etiological agent of a corneal ulcer with hypopyon formation (45).

Antimicrobial susceptibilities of the two species differ. *Pseudomonas testosteroni* is usually susceptible to all antimicrobials except ampicillin and, sometimes, carbenicillin. *Pseudomonas acidovorans* is usually susceptible only to cefoxitin and the third-generation cephalosporins, the ureidopenicillins, tetracycline, the quinolones, and trimethoprim—sulfamethoxazole (126, 128, 164, 221).

The VA Group: VA-1, Pseudomonas pickettii, Pseudomonas thomasii, and IVD

The biovar VA-1 has been reported as an agent of septicemia (218, 225). Six cases were related to the use of contaminated 0.05% chlorhexidine for skin disinfection (225). Furthermore, one patient has acute meningitis (127). *Pseudomonas pickettii* (VA-2) has been found in many clinical and environmental sources (361, 373). Three cases of septicemia are extant (150, 185). The source may have been intravenous catheter lines. The organism has also been isolated as a colonizer of the respiratory tract following endotracheal toilette with contaminated 0.9% NaCl (308). Finally, a report of a wound infection exists (185). *Pseudomonas thomasii*, closely related to if not identical with *P. pickettii* (232) contaminated the distilled water supply of a hospital (20) and caused nosocomial bacteremia, urinary and respiratory infections (345) through insufficiently sterilized fluids (105, 345). The group IVD, also closely related if not identical, has caused endocarditis (172) and was found in the sputum of 16 patients with chronic respiratory disease in one series (hospital epidemic?) (234).

The biovar VA-1 is generally sensitive to more antimicrobials than *P. pickettii*. Both have been susceptible to chloramphenicol, tetracycline, nalidixic acid, trimethoprim—sulfamethoxazole, cephalosporins of the second and third generation, and the ureidopenicillins, and resistant to ampicillin and polymyxin B. The effects of cephalothin and carbenicillin vary, although VA-1 is more often sensitive than *P. pickettii*. The latter is always resistant to aminoglycosides, VA-1 only irregularly so (126, 128, 154, 163).

Pseudomonas paucimobilis

This species is widely distributed in the environment, including hospital water sources (173, 203, 252, 300, 374, 479). Disease has only been described recently: nosocomial septicemia (416, 424), meningitis in a "normal" host (181), and infection of a leg ulcer (341). The author described nine clinically insignificant strains from blood cultures in 1972 (479). Water sources (saline for irrigation, tap water used for rinsing) have led to colonization in patients (80, 125). A pseudoepidemic of bacteremia proved to be due to water in a shaking incubator (117). Strains are usually susceptible to erythromycin, tetracycline, chloramphenicol, trimethoprim—sulfamethoxazole, and the aminoglycosides, and resistant to the ureidopenicillins, the first- and second-generation cephalosporins, cefoperazone, and moxalactam. Susceptibility to other drugs varies (126, 128, 417).

The Group VE: VE-1 and VE-2

These bacteria have been isolated from a variety of environmental and human sources (165, 340, 480). In one series of eight strains, none was judged clinically significant. The authors also mention three earlier papers describing human isolates called *Chromobacterium typhiflavum* which are at least similar to VE bacteria (340). Recently, septicemia due to VE-2 in two severely traumatized postsurgical patients was published (346, 405). Septicemia with VE-1 was seen in a patient under steroid treatment for lupus erythematosus who underwent treatment for pancreatitis (29). The VE organisms have been susceptible to many antimicrobials, except the first- and second-generation cephalosporins; susceptibility to ampicillin, chloramphenicol, and trimethoprim—sulfamethoxazole varies (126, 128).

ALCALIGENES SPECIES

Alcaligenes faecalis

The habitat and survival of *Alcaligenes faecalis*, first isolated in 1896 from stale beer and from feces (343), seem to be similar to those in *Pseudomonas* species. Besides soil and water, sources include moist items in hospitals, such as nebulizers and respirators (296), intravenous drugs (294), hemodialysis systems (66, 130), tissue cultures (458), and 0.1% chlorhexidine solution (107). *Alcaligenes faecalis* may multiply in tap water and in certain mineral salt solutions as well (40).

Alcaligenes faecalis has probably been isolated quite frequently and from numerous human sources. Earlier reports of isolation, however, are by and large unsupported by evidence that would exclude

Achromobacter species and nonoxidizing *Pseudomonas* species (which would call for the use of oxidative—fermentative carbohydrate media and flagellar strains for differentiation from *Alcaligenes*). Thus the approximately 50 reports of "*Alcaligenes*" infections which have been summarized elsewhere (2, 154, 157a, 415, 495) will not be discussed here. Only papers with sufficient evidence of a correct bacteriological diagnosis will be covered. The organism may perhaps occasionally be part of the normal skin flora (34).

Nonfatal septicemia due to *A. faecalis* has been described following appendectomy (415), use of contaminated hemodialysis fluid (66), intravenous injection of contaminated succinylcholine (294), and urological instrumentation possibly involving contaminated water (245). Mixed septicemia, with *A. faecalis* as one component, followed transurethral resection of the prostate (196), cardiac surgery (196), and war injuries (465). In the latter instance the organism was also isolated from wounds (465). Of 11 patients with *Alcaligenes* septicemia, 4 had mixed infection and 3 died. Each one of the patients showed a different predisposing factor (110). On the other hand, blood isolations of *A. faecalis* have been made from patients without signs of septicemia (154). In the same series 15 strains (14 of them mixed) were recovered from sputum, 5 from urine (3 mixed), and 2 (mixed) from osteomyelitides. They were usually hospital acquired, and none was considered etiologically significant (154).

Animal pathogenicity of *A. faecalis* (tested by intraperitoneal or subcutaneous injection into guinea pigs and rabbits, intravenous injection into rabbits, intraperitoneal injection into mice) was found lacking, but mice did die if mucin was added to the inoculum (176, 394).

Except for sensitivity to polymyxin B and trimethoprim—sulfamethoxazole, susceptibility is erratic (126, 128). Ceftazidime and ceftriaxone have been effective against most strains, however (482).

Alcaligenes odorans

The ecology of this organism, first described in the 1920s (445), has not been studied. Most human strains have been cultured from urine (8, 52, 159, 292, 340, 455). Ear discharge, wounds, sputum (8, 52, 159, 340, 455), and feces (445) constitute further sources. Mixed cultures are frequent. Clinical data are available from two patients with urinary infection (8). A hospital epidemic has been reported, although the mode of spread was not documented (52). There is no pathogenicity for rabbits and guinea pigs by various routes (52). The organism is sensitive to most antibiotics, except for frequent resistance to cefuroxime and chloramphenicol (126, 128, 164).

Alcaligenes denitrificans

This species has been recovered from blood collection tubes (480), blood, ear, spinal fluid, and urine (455). Its susceptibility resembles that of *A. faecalis,* except that it is more often susceptible to cefamandole, cefoperazone, carbenicillin, and the ureidopenicillins but less often to the aminoglycosides (126, 128, 164).

Group IVE

Bacteria of this group, which seems to be related to *Alcaligenes* (455), have been most often isolated from human urines (455, 498). Further clinical data are lacking. These bacteria are susceptible to many antibiotics, except ampicillin and carbenicillin; many strains are also resistant to tetracycline (126, 128, 164, 498).

ACHROMOBACTER SPECIES

Achromobacter xylosoxidans

Achromobacter xylosoxidans has been found in moist hospital areas. Outbreaks of septicemia (286), cerebral ventriculitis (408), and various other infections (177, 467) have been traced to contamination of saline used in diagnostic tracer procedures (286), 0.1% chlorhexidine (407), and neomycin-containing preparations (177). The organism has been isolated, often in mixed culture, from many human sources (71, 208, 214, 357, 467, 507). Remarkable sources are a pancreatic abscess (11), the external ear (357, 507), and sputum in a case of cystic fibrosis (21) and in a case of pneumonia associated with IgM deficiency (111). Of the strains of Holmes et al. (208), five possibly played a pathogenic role; two were also isolated from chlorhexidine solutions. One paper has reported recurrent purulent meningitis in a 9-year-old girl due to *A. xylosoxidans,* with agglutination of the organism by the patient's serum (407). Another clinically less well characterized case (septicemia and cervical lymphadenopathy with positive culture results) was related to Yabuuchi et al. (509). Cases of septicemia (143) and meningitis (250, 413) in premature infants due to *Achromobacter* will not be discussed here, since the organisms isolated have not been described well enough to warrant their classification as *Achromobacter* species.

Achromobacter xylosoxidans strains are susceptible to minocycline, doxycycline (but not tetracycline!), carbenicillin, polymyxin B, trimethoprim—sulfamethoxazole, the ureidopenicillins, ceftazidime, and cefoperazone but resistant to other antimicrobials (126, 128, 164, 214).

Achromobacter VD-1 and VD-2

Achromobacter strains of group VD have been isolated from the respiratory tract, blood, urine, wounds, and feces (455) and from hospital water sources (296). No data on diseases due to these organisms are extant. Strains were susceptible to aminoglycosides, trimethoprim-sulfamethoxazole, nalidixic acid, tetracycline, and carbenicillin, but resistant to other antibiotics (164).

BORDETELLA BRONCHISEPTICA

Bordetella bronchiseptica is strictly parasitic and is found mainly in animals, rarely in man. Dogs, cats, guinea pigs, rats, rabbits, pigs, horses, monkeys, goats, ferrets, opossums, raccoons, foxes, skunks, and turkeys may carry the organism or may acquire a respiratory infection due to it (76, 154, 171). Epidemics in laboratory animals are well known (76, 285). In one laboratory epidemic of dogs and cats purulent conjunctivitis, keratitis, tracheitis, bronchopneumonia, and occasional otitis and meningitis were observed (285). In swine spontaneous atrophic rhinitis, pneumonia, and otitis have been described (171). Experimentally, intraperitoneal inoculation of guinea pigs causes fatal peritonitis, subcutaneous inoculation produces only a local lesion, and feeding does not result in infection (171). In mice [who do not have natural disease from *B. bronchiseptica* (76, 121, 285)] intracerebral inoculation gives rise to ventriculitis and intranasal infection to pulmonary (mainly vascular) damage (76). Rats are known to become naturally infected and to develop bronchopneumonia after intranasal instillation of *B. bronchiseptica* (76). For a further discussion of animal disease due to *B. bronchiseptica*, as well as the role of toxins, a recent review should be consulted (171).

Carriers of *B. bronchiseptica* in humans (450), as well as human disease, are rare. Most isolates came from the respiratory tracts of persons in close contact with animals (e.g., caretakers) who either were asymptomatic or had symptoms of a mild upper respiratory infection (154, 285). In a few cases with repeated isolation of *B. bronchiseptica* from sputum, a pertussis-like syndrom was seen (47, 48, 116, 240, 272). Some of these patients had pets (cats, rabbits) which were carriers or had overt pulmonary disease (47, 48, 240). A long-lasting, febrile bronchopneumonia in a 14-year-old child without obvious animal contact was observed in 1958. It resisted penicillin and nystatin treatment but responded to tetracycline (239). The serum of the patient agglutinated the *B. bronchiseptica* strain in periodically increasing titers. Recently two cases of fatal pneumonia in immunocompromised patients (Hodgkin's disease, malnutrition, and alcoholism) were described (158, 437); one patient had a sick pet dog which had died shortly before her hospital admission. We found one strain in a patient with cystic fibrosis (21). Gardner et al. (154) isolated *B. bronchiseptica* from the sputum of 16 hospitalized patients (15 in mixed culture) and one strain each from urine and blood. None of the patients had a whooping cough-like illness, and in only one patient with terminal tracheobronchitis did an etiological role of *B. bronchiseptica* seem likely. Prior antibiotic treatment and tracheal manipulation were common in the history of these patients. One patient from whose blood *B. bronchiseptica* was isolated, had a mixed septicemia (with *Streptococcus pneumoniae* as the second organism), presumably as an extension of a terminal pneumonia following *Staphylococcus aureus* endocarditis. The urine isolate came from the catheter specimen of a patient showing urinary tract infection with *B. bronchiseptica*, *Providencia* species, and nonhemolytic streptococci.

Systemic infections are rare. A case of endocarditis (157a) as well as of mixed subacute endocarditis with *B. bronchiseptica* and *Staphylococcus epidermidis* was seen; in the latter treatment with penicillin, streptomycin, and tetracycline was successful (85). Meningitis following reduction of an orbital fracture (due to the kick of a horse) has been reported (70). It responded to chloramphenicol, but the source for the *B. bronchiseptica* was not determined. Finally, a canine-related peritonitis in a patient on continuous ambulatory peritoneal dialysis has been treated successfully with chloramphenicol (61).

Most strains are resistant to ampicillin, cefamandole, cefoxitin, ceftizoxime, and azthreonam but sensitive to other antimicrobials used nowadays (12, 13, 126, 128, 164, 191).

FLAVOBACTERIUM SPECIES

Ecology, Animal Pathogenicity, and Occurrence in Human Specimens

The natural habitats of flavobacteria are soil, water, plants, and foodstuffs (190, 327, 406). In hospitals water systems are a significant source: incubators, nebulizers (heated and unheated), water baths, drinking fountains, sink faucets, distilled water lines, dental chair spray units, cold humidifiers (195, 296, 426), saline solutions for irrigation (355) and hemodialysis systems (130) have been found contaminated with flavobacteria. Tap water has been found to harbor flabobacteria by some (195) but not by others (327). Unlike *P. aeruginosa*, flavobacteria may resist 10 ppm of chlorine for 10 min in tap water (195). Survival of flavobacteria in intravenous anesthetics has also been observed (324). One sample of Hibitane (British Pharmacopoeia, 1:1000 to 1:5000 aqueous chlorhexidine) contained more than 1000 organisms per milliliter (78).

The experience of all authors who have encountered flavobacteria is that they are rare isolates from human material, occurring in 0.1– 1% of all positive blood cultures (279, 300), in 0.01% of all positive urine cultures (279), in 3 of 4430 (42) and in 88 of 27600 (329) genital tract specimens. Epidemic situations constitute exceptions. Environmental samples contain flavobacteria much more often than clinical samples (326).

Flavobacterium meningosepticum is not pathogenic for mice or rabbits intravenously or intraperitoneally or for hamsters subcutaneously or intravenously; intracerebral pathogenicity for mice is doubtful (60, 175, 231).

Neonatal Meningitis Due to Flavobacterium meningosepticum

This is the best-known instance of *Flavobacterium* pathogenicity. Since the first description of the disease in 1944 (409), around 120 cases have been published (2, 3, 46, 60, 62, 69, 72, 78, 89, 104, 110, 114, 124, 135, 152, 157, 189, 191, 228, 231, 243, 249, 258, 265, 268, 310, 327, 355, 369, 370, 377, 389, 401, 421, 446, 455, 457, 463, 466, 473, 475, 481, 508, 515). More than half of the infants reported were of low birth weight (2500 g); the same percentage showed an onset of symptoms at an age of less than 8 days [the total range was 1—60 days, but 6-, 7-, (389), and 19-month-old patients (515) may also belong to this group]. The most frequent serotype found was C (which was also involved in several epidemics), followed by B, F, A, M, and E. Symptoms were typical of neonatal meningitis (failure to feed, respiratory distress, convulsions or lethargy, failure to gain weight, discrete localizing signs). The spinal fluid showed polymorphonuclear pleocytosis, increase in the protein and decrease in the glucose levels, and a positive direct smear. Blood cultures were often positive as well, and serum agglutinins rose in some instances. Few hematological data are extant.

The overall case fatality rate was about 55%; and in most of the survivors a hydrocephalus developed. Ventriculitis has been blamed for this as well as for the tendency to relapse (466). Treatment was correlated with survival only if intrathecal (and/or intravenous) (189) antibiotics were administered whose concentrations in the spinal fluid surpassed the minimal inhibitory concentrations (MICs) of the isolates: for example, rifampin (69, 249, 355, 377), sulfamethoxazole—trimethoprim (135, 243), erythromycin (265, 355, 377), and vancomycin (157, 189, 355), although MICs for the latter two drugs were often in the "resistant" range. In one case, the intravenous and intrathecal combination of rifampicin plus erythromycin failed to sterilize the cerebrospinal fluid from *Flavobacterium* IIB (!) and had to be followed by the successful combination of mezlocillin plus cefoxitin (228). In most of these cases, however, evaluation of treatment results is clouded by the fact that infants had received several antibiotics by various routes.

Whether the causative organisms had an environmental or a maternal origin is unknown. In two epidemics (62, 463) and possibly another one (355), *identical* serotypes were isolated from human specimens and environmental sources, that is, a faulty sink trap (62) or saline solution (355), but other serotypes were found in the environment as well. Maternal complications of delivery were reported in only a few instances, and in cases of multiple births only one infant has fallen ill. The likely portal of entry was the pharynx. About one-half of the reported cases were associated with epidemic situations. Colonization of cohorts, but not of personnel, was common (see below). In one series the upper respiratory tract was colonized with *F. meningosepticum* in 25% of the premature infants on an epidemic ward, but only in 4% of the full-term babies (402). In another series 14 overt infections and 44 colonizations in babies were recorded (157). The organisms persisted in the upper respiratory tract for a mean period of 17 days, and duration of colonization was more prolonged in infants receiving antibiotics (191). In a third series, however, there were no positive throat cultures of cohorts (463).

Adult Meningitis Due to Flavobacterium Species

Only seven cases are extant (18, 188, 231, 243, 268, 274, 377). They have occurred in patients between the ages of 19 and 69 years. All of them had underlying diseases predisposing to gram-negative meningitis (hematological malignancies, cranial carcinoma, malnutrition, chronic renal failure) and showed typical signs and symptoms of such a disease. One case was due to *F. meningosepticum* type A (243), two to type C (231, 274), and one involved *Flavobacterium* IIB (18); the other three *F. meningosepticum* strains were not typed. Four patients died. All patients received multiple antibiotics by various routes.

Pneumonitis and Airway Colonization with Flavobacterium meningosepticum

In one of the nursery eipdemics due to *F. meningosepticum* type C, 5 of 14 infants were reported to have had pneumonia with bacteremia (62); two also had meningitis. Four of the five died, but signs and symptoms were not described. Colonization of the upper airways in cohorts of newborns with *F. meningosepticum* meningitis was common (46, 62, 190, 402); it was more frequent in premature infants (402) and persisted for about 1½ weeks in those who did not receive specific therapy and for 3 weeks in those who received antibiotics (190). Colonization of personnel was never observed (190). Nosocomial *F. meningosepticum* (type D) pneumonia was observed in a tracheotomized 14-year-old patient after ampicillin therapy. Chest roentgenogram and sputum smear were in line with the diagnosis. Therapy with chloramphenicol (to which the organism was sensitive in the disk test) was apparently successful (456). A case of nosocomial pneumonia in an adult on a respirator was reported in 1979 (310). In another study a polymyxin B aerosol was administered to the upper airways of 292 patients in a respiratory—surgical intensive care unit with the aim of preventing *P. aeruginosa* infection (131). A total of 18 patients became colonized with (polymyxin-resistant) flavobacteria, and 1 developed fatal pneumonia from it. A prospective study from the same unit showed that of 2329 patients, 195 (8.4%) became colonized with a single strain of *Flavobacterium* IIB during their hospitalization; the strain was also recovered from numerous water sources in the hospital and from drinking water (108). Thus, in line with observations on other gram-negative rods, colonization with flavobacteria is more frequent than infection and usually prefers debilitated individuals.

Etiologically indeterminate strains of flavobacteria have been cultured from the sputum of debilitated individuals (327, 340, 390). No clinical data exist on 43 strains of *F. meningosepticum* type G and 9 strains of *Flavobacterium* IIB which were isolated from the respiratory tracts of patients in a medical intensive care unit in Strasbourg, France (370, 371). We isolated one strain of *Flavobacterium* IIB from a child with cystic fibrosis (21).

Isolates of Flavobacterium Species from the Blood

In the experience of this author and of others (180, 279, 300, 326), flavobacteria isolated from blood cultures are more often etiologically insignificant than significant. In fact, five isolates reported were associated with the use of contaminated chlorhexidine gluconate (78). There are, however, reports of significant isolates associated with meningitis (see above), with isoimmunization with anti-Rh (390), with tooth extraction (mixed culture) (349), with implantation of transvenous pacemekers (262), and with a variety of immunocompromising disorders (110, 188, 257, 447). In one series of 20 cases (110) there were only 3 fatalities. A relatively benign course was also evident in three other series: Fever and systemic symptoms of *Flavobacterium* sp. septicemia in three postcardiotomy patients reacted favorably to apparently appropriate antimicrobial treatment (33). Strains identical to the infecting one were cultured from a water bath used for warming donor blood and from residual water in hoses supplying the heat exchanger of a heart—lung machine (33). Fever as the only sign of *F. meningosepticum* (type F) septicemia abated under *in*appropriate chemotherapy (324, 328); contaminated injectable drugs seemed to have been the source (324). Fever persisted, however, even under appropriate antimicrobial treatment until arterial catheters infected with *Flavobacterium* IIB were removed (430). These infections could be traced to contaminated syringes used to obtain arterial blood gas specimens. No clinical data are available on type H to L strains isolated from blood in France (370).

The first isolates of "*Flavobacterium*" from a case of subacute bacterial endocarditis (398) turned out to be *P. cepacia*. Two instances of the disease due to *F. meningosepticum* were superimposed on rheumatic heart disease. Apparent cure resulted from valvular surgery in one case, that of an addict (499), while the other patient died (510). Both had been treated with antimicrobials.

Isolates of Flavobacterium Species from Other Sources

Flavobacterium strains from wounds or ulcers have usually occurred in mixed culture and have been etiologically indeterminate (168, 173, 206, 208, 340) or were associated with contaminated hibitane (78). Two strains of *Flavobacterium odoratum*, isolated in pure culture, were deemed significant: one from an amputation site on a foot repeatedly recovered following gentamicin treatment for a *P. aeruginosa* infection (91a), and one cultured from a gangrenous foot (208).

One patient from whose eye a flavobacterium was isolated had purulent conjunctivitis (173). Endophathalmitis due to *F. meningosepticum* type E occurred following keratoplasty. The cornea had ben pretreated in a neomycin—gramicidin—polymyxin B solution. The strain was resistant to these drugs and also grew from the culture medium used for storage of the cornea (251). In this connection it is of interest to note that in one series flavobacteria were recovered from 10.8% of 240 bank eyes (357), but not from normal eyes. One strain of *Flavobacterium breve* (207) from the conjunctiva is reported.

Isolates of *Flavobacterium* sp. from urines have been reported without clinical data (206, 207, 209, 226, 326, 370). Strains of IIF and *F. odoratum* have been the most common ones (98, 206, 208, 236), but only a few were deemed clinically significant.

Early on *F. odoratum* had been isolated from the stools of individuals with diarrhea due to known causes (445a). One case of spontaneous peritonitis in a cirrhotic due to *Flavobacterium multivorum* has been reported (97a); there was improvement with gentamicin and carbenicillin. The relative occurrence of flavobacteria in the genital tract was mentioned before. Group IIF is an occasional inhabitant of the human cervix (329, 455).

The related organism IIJ has been isolated as the most frequent obligate aerobe from the oral and nasal fluid of dogs (19, 393). It is thus not surprising that human isolations were made most frequently from dog or cat bites (455).

Holmes et al. (200, 204) have collected isolates of *Flavobacterium spiritivorum* and *Flavobacterium thalpophilum* from many human sources. Clinical data have not been available until now. One case of neonatal meningitis due to IIE became known recently (492); it responded to chloramphenicol plus penicillin.

Antimicrobial Susceptibility of Flavobacterium Species

From the bacteriological data submitted, it can be assumed that most collections that were tested represented mixtures of *F. meningosepticum* and *Flavobacterium* IIB. The earliest systematic study (1964) on 27 strains (367) yielded results that contrasted to later ones insofar as the majority of strains showed minimal inhibitory concentrations (MICs) in the "sensitive" range ($\leqslant 10$ mg/liter) to ampicillin, chloramphenicol, and tetracycline. In line with other investigations, resistance ($\geqslant 20$ mg/liter) was observed to cephalothin, colistin, and penicillin, while most strains were sensitive ($\leqslant 1.25$ mg/liter) to erythromycin (367). A 1971 paper (7) on 11 *F. meningosepticum* strains found all of them resistant ($\geqslant 25$ mg/liter) to cephalothin, ampicillin, tetracycline, chloramphenicol, as well as polymyxin B, to methicillin, gentamicin, vancomycin, and carbenicillin ($\geqslant 100$ mg/liter), but again sensitive to nalidixic acid ($\leqslant 50$ mg/liter). All were sensitive or moderately so ($\leqslant 10$ mg/liter) to erythromycin and rifampin, but only the latter drug was bactericidal to seven of the strains. Sulfamethoxazole was inhibitory ($\leqslant 40$ mg/liter) for all strains, and trimethoprim for none ($\geqslant 20$ mg/liter) but the combination of the two inhibited all strains at 5 mg/liter. There was sensitivity to minocycline (4 mg/liter) and to piperacillin (128 mg/liter, MIC_{90} = 16 mg/liter) (128). *Flavobacterium* IIB was, in addition, usually sensitive to ceftazidime, cefoperazone, azlocillin, and mezlocillin (126, 128, 482). Resistance to gentamicin, tobramycin, and amikacin in flavobacteria was first described in 1976 (105, 295). Strains tested against other second- and third-generation cephalosporins were found variably susceptible (128, 434, 482). Finally, development of higher MICs to sulfamethoxazole, clindamycin, vancomycin, and erythromycin has been observed during therapy of meningitis with intravenous sulfamethoxazole, erythromycin, and vancomycin (190).

Kirby-Bauer disk tests (26) usually show flavobacteria to be sensitive to erythromycin, sulfamethoxazole—trimethoprim, nalidixic acid and vancomycin, variably susceptible to chloramphenicol, tetracycline and cefoxitin, and resistant to most aminoglycosides (with the exception of gentamicin in some strains), to the polymyxins, to the first generation of cephalosporins, and to the penicillins (2, 157, 164). Comparisons between MIC values and Kirby—Bauer "assignments," however, have led to unexpected results. A total of 6 strains of *F. meningosepticum* plus 22 strains of "flavobacteria" (1) and 20 strains of *Flavobacterium* IIB (486) were tested using customary MIC breakpoints (313). All strains proved to be resistant to ampicillin, carbenicillin, penicillin, cephalothin, chloramphenicol, tetracycline, colistin, vancomycin, gentamicin, tobramycin, and amikacin. There was variable susceptibility to clindamycin and erythromycin (confirmed in Ref. 99) and sensitivity of rifampin ($\leqslant 1$ mg/liter), nalidixic acid, and sulfamethoxazole—trimethoprim ($\leqslant 25$ mg/liter). A consistently bactericidal effect was found only with rifampin. In the Kirby—Bauer tests, however, there were many strains "sensitive" or "intermediate" to vancomycin, chloramphenicol, tetracycline, gentamicin, and cefoxitin. False susceptibilities were interpreted from Kirby—Bauer data in all strains "sensitive" by disk to gentamicin, chloramphenicol, and tetracycline, since they actually gave "resistant" MICs. In the cases of clindamycin and erythromycin, some disk-sensitive strains showed "intermediate" MICs. Contrariwise, strains sensitive to cefoxitin by MIC exceeded those sensitive by disk, but that discrepancy disappeared if sensitive and intermediate strains were combined (1). Thus it seems important to also perform dilution susceptibility tests for drugs that display "sensitivity" in the disk test.

So far, comparative studies between disk and dilution tests have not been done on other species. Strains of IIE, IIF and IIJ resemble each other in their disk sensitivity to penicillin, ampicillin, carbenicillin, cephalothin, chloramphenicol, tetracycline, nalidixic acid, erythromycin, sulfamethoxazole—trimethoprim, and clindamycin; their variable susceptibility to gentamicin; and their resistance to tobramycin. The group IIJ is reported resistant to polymyxin B, while IIF is sensitive (19, 164, 329, 393, 481). *Flavobacterium breve* behaves like *F. meningosepticum* except for sensitivity to erythromycin, chloramphenicol, and tetracycline; *F. odoratum* is also sensitive to erythromycin, clindamycin, and thienamycin (98, 206—208, 482). *Flavobacterium spiritivorum* and *F. thalpophilum* resemble *F. meningosepticum* (200, 204). *Flavobacterium multivorum* resembles *F. breve* and is also sensitive to ceftriaxone, cefotaxime, and ceftazidime (482).

AGROBACTERIUM SPECIES (GROUP VD-3)

These organisms, first described as plant pathogens, have been isolated from human specimens, mostly blood, sputum, urine, and other fluids (233, 247, 375, and own observations). No strain has thus far been described as etiologically significant. Susceptibility tests disclose a unique pattern of sensitivity to polymyxin B, trimethoprim—sulfamethoxazole, gentamicin, tetracycline, and carbenicillin and resistance to tobramycin and amikacin (164). Susceptibilities to other drugs vary.

ACKNOWLEDGMENTS

This is an update of the article "Clinical Role of Infrequently Encountered Nonfermenters" by the author which appeared in *Glucose Nonfermenting Gram-Negative Bacteria in Clinical Microbiology* (G. L. Gilardi, ed., CRC Press, Boca Raton, Fla., 1978). Parts of the first two main sections had first appeared in two publications, both entitled "Clinical Microbiology of Unusual Pseudomonas Species." The first publication appeared in *Progress in Clinical Pathology*, Vol. 5 (M. Stefanini, ed., Grune and Stratton, New York, 1973 pp. 185–218), and the second in *Unusual Organisms of Clinical Significance* (Committee on Continuing Education, Board of Education and Training, American Society for Microbiology, Washington, D.C., 1976). Parts of the section on *Flavobacterium* species appeared in *The Flavobacterium–Cytophaga Group* (H. Reichenbach and O. B. Weeks, eds., Verlag Chemie, Weinhein, 1981). Permission for reprinting certain passages has been obtained from the publishers Grune and Stratton, New York; the American Society for Microbiology, Washington D.C.; CRC Press, Boca Raton, Fla; and Verlag Chemie, Weinheim.

REFERENCES

1. R. C. Aber, C. Wennersten, and R. C. Moellering, *Antimicrob. Agents Chemother. 14*:483 (1978).
2. E. Abrutyn and S. Plotkin, in *Pathogenic Microorgnaisms from Atypical Clinical Sources* (A. von Graevenitz and T. Sall, eds), Marcel Dekker, New York, 1975, p. 113.
3. K. C. Agarwal and M. Ray, *Indian J. Med. Res. 59*:1006 (1971).
4. M. Alain, J. Saint-Etiene, and V. Reynes, *Med. Trop. Marseille 9*:119 (1949).
5. A. D. Alexander, D. L. Huxsoll, A. R. Warner, V. Shepler, and A. Dorsey, *Appl. Microbiol. 20*:825 (1970).
6. A. D. Alexander and L. C. Williams, *Appl. Microbiol. 22*:11 (1971).
7. G. Altmann and B. Bogokovsky, *J. Med. Microbiol. 4*:296 (1971).
8. A. B. Amin and A. Pendse, *Indian J. Med. Sci. 27*:768 (1973).
9. K. Anderson and R. Keynes, *Br. Med. J. 2*:274 (1958).
10. P. C. Appelbaum and A. J. Bowen, *Br. J. Dermatol. 98*:229 (1978).
11. P. C. Appelbaum and D. B. Campbell, *J. Clin. Microbiol. 12*:282 (1980).
12. P. C. Appelbaum, J. Tamim, J. Stavitz, R. C. Aber, and G. A. Pankuch, *Eur. J. Clin. Microbiol. 1*:159 (1982).
13. P. C. Appelbaum, J. Tamim, G. A. Pankuch, and R. C. Aber, *Chemotherpay 29*:337 (1983).
14. L. R. Ashdown, *Rev. Infect. Dis. 1*:891 (1979).
15. L. R. Ashdown, *J. Clin. Microbiol. 14*:361 (1981).
16. L. R. Ashdown, V. A. Duffy, and R. A. Douglas, *Med. J. Aust. 1*:314 (1980).
17. B. E. Atkinson, D. L. Smith, and W. R. Lockwood, *Ann. Intern. Med. 83*:369 (1975).

18. D. H. Bagley, J. C. Alexander, V. J. Gill, and A. S. Ketcham, *Arch. Intern. Med.* 136:229 (1976).
19. W. E. Bailie, E. C. Stowe, and A. M. Schmitt, *J. Clin. Microbiol.* 7:223 (1978).
20. R. M. Baird, K. M. Elhag, and E. J. Shaw, *J. Med. Microbiol.* 9:493 (1976).
21. R. S. Baltimore, K. Radnay-Baltimore, A. von Graevenitz, and T. F. Dolan, *Helv. Pediatr. Acta* 37:547 (1982).
22. D. C. J. Bassett, *J. Clin. Pathol.* 24:708 (1971).
23. D. C. J. Bassett, *J. Clin. Pathol.* 24:798 (1971).
24. D. C. J. Bassett, J. A. S. Dickson, and G. H. Hunt, *Lancet* 1:1263 (1973).
25. D. C. J. Bassett, K. J. Stokes, and W. R. G. Thomas, *Lancet* 1:1188 (1970).
26. A. W. Bauer, W. M. M. Kirby, J. C. Sherris, and M. Turck, *Am. J. Clin. Pathol.* 45:493 (1966).
27. D. A. Bemis, J. A. Greisen, and M. J. G. Appel, *J. Clin. Microbiol.* 5:471 (1977).
28. T. Ben-Tovim, E. Eylan, A. Romano, and R. Stein, *Infection* 2:162 (1974).
29. S. A. Berger, Y. Sigman-Igra, J. Stadler, and A. Campus, *J. Clin. Microbiol.* 17:926 (1983).
30. U. Berger and H. D. Piotrowski, *Med. Microbiol. Immunol.* 165:169 (1978).
31. R. L. Berkelman, S. Lewin, J. R. Allen, R. L. Anderson, L. D. Budnick, S. Shapiro, S. M. Friedman, P. Nicholas, R. S. Holzman, and R. W. Haley, *Ann. Intern. Med.* 95:32 (1981).
32. R. L. Berkelman, J. Godley, J. A. Weber, R. L. Anderson, A. M. Lerner, N. J. Petersen, and J. R. Allen, *Ann. Intern. Med.* 96:456 (1982).
33. W. B. Berry, A. G. Morrow, D. C. Harrison, H. D. Hochstein, and C. K. Himmelsbach, *J. Thorac. Cardiovasc. Surg.* 45:476 (1963).
34. D. J. Bibel and J. R. LeBrun, *J. Invest. Dermatol.* 64:119 (1975).
35. J. Z. Biegeleisen, R. Mosquera, and W. B. Cherry, *Am. J. Trop. Med. Hyg.* 13:89 (1964).
36. D. J. Blazevic, M. H. Koepcke, and J. M. Matsen, *Appl. Microbiol.* 25:197 (1973).
37. J. Blessing, J. Walker, B. Maybury, A. Yeager, and N. Lewiston, *Am. Rev. Respir. Dis.* 119:262 (1979).
38. J. G. A. Borghans, M. T. C. Hosli, H. Olsen, E. M. Ravn, K. Siboni, and P. Sogaard, *Acta Pathol. Microbiol. Scand 87B*:15 (1979).
39. E. J. Bottone, S. D. Douglas, A. R. Rausen, and G. T. Keusch, *J. Clin. Microbiol.* 1:425 (1975).

40. K. Botzenhart, *Zentralbl. Bakteriol. I Abt. Orig. B* 163:470 (1976).
41. M. Bourgain, P. H. Bonnet, and D. Raby, *Ann. Inst. Pasteur* 95:361 (1958).
42. K. Bovre, N. Hagen, B. P. Berdal, and E. Jantzen, *Acta Pathol. Microbiol. Scand.* 85B:27 (1977).
43. A. L. Braude, F. J. Carey, and J. Siemienski, *J. Clin. Invest.* 34:311 (1954).
44. J. Bret and R. Durieux, *Gynecol. Obstet.* 61:55 (1962).
45. J. H. Brinser and E. Torczynski, *Am. J. Opthalmol.* 84:462 (1977).
46. J. A. Brody, H. Moore, and E. O. King, *Am. J. Dis. Child.* 96:1 (1958).
47. F. Brooksaler and J. D. Nelson, *Am. J. Dis. Child.* 114:389 (1967).
48. J. H. Brown, *Bull. Johns Hopkins Hosp.* 38:147 (1926).
49. C. W. Bruch, *Am. Perfum. Cosmet.* 86:46 (1971).
50. W. G. Brundage, C. J. Thuss, and D. C. Walden, *Am. J. Trop. Med. Hyg.* 17:183 (1968).
51. E. T. Brygoo, *Bull. Soc. Pathol. Exot.* 46:347 (1953).
52. B. Brzin, *Zentralbl. Bakteriol. I Abt. Orig. A* 218:56 (1971).
53. B. Brzin, *Zentralbl. Bakteriol. I Abt. Orig. A* 248:579 (1981).
54. D. H. Buchholz, V. M. Young, N. R. Friedman, J. A. Reilly, and M. R. Mardiney, *N. Engl. J. Med.* 285:429 (1971).
55. R. J. Buchman, J. E. Kmiecik, and A. M. La Noue, *Am. J. Surg.* 125:324 (1973).
56. D. W. Burdon and J. L. Whitby, *Br. Med. J.* 2:153 (1967).
57. W. H. Burkholder, *Phytopathology* 40:115 (1950).
58. R. Burri and A. Stutzer, *Zentralbl. Bakteriol. 2 Orig.* 1:251, 350, 392, 422 (1895).
59. K. J. Büsing and K. Freytag, *Zentralbl. Bakteriol. I. Abt. Orig. A* 160:577 (1953).
60. R. Buttiaux and J. Vandepitte, *Ann. Inst. Pasteur* 98:398 (1960).
61. L. H. Byrd, L. Anama, M. Gutkin, and H. Chmel, *J. Clin. Microbiol.* 14:232 (1981).
62. H. A. Cabrera and G. H. Davis, *Am. J. Dis. Child.* 101:289 (1961).
63. H. A. Cabrera and M. A. Drake, *Am. J. Clin. Pathol.* 64:700 (1975).
64. E. M. Carpenter and D. Dicks, *J. Clin. Pathol.* 35:581 (1982).
65. L. A. Carson, M. S. Favero, W. W. Bond, and N. J. Peterson, *Appl. Microbiol.* 25:476 (1973).
66. R. Y. Cartwright and B. L. Radford, *Br. Med. J.* 4:711 (1972).
67. L. Chambon, *Ann. Inst. Pasteur* 89:229 (1955).
68. L. Chambon and J. Fournier, *Ann. Inst. Pasteur* 91:472 (1956).

69. T. Chandrika and S. P. Adler, *Pediatr. Infect. Dis.* 1:40 (1982).
70. K. C. Chang, R. M. Zakheim, C. T. Cho, and J. C. Montgomery, *J. Pediatr.* 86:639 (1975).
71. B. Chester and L. H. Cooper, *J. Clin. Microbiol.* 9:425 (1979).
72. C. Cintado, R. Garcia, M. Menéndez, P. Macias, and R. Torconteras, *Am. Esp. Pediatr.* 11:147 (1978).
73. A. J. Clayton, R. S. Lisella, and D. G. Martin, *Mil. Med.* 138:24 (1973).
74. J. A. Cook, *Med. J. Aust.* 2:627 (1962).
75. E. G. Cooper, *J. Am. Med. Assoc.* 200:452 (1967).
76. C. Coronini, H. Flamm, and W. Kovac, *Zentralbl. Bakteriol. I Abt. Orig.* 172:437 (1958).
77. W. A. Cowlishaw, M. E. Hughes, and H. C. R. Simpson, *J. Clin. Pathol.* 29:1088 (1976).
78. M. M. Coyle-Gilchrist, P. Crewe, and G. Roberts, *J. Clin. Pathol.* 29:824 (1976).
79. J. Cragg and A. V. Andrews, *Br. Med. J.* 3:54 (1969).
80. L. R. Crane, L. C. Tagle, and W. A. Palutke, *J. Am. Med. Assoc.* 246:985 (1981).
81. D. E. Craven, B. Moody, M. G. Connolly, N. R. Kollisch, K. D. Stottmeier, and W. S. McCabe, *N. Engl. J. Med.* 305:621 (1981).
82. L. Cravitz and W. R. Miller, *J. Bacteriol.* 86:52 (1950).
83. L. Cravitz and W. R. Miller, *J. Infect. Dis.* 86:46 (1950).
84. R. H. Dailey and E. J. Benner, *N. Engl. J. Med.* 279:361 (1968).
85. A. J. D. Dale and J. E. Geraci, *Proc. Staff Meet. Mayo Clin.* 36:288 (1961).
86. A. M. Dannenberg and E. M. Scott, *J. Exp. Med.* 107:153 (1958).
87. A. M. Dannenberg and E. M. Scott, *Am. J. Pathol.* 34:1099 (1958).
88. A. M. Dannenberg and E. M. Scott, *J. Immunol.* 84:223 (1960).
89. F. Daoulas-Le Bourdelles, P. Berche, J. L. Avril, and M. Véron, *Med. Mal. Infect.* 7:456 (1977).
90. F. Daschner, H. Langmaack, M. Grehn, A. Steffens, and M. Just, *Chemotherapy* 27:39 (1981).
91. F. Daschner, A. Steffens, M. Just, and M. Metzger, *Infection* 8:S433 (1980).
91a. J. M. Davis, M. M. Peel, and J. A. Gillians, *Med. J. Aust.* 2:703 (1979).
92. J. Debette and R. Blondeau, *Can. J. Microbiol.* 26:460 (1980).
93. J. Debois, H. Degreef, J. Vandepitte, and J. Spaepen, *J. Clin. Pathol.* 28:993 (1975).
94. H. Degreef, J. Debois, and J. Vandepitte, *Dermatologica* 151:296 (1975).

95. F. Denis, A Sow, M. David, J. P. Chiron, A. Samb, and I. Diop Mar, *Med. Mal. Infect.* 7:228 (1977).
96. D. Denney, R. H. Bigley, A. D. Rashad, W. J. MacDonald, and M. J. Miller, *West J. Med.* 122:160 (1975).
97. H. A. Derby and B. W. Hammer, *Iowa Agric. Stn. Res. Bull.* 145:387 (1931).
97a. V. K. Dhawan, K. R. Rajashekaraiah, W. I. Metzger, T. W. Rice, and C. A. Kallick, *J. Clin. Microbiol.* 11:492 (1980).
98. J. D. Dick, S. B. Wee, D. A. Noe, and P. Charache, *Abstr. Annu. Meet. ASM* 344 (1979).
99. J. R. DiPersio and T. L. Krafczyk, *Antimicrob. Agents Chemother.* 14:274 (1978).
100. J. R. DiPersio and T. L. Krafczyk, *Chemotherapy* 26:323 (1980).
101. W. E. Dismukes, A. W. Karchmer, M. J. Buckley, W. G. Austen, and M. N. Swartz, *Circulation* 48:365 (1973).
102. A. Dodin and J. Fournier, *Ann. Inst. Pasteur* 119:211, 738 (1970).
103. A. Dodin, M. Galimand, M. A. Chove, and R. Sanson, *Med. Mal. Infect.* 6:395 (1976).
104. J. R. Dooley, L. J. Nims, V. H. Lipp, A. Beard, and L. T. Delaney, *J. Trop. Pediatr.* 26:24 (1980).
105. E. Dowsett, *Lancet* 1:1338 (1972).
106. F. A. Drasar, W. Farrell, J. Maskell, and J. D. Williams, *Br. Med. J.* 2:1284 (1976).
107. C. Dulake and E. Kidd, *Lancet* 1:980 (1966).
108. G. DuMoulin, *J. Clin. Microbiol.* 10:155 (1979).
109. M. E. Dunne, S. Polakavetz, W. G. Armiger, A. Raneri, M. J. Snyder, and J. D. Stafford, *Morbid. Mortal. Week.* 22:265, 284 (1973).
110. H. L. DuPont and W. W. Spink, *Medicine* 48:307 (1969).
111. D. L. Dworzack, C. M. Murray, G. H. Hodges, and W. G. Barnes, *Am. J. Clin. Pathol.* 70:712 (1978).
112. P. H. Dyte and J. A. Gillians, *Med. J. Aust.* 1:444 (1977).
113. G. M. Ederer and J. M. Matsen, *J. Infect. Dis.* 125:613 (1972).
114. R. Eeckels, J. Vandepitte, and V. Seynhave, *Belg. Tijdschr. Geneeskd.* 21:244 (1965).
115. T. C. Eickhoff, J. V. Bennett, P. S. Hayes, and J. Feeley, *J. Infect. Dis.* 121:95 (1970).
116. G. Eldering, J. Holwerda, A. Davis, and J. Baker, *Appl. Microbiol.* 18:618 (1969).
117. S. R. Ell, P. O'Keefe, K. D. Thompson, and A. Earl, *Abstr. ICAAC* 216 (1979).
118. C. Eller, *Appl. Microbiol.* 17:26 (1969).
119. C. Ertug, *Dis. Chest* 40:693 (1961).
120. M. L. Eschete, F. Williams, and B. C. West, *Arch. Intern. Med.* 140:1533 (1980).

121. D. W. Evans and H. B. Maitland, *J. Pathol. Bacteriol. 48*:67 (1939).
122. E. D. Everett and R. A. Kishimoto, *J. Infect Dis. 128S*:107 (1973).
123. E. D. Everett and R. A. Nelson, *Am. Rev. Respir. Dis. 112*: 331 (1975).
124. A. Eykens, E. Eggermont, R. Eeckels, J. Vandepitte, and J. Spaepen, *Helv. Paediatr. Acta 28*:421 (1973).
125. H. Faden, M. Britt, and B. Epstein, *Infect. Control 2*:233 (1981).
126. R. J. Fass, *Antimicrob. Agents Chemother. 18*:483 (1980).
127. R. J. Fass and J. Barnishan, *Ann. Intern. Med. 84*:51 (1976).
128. R. J. Fass and J. Barnishan, *Rev. Infect. Dis. 2*:841 (1980).
129. M. S. Favero and C. H. Drake, *Appl. Microbiol. 14*:627 (1966).
130. M. S. Favero, N. J. Petersen, L. A. Carson, W. W. Bond, and S. H. Hindman, *Health Lab. Sci. 12*:321 (1975).
131. T. W. Feeley, G. C. DuMoulin, J. Hedley-Whyte, L. D. Bushell, J. P. Gilbert, and D. S. Feingold, *N. Eng. J. Med. 293*:471 (1975).
132. T. P. Felegie, V. L. Yu, L. W. Rumans, and R. B. Yee, *Antimicrob. Agents Chemother. 16*:833 (1979).
133. M. Felsby, G. Munk-Anderson, and K. Siboni, *J. Med. Microbiol. 6*:413 (1973).
134. S. K. Felts, W. Schaffner, M. A. Melly, and M. G. Koenig, *Ann. Intern Med. 77*:881 (1972).
135. J. J. Ferlauto and D. H. Wells, *South. Med. J. 74*:757 (1981).
136. J. J. Fischer, *J. Infect. Dis. 128S*:771 (1973).
137. M. C. Fisher, S. S. Long, E. M. Roberts, J. M. Dunn, and R. H. Balsara, *J. Am. Med. Assoc. 246*:1571 (1981).
138. M. W. Fisher, A. B. Hillegas, and P. L. Nazeeri, *Appl. Microbiol. 22*:13 (1971).
139. P. C. Fleming and B. Knie, *J. Antimicrob. Chemother. 8B*:169 (1981).
140. J. R. Flemma, F. C. DiVincenti, L. N. Dotin, and B. A. Pruitt, *Ann. Thorac. Surg. 7*:491 (1969).
141. T. Floch, J. C. Vergos, Y. Pidoux, and H. A. Floch, *Med. Mal. Infect. 9*:230 (1979).
142. C. Fluegge, *Die Mikroorganismen*, 2nd ed., F. C. W. Vogel, Leipzig, 1886.
143. J. F. Foley, C. R. Gravella, W. E. Englehard, and T. D. Y. Chin, *Am. J. Dis. Child. 101*:279 (1961).
144. W. Foulon, A. Naessens, S. Lauwers, M. Volckaert, P. Devroey, and J. J. Amy, *Lancet 2*:358 (1981).
145. J. Fournier and L. Chambon, *La Mélioidose: Maladie d'actualité et le bacille de Whitmore*, Editions Médicales Flammarion, Paris, 1958.
146. M. J. Frank and W. Schaffner, *J. Am. Med. Assoc. 236*:2418 (1976).

147. M. Franklin, *Bacteriol. Proc.* p. 85 (1969).
148. M. Franklin, *J. Infect. Dis.* *124S*:530 (1971).
149. D. Fritsche, R. Lütticken, and H. Böhmer, *Zentralbl. Bakteriol. I Abt. Orig. A* *229*:89 (1974).
150. S. Fujita, T. Yoshida, and F. Matsubara, *J. Clin. Microbiol.* *13*:781 (1981).
151. P. B. Fuller, E. Fisk, R. B. Byrd, G. A. Griggs, and M. R. Smith, *Chest* *74*:222 (1978).
152. I. Furuta, H. Kaya, and T. Tsuchiya, *J. Jpn. Assoc. Infect. Dis.* *48*:313 (1974).
153. H. Gamerman, H. H. Mollaret, A. Dodin, J. M. Delecloy, and R. Romeo, *Med. Mal. Infect.* *7*:395 (1977).
154. P. Gardner, W. B. Griffin, M. N. Swartz, and L. Kunz, *Am. J. Med.* *48*:735 (1970).
155. S. G. Geftic, H. Heymann, and F. W. Adair, *Appl. Environ. Microbiol.* *37*:505 (1979).
156. S. M. Gelbart, G. F. Reinhardt, and H. B. Greenlee, *J. Clin. Microbiol.* *3*:62 (1976).
157. R. M. George, C. P. Cochran, and W. E. Whealer, *Am. J. Dis. Child.* *101*:296 (1961).
157a. J. E. Geraci and W. R. Wilson, *Mayo Clin. Proc.* *57*:145 (1982).
158. H. K. Ghosh and J. Tranter, *J. Clin. Pathol.* *32*:546 (1979).
159. G. L. Gilardi, *Can. J. Microbiol.* *13*:895 (1967).
160. G. L. Gilardi, *Am. J. Clin. Pathol.* *51*:58 (1969).
161. G. L. Gilardi, *Appl. Microbiol.* *20*:521 (1970).
162. G. L. Gilardi, *Ann. Intern. Med.* *77*:211 (1972).
163. G. L. Gilardi, in *The Clinical Laboratory as an Aid in Chemotherapy of Infectious Disease* (A. Bondi, J. T. Bartola, and J. E. Prier, eds.), University Park Press, Baltimore, 1977, p. 121.
164. G. L. Gilardi, personal communication (1980).
165. G. L. Gilardi, S. Hirschl and M. Mandel, *J. Clin. Microbiol.* *1*:384 (1975).
166. G. L. Gilardi and H. J. Mankin, *N.Y. State J. Med.* *73*:2789 (1973).
167. D. N. Gilbert, W. L. Moore, C. L. Hedberg, and J. P. Sanford, *Ann. Intern. Med.* *68*:662 (1968).
168. D. N. Gilbert, J. P. Sanford, E. Kutscher, C. V. Sanders, J. P. Luby, and J. A. Barnett, *Arch. Environ. Health* *26*:125 (1973).
169. A. Goetz, V. L. Yu, J. E. Hanchett, and J. D. Rihs, *Arch. Intern. Med.* *143*:1909 (1983).
170. R. Gold, E. Jur, H. Levison, A. Isles, and P. C. Fleming, *J. Antimicrob. Chemother.* *12A*:331 (1983).
171. R. A. Goodnow, *Microbiol. Rev.* *44*:722 (1980).

172. C. D. Graber, L. P. Jervey, W. E. Ostrander, L. H. Dalley, and R. E. Weaver, *Am. J. Clin. Pathol.* 49:220 (1968).
173. P. W. Greaves, *J. Med. Lab. Technol.* 23:115 (1966).
174. R. N. Green and P. G. Tuffnell, *Am. J. Med.* 44:599 (1968).
175. R. A. Greenwald, G. Nash, and F. D. Foley, *Am. J. Clin. Pathol.* 52:188 (1969).
176. M. Grehn, *Zentralbl. Bakteriol. I Abt. Orig. A* 235:84 (1976).
177. D. Gröschel, L. D. Cody, and C. Tiemann, *Abstr. ICAAC* 378 (1979).
178. E. Grunberg, G. Beskind, W. F. DeLorenzo, and E. Titsworth, *Am. Rev. Respir. Dis.* 101:623 (1970).
179. G. Guenther, *Hyg. Rundsch.* 4:721 (1894).
180. L. Gutmann et al., *Med. Mal. Infect.* 7:482 (1977).
181. V. Hajiroussou, B. Holmes, J. Bullas, and C. A. Pinning, *J. Clin. Pathol.* 32:953 (1979).
182. W. H. Hall and R. E. Manion, *Antimicrob. Agents Chemother.* 4:193 (1973).
183. E. A. Hambie, S. A. Larsen, M. Felker, W. L. Jones, and J. C. Feeley, *J. Clin. Microbiol.* 5:167 (1977).
184. J. Hamilton, W. Burch, G. Grimmet, K. Orme, D. Brewer, R. Frost, and C. Fulkerson, *Antimicrob. Agents Chemother.* 4:551 (1973).
185. W. Hansen, G. Glupczynski, and E. Yourassowsky, *Med. Mal. Infect.* 12:507 (1982).
186. P. C. Hardy, G. M. Ederer, and J. M. Matsen, *N. Engl. J. Med.* 283:33 (1970).
187. H. D. Harlowe, *Laryngoscope* 82:882 (1972).
188. S. P. Harrington and C. A. Perlino, *South. Med. J.* 74:764 (1981).
189. H. B. Hawley and D. Gump, *Am. J. Dis. Child.* 126:261 (1973).
190. P. R. Hayes, *J. Appl. Bacteriol.* 43:345 (1977).
191. B. T. Hazuka, A. S. Dajani, K. Talbot, and B. M. Keen, *J. Clin. Microbiol.* 6:450 (1977).
192. J. R. Heckly, and M. N. Klumpp, *Bacteriol. Proc.* 81 (1964).
193. J. R. Heckly, and C. Nigg, *J. Bacteriol.* 76:427 (1958).
194. A. Heidt, H. Monteil, and C. Richard, *J. Clin. Microbiol.* 18:738 (1983).
195. L. G. Herman, *Health Lab. Sci.* 13:5 (1976).
196. P. E. Hermans, and J. A. Washington, *Ann. Intern. Med.* 73:387 (1970).
197. M. N. Hezebicks, and C. Nigg, *Antibiot. Chemother.* 8:543 (1958).
198. G. L. Hobby, T. F. Lenert, J. Maier-Engallena, and E. DeNoia-Cicenia, *Am. Rev. Respir. Dis.* 99:952 (1969).
199. G. L. Hobby, and T. F. Lenert, *Am. Rev. Respir. Dis.* 103:569 (1971).

200. B. Holmes, D. G. Hollis, A. G. Steigerwalt, M. J. Pickett, and D. J. Brenner, *Int. J. Syst. Bacteriol.* 33:667 (1983).
201. B. Holmes, S. P. Lapage, and B. G. Easterling, *J. Clin. Pathol.* 32:66 (1979).
202. B. Holmes, S. P. Lapage, and H. Malnick, *J. Clin. Pathol.* 28:149 (1975).
203. B. Holmes, R. J. Owen, A. Evans, H. Malnick, and W. R. Willcox, *Int. J. Syst. Bacteriol.* 27:133 (1977).
204. B. Holmes, R. J. Owen, and D. G. Hollis, *Int. J. Syst. Bacteriol.* 32:157 (1982).
205. B. Holmes, R. J. Owen, and R. E. Weaver, *Int. J. Syst. Bacteriol.* 31:21 (1981).
206. B. Holmes, J. J. S. Snell, and S. P. Lapage, *Int. J. Syst. Bacteriol.* 27:330 (1977).
207. B. Holmes, J. J. S. Snell, and S. P. Lapage, *Int. J. Syst. Bacteriol.* 28:201 (1978).
208. B. Holmes, J. J. S. Snell, and S. P. Lapage, *J. Clin. Pathol.* 30:595 (1977).
209. B. Holmes, J. J. S. Snell, and S. P. Lapage, *J. Clin. Pathol.* 32:73 (1979).
210. C. Howe and W. R. Miller, *Ann. Intern. Med.* 26:93 (1947).
211. C. Howe, A. Sampath, and M. Spotnitz, *J. Infect. Dis.* 124:598 (1971).
212. R. Hugh, and G. L. Gilardi, in *Manual of Clinical Microbiology* (E. H. Lennette, A. Balows, W. J. Hausler, and J. P. Truant, eds.), American Society for Microbiology, Washington, D.C., 1980, p. 288.
213. R. Hugh and E. Ryschenkow, *J. Gen. Microbiology* 26:1 (1961).
214. Y. Igra-Siegman, H. Chmel, and C. Cobbs, *J. Clin. Microbiol.* 11:141 (1980).
215. H. Iizuka and K. Komagata, *J. Gen. Appl. Microbiol.* 10:207 (1964).
216. P. Ikari and R. Hugh, *Bacteriol. Proc.* 41 (1963).
217. S. Z. Ileri, *Br. Vet. J.* 121:164 (1965).
218. H. Japp, A. von Graevenitz, J. Wüst, and G. L. Gilardi, *Clin. Microbiol. Newslett.* 3:124 (1981).
219. O. Jessen, *Pseudomonas aeruginosa and Other Green Fluorescent Pseudomonads. A Taxonomic Study,* E. Munksgaard, Copenhagen, 1965.
220. J. J. John, *Am. Rev. Respir. Dis.* 114:1021 (1976).
221. R. N. Jones and A. L. Barry, *Diagn. Microbiol. Infect Dis.* 1:165 (1983).
222. H. S. Juffs, *J. Appl. Bacteriol.* 36:585 (1973).
223. H.-M. Just, A. Beckert, M. Bassler, and F. D. Daschner, *Chemotherapy* 28:397 (1982).
224. H.-M. Just, E. Phillips, M. Bassler, and F. D. Daschner, *Eur. J. Clin. Microbiol.* 1:371 (1982).

225. A. Kahan, A. Philippon, G. Paul, S. Weber, C. Richard, G. Hazebroucq, and M. Degeorges, *J. Infect.* 7:256 (1983).
226. N. Karabatsos and R. D. Herrold, *J. Urol.* 84:187 (1960).
227. R. A. Kaslow, D. C. Mackel, and G. F. Mallison, *J. Am. Med. Assoc.* 236:2407 (1976).
228. M. C. Kelsey, A. M. Emmerson, and Y. Drabu, *Eur. J. Clin. Microbiol.* 1:138 (1982).
229. T. F. Keys, L. J. Melton, M. D. Maker, and D. M. Ilstrup, *J. Infect. Dis.* 147:489 (1983).
230. G. Kielwein, *Arch. Lebensmittelhyg.* 22:15 (1971).
231 E. O. King, *Am. J. Clin. Pathol.* 31:241 (1959).
232. A. King, B. Holmes, I. Phillips, and S. P. Lapage, *J. Gen. Microbiol.* 114:137 (1979).
233. M. Kiredjian, *Med. Mal. Infect.* 9:233 (1979).
233a. G. C. Klein, *J. Clin. Microbiol.* 11:27 (1980).
234. B. D. Knuth, M. R. Owen, and R. Latorraca, *Am. J. Med. Technol.* 35:227 (1969).
235. P. Kohut, *Bratisl. Lek. Listy* 71:65 (1979).
236. P. Kohut and M. Rusinko, *Cesk. Epid. Mikrobiol. Immun.* 27:151 (1978).
237. E. A. Konopka, S. C. Jones, A. Stieglitz, and H. C. Zogonas, *Antimicrob. Agents Chemother.* 503 (1970).
238. T. Kothari, M. P. Reyes, N. Brooks, W. J. Brown, and A. M. Lerner, *Can. Med. Assoc. J.* 116:1231 (1977).
239. P. Krepler and H. Flamm, *Wien. Klin. Wochenschr.* 35:641 (1958).
240. K. H. Kristensen and H. Lautrop, *Ugeskr. Laeg.* 124:303 (1962).
241. E. Kuehnel and H. Lundh, *Dial. Transplant.* 5:44 (1976).
242. S. P. Lapage, L. R. Hill, and J. D. Reeve, *J. Med. Microbiol.* 1:195 (1968).
243. S. P. Lapage and R. J. Owen, *J. Clin. Pathol.* 26:747 (1973).
244. L. R. Laraya-Luasay, M. Lipstein, and N. N. Huang, *Pediatr. Res.* 11:502 (1977).
245. P. M. Last, P. A. Harbison, and J. A. Marsh, *Lancet* 1:74 (1966).
246. P. Laudat, A. Audurier, J. Loulergue, B. Legros, and F. Lapierre, *J. Infect* 7:281 (1983).
247. H. Lautrop, *Acta Pathol. Microbiol. Scand.* 187S:63 (1967).
248. W. J. Ledger and J. T. Headington, *Int. J. Gynaecol. Obstet.* 10:87 (1972).
249. E. L. Lee, M. J. Robinson, M. L. Thong, and S. D. Puthucheary, *Arch. Dis. Child.* 51:209 (1976).
250. S. L. Lee and K. L. Tan, *Singapore Med. J.* 13:261 (1972).
251. M. LeFrancois and J. L. Baum, *Arch. Ophthalmol.* 94:1907 (1976).
252. E. Leifson, *Int. Bull. Bacteriol. Nomencl. Taxon.* 12:133 (1962).

253. E. Leifson and R. Hugh, *J. Gen. Microbiol.* 10:68 (1954).
254. R. E. Levin, *Appl. Microbiol.* 16:1734 (1968).
255. R. E. Levin, *Antonie van Leeuwenhoek J. Microbiol. Serol.* 38:121 (1972).
256. H. B. Levine and R. L. Maurer, *J. Immunol.* 81:433 (1958).
257. J. Lewis and F. R. Fekety, *Johns Hopkins Med. J.* 124:106 (1969).
258. R. Lietz, W. Handrick, V. Steinert, and H. Uhlig, *Paediatr. Grenzgeb.* 16:363 (1977).
259. P. V. Liu, *Am. J. Clin. Pathol.* 41:150 (1964).
260. M. -L. Loiseau-Marolleau and N. Malarre, *Pathol. Biol.* 25:637 (1977).
261. H. F. Long and B. W. Hammer, *J. Bacteriol.* 41:100 (1941).
262. P. Ma, W. E. Delaney, and W. J. Grace, *Crit. Care Med.* 2:135 (1974).
263. L. L. Mackenzie and P. H. Gilligan, *Abstr. Annu. Meet. Am. Soc. Microbiol.* 317 (1983).
264. P. A. Mackowiak and J. W. Smith, *J. Am. Med. Assoc.* 240:764 (1978).
265. E. Maderazo, H. P. Bassaris, and R. Quintiliani, *J. Pediatr.* 85:675 (1974).
266. T. Madhavan, *Ann. Intern. Med.* 80:670 (1974).
267. T. Madhavan, E. J. Fisher, F. Cox, and E. L. Quinn, *Ann. Intern. Med.* 78:971 (1973).
268. M. Madruga, U. Zanon, G. M. N. Pereira, and A. C. Galvao, *J. Infect. Dis.* 121:328 (1970).
269. C. R. Magnussen, M. T. Sammartino, and K. D. Ernest, *Antimicrob. Agents Chemother.* 22:154 (1982).
270. J. I. Maiztegui, J. Z. Biegeleisen, W. B. Cherry, and E. M. Kass, *N. Engl. J. Med.* 272:222 (1965).
271. W. F. Malizia, G. A. West, W. G. Brundage, and D. C. Walden, *Health Lab. Sci.* 6:27 (1969).
272. C. S. Man, *Pediatrics* 6:277 (1950).
273. I. N. Mandell, H. D. Feiner, N. M. Price, and M. Simberkoff, *Arch. Dermatol.* 113:199 (1977).
274. R. M. Mani, K. C. Kuruvila, P. M. Batlivala, P. N. Damle, G. V. Shirgaonkar, R. P. Soni, and P. R. Vyas, *J. Clin. Pathol.* 31:220 (1978).
275. W. Mannheim, and H. Buerger, *Z. Med. Mikrobiol. Immunol.* 152:249 (1966).
276. C. Marne, R. Pallarés, and A. Sitges-Sierra, *J. Clin. Microbiol.* 17:1173 (1983).
277. W. J. Martin, M. D. Maker, and J. A. Washington, *Am. J. Clin. Pathol.* 60:831 (1973).
278. W. J. Martone, C. A. Osterman, K. A. Fisher, and R. P. Wenzel, *Rev. Infect. Dis.* 3:708 (1981).
279. J. M. Matsen, *Health Lab. Sci.* 12:305 (1975).

280. R. J. Mathews, G. M. Ederer, L. V. Cunningham, and J. M. Matsen, *Abstr. Annu. Meet. Am. Soc. Microbiol.* 41 (1975).
281. J. J. Mathewson and R. B. Simpson, *J. Clin. Microbiol.* 15: 1016 (1982).
282. E. E. Mays and E. A. Ricketts, *Chest* 68:261 (1975).
283. J. B. McCormick, D. J. Sexton, J. G. McMurray, E. Carey, P. Hayes, and R. A. Feldman, *Ann. Intern. Med.* 83:512 (1975).
284. J. B. McCormick, R. E. Weaver, P. S. Hayes, J. M. Boyce, and R. A. Feldman, *J. Infect. Dis.* 135:103 (1977).
285. J. P. McGowan, *J. Pathol. Bacteriol.* 15:372 (1911).
286. M. B. McGuckin, R. J. Thorpe, K. M. Koch, A. Alavi, M. Staum, and E. Abrutyn, *Am. J. Epidemiol.* 115:785 (1982).
287. K. T. McKee, M. A. Melly, H. L. Greene, and W. Schaffner, *Am. J. Dis. Child.* 133:649 (1979).
288. G. W. Meyer, *Calif. Med.* 119:15 (1973).
289. W. R. Miller, L. Pannell, L. Cravitz, W. A. Tanner, and M. A. Ingalls, *J. Bacteriol.* 65:115 (1948).
290. W. R. Miller, L. Pannell, L. Cravitz, W. A. Tanner, and T. Rosebury, *J. Bacteriol.* 65:127 (1948).
291. W. B. Miller, L. Pannell, and M. S. Ingalls, *Am. J. Hyg.* 47: 205 (1948).
292. R. G. Mitchell and S. K. R. Clarke, *J. Gen. Microbiol.* 40: 343 (1965).
293. R. G. Mitchell and A. C. Hayward, *Lancet* 1:793 (1966).
294. J. H. Modell, *Anesthesiology* 27:329 (1966).
295. R. C. Moellering, C. Wennersten, L. J. Kunz, and J. W. Poitras, *Am. J. Med.* 62:873 (1977).
296. H. L. Moffet and T. Williams, *Am. Dis. Child.* 114:7 (1967).
297. H. L. Moffet, D. Allan, and T. Williams, *Am. J. Dis. Child.* 114:13 (1967).
298. H. H. Mollaret, *Med. Mal. Infect.* 7:391 (1977).
299. B. L. Monias, *J. Infect. Dis.* 43:330 (1928).
300. H. Monteil, V. Heinrich, and C. Richard, *Med. Mal. Infect.* 6:180 (1976).
301. H. Monteil, C. Richard, and A. Heidi, *Med. Mal. Infect.* 11: 544 (1981).
302. M. R. Moody and V. M. Young, *Antimicrob. Agents Chemother.* 7:836 (1975).
303. M. R. Moody and V. M. Young, *J. Antimicrob. Chemother.* 5:143 (1979).
304. M. R. Moody, V. M. Young, and D. M. Kenton, *Antimicrob. Agents Chemother.* 2:344 (1972).
305. H. B. Moore and M. J. Pickett, *Can. J. Microbiol.* 6:43 (1960).
306. *Morbidity and Mortality Weekly Report* 28:289, 409 (1979).
307. *Morbidity and Mortality Weekly Report* 30:610 (1981).
308. *Morbidity and Mortality Weekly Report* 32:495 (1983).

309. M. B. Morris and J. B. Roberts, *Nature London* 183:1538 (1959).
310. M. Nadarajah and T. H. Tan, *Singapore Med. J.* 20:293 (1979).
311. S. I. Narasimhan, D. L. Gopaul, and L. A. Hatch, *Am. J. Clin. Pathol.* 68:304 (1977).
312. *National Nosocomial Infection Study — Quarterly Reports,* Center for Disease Control, Atlanta, May 1972, p. 17.
313. National Committee for Clinical Laboratory Standards, *Performance Standards for Antimicrobial Disk Susceptibility Tests* 3/14:469 (1983).
314. H. C. Neu, C. J. Garvey, and M. P. Beach, *J. Infect. Dis. Suppl.* 128:336 (1973).
315. C. Nigg, *J. Immunol.* 91:18 (1963).
316. C. Nigg, R. J. Heckly, and M. Colling, *Proc. Soc. Exp. Biol. Med.* 89:17 (1955).
317. C. Nigg and M. M. Johnston, *J. Bacteriol.* 82:159 (1961).
318. C. Nigg, J. Ruch, E. Scott, and K. Noble, *J. Bacteriol.* 71:530 (1956).
319. C. E. Nord, T. Wadström, and B. Wretlind, *Med. Microbiol. Immunol.* 160:1 (1974).
320. C. E. Nord, T. Wadström, and B. Wretlind, *Antimicrob. Agents. Chemother.* 6:521 (1974).
321. C. E. Nord, T. Wadström, and B. Wretlind, *Med. Microbiol. Immunol.* 161:89 (1975).
322. J. J. Nussbaum, D. S. Hull, and M. J. Carter, *Arch. Ophthalmol.* 98:1224 (1980).
323. H. Olsen, *Dan. Med. Bull.* 14:1 (1967).
324. H. Olsen, *Dan. Med. Bull.* 14:6 (1967).
325. H. Olsen, *Acta Pathol. Microbiol. Scand.* 70:601 (1967).
326. H. Olsen, *Acta Pathol. Microbiol. Scand.* 75:313 (1969).
327. H. Olsen, *Acta Pathol. Microbiol. Scand.* 67:291 (1966).
328. H. Olsen, W. C. Frederiksen, and K. E. Siboni, *Lancet 2*: 1294 (1965).
329. H. Olsen and T. Ravn, *Acta Pathol. Microbiol. Scand. Sec. B.* 79:106 (1971).
330. G. R. Osteraas, J. M. Hardman, J. W. Bass, and C. Willson, *Am. J. Dis. Child.* 122:446 (1971).
331. L. A. Otto, B. S. Deboo, E. L. Capers, and M. J. Pickett, *J. Clin. Microbiol.* 7:341 (1978).
332. R. J. Owen and S. P. Lapage, *Antonie van Leeuwenhoek J. Microbiol. Serol.* 40:255 (1974).
333. R. J. Owen and J. J. S. Snell, *Antonie van Leeuwenhoek J. Microbiol. Serol.* 39:473 (1973).
334. G. Pappalardo, D. Roussianos, and F. Tanner, *Experientia* 36:493 (1980).
335. K. Parent and P. D. Mitchell, *Gastroenterology* 71:365 (1976).

336. K. Parent and P. D. Mitchell, *Gastroenterology* 72:1111 (1977).
337. S. Patrick, J. M. Hindmarch, R. V. Hague, and D. M. Harris, *J. Clin. Pathol.* 28:741 (1975).
338. P. Patamasucon, C. Pichyangkura, and G. W. Fischer, *J. Pediatr.* 87:133 (1975).
339. P. Patamasucon, U. B. Schaad, and J. D. Nelson, *J. Pediatr.* 100:175 (1982).
340. M. M. Pedersen, E. Marso, and M. J. Pickett, *Am. J. Clin. Pathol.* 54:178 (1970).
341. M. M. Peel, J. M. Davis, W. L. H. Armstrong, J. R. Wilson, and B. Holmes, *J. Clin. Microbiol.* 9:561 (1979).
342. F. A. Perryman and D. J. Flournoy, *J. Clin. Microbiol.* 12:79 (1980).
343. J. Petruschky, *Zentralbl. Bakteriol. I Orig.* 19:187 (1896).
344. I. Phillips, *Lancet* 1:375 (1971).
345. I. Phillips, D. Eykyn, and M. Laker, *Lancet* 1:1258 (1972).
346. F. D. Pien, *J. Clin. Microbiol.* 6:435 (1977).
347. F. D. Pien and H. Y. Higa, *J. Clin. Microbiol.* 7239 (1978).
348. J. A. Piggott and L. Hochholzer, *Arch. Pathol.* 90:101 (1970).
349. M. Pintér and J. Ivanyi, *Br. Med. J.* 2:1555 (1965).
350. M. Pintér and M. Kantor, *Orv. Hetil.* 115:254 (1974).
351. S. D. Pitlik, K. Bolivar, D. H. Ho, and G. P. Bodey, *Abstr. Annu. Meet. Am. Soc. Microbiol.* 50 (1983).
352. M. Pittman, *J. Lab. Clin. Med.* 42:273 (1953).
353. H. Pivnick, *Bacteriol. Proc.* 42 (1954).
354. A. A. Plotkin and R. Austrian, *Am. J. Med. Sci.* 235:621 (1958).
355. S. Plotkin and J. C. McKittrick, *J. Am. Med. Assoc.* 198:662 (1966).
356. R. H. Poe, H. R. Marcus, and G. L. Emerson, *Am. Rev. Respir. Dis.* 115:861 (1977).
357. F. M. Polack, D. Locatcher-Khorazo, and E. Gutierrez, *Arch. Ophthalmol.* 78:219 (1967).
358. K. E. Price, *J. Antimicrob. Chemother.* 8A:89 (1981).
359. K. E. Price, M. D. DeFuria, and T. A. Pursiano, *J. Infect. Dis. Suppl.* 134:249 (1976).
360. J. J. Rahal, M. S. Simberkoff, and P. J. Hyams, *J. Infect. Dis. Suppl.* 128:762 (1973).
361. E. Ralston, N. J. Palleroni, and M. Doudoroff, *Int. J. Syst. Bacteriol.* 23:15 (1973).
362. C. Randall, *Can. J. Publ. Health* 71:119 (1980).
363. F. T. Rapaport, J. W. Millar, and J. Ruch, *Arch. Pathol.* 71:429 (1961).
364. R. H. Rapkin, *Pediatrics* 57:239 (1976).
365. J. Redding and P. W. McWalter, *Br. Med. J.* 2:275 (1980).
366. M. S. Redfearn, N. J. Palleroni and R. Y. Stanier, *J. Gen. Microbiol.* 43:293 (1966).

367. J. Reinarz, B. Mays, and J. Sanford, *Antimicrob. Agents Chemother.* 451 (1964).
368. D. J. Reinhardt, D. Adams, P. Dickson, and V. Traina, *Dev. Ind. Microbiol.* 20:705 (1979).
369. B. Remion, *Pediatr. Pol.* 48:887 (1973).
370. C. Richard, H. Monteil, and B. Laurent, *Ann. Inst. Pasteur* 130B:141 (1979).
371. C. Richard, H. Monteil, A. Le Faou, and B. Laurent, *Med. Mal. Infect.* 9:124 (1979).
372. P. S. Riley, H. W. Tatum, and R. E. Weaver, *Appl Microbiol.* 24:798 (1972).
373. P. S. Riley and R. E. Weaver, *J. Clin. Microbiol.* 1:61 (1975).
374. P. S. Riley and R. E. Weaver, *Abstr. Annu. Meet. Am. Soc. Microbiol.* 122 (1972).
375. P. S. Riley and R. E. Weaver, *J. Clin. Microbiol.* 5:172 (1977).
376. R. A. Rimington, *Med. J. Aust.* 1:50 (1962).
377. I. Rios, J. J. Klimek, E. Maderazo, and R. Quintiliani, *Antimicrob. Agents Chemother.* 14:444 (1978).
378. T. Risko and I. Nikodemusz, *Nepegeszseguǵy* 31:106 (1950).
379. J. B. M. Roberts and D. C. E. Speller, *Lancet* 2:1099 (1973).
380. K. B. Rogers, *J. Appl. Bacteriol.* 23:53 (1960).
381. T. Rosebury, *Microorganisms Indigenous to Man*, McGraw-Hill, New York, 1962.
382. B. J. Rosenstein and D. E. Hall, *Johns Hopkins Med. J.* 147:188 (1980).
383. S. L. Rosenthal, *Am. J. Clin. Pathol.* 62:807 (1974).
384. S. L. Rosenthal, J. H. Zuger, and E. Apollo, *Am. J. Clin. Pathol.* 64:382 (1975).
385. H. L. Rubin, A. D. Alexander, and K. H. Vager, *Mil. Med.* 128:538 (1963).
386. K. A. Rudell and C. R. Anselmo, *Antimicrob. Agents Chemother.* 7:400 (1975).
387. A. D. Russell and A. P. Mills, *J. Clin. Pathol.* 27:63 (1974).
388. A. M. Rutenburg, G. M. Koota, and F. B. Schweinburg, *Ann. N.Y. Acad. Sci.* 76:348 (1958).
389. A. Samb, J.-P. Chiron, F. Denis, A. Sow, and I. Diop Mar, *Bull. Soc. Med. Afr. Noire Lang. Fr.* 22:84 (1977).
390. V. Sanchis-Bayarri, R. Borras, and M. S. Mari, *Rev. Clin. Esp.* 133:455 (1974).
391. D. C. Sands, M. D. Schroth, and D. C. Hildebrand, *J. Bacteriol.* 101:9 (1970).
392. J. O. Sanford, *Annu. Rev. Respir. Dis.* 104:452 (1971).
393. D. A. Saphir and G. R. Carter, *J. Clin. Microbiol.* 3:344 (1976).
394. J. K. Sarkar, B. Choudhury, and B. P. Tribedi, *Indian J. Med. Res.* 47:1 (1959).
395. T. K. Sarkar, G. L. Gilardi, A. S. Aguam, J. Josephson, and G. Leventhal, *Postgrad. Med.* 65:253 (1979).

396. W. Schaffner, G. Reisig, and R. A. Verall, *Lancet* 1:1050 (1973).
397. H. H. Schassan, R. Hörning, R. Malottke, and J. Potel, *Curr. Chemother. Immunother.* 1:736 (1982).
398. J. Schiff, L. S. Suter, R. D. Gourley, and W. D. Sutliff, *Ann. Intern. Med.* 55:499 (1961).
399. W. F. Schlech, J. B. Turchik, R. E. Westlake, G. C. Klein, J. D. Band, and R. E. Weaver, *N. Engl. J. Med.* 305:1133 (1981).
400. V. Schmidt, R. Kapila, Z. Kaminsky, and D. Louria, *J. Clin. Microbiol.* 10:385 (1979).
401. S. J. Seligman, T. Madhavan, and D. Alcid, *J. Infect. Dis. Suppl.* 128:322 (1973).
402. R. Seligmann, M. Komarov, and R. Reitler, *Br. Med. J.* 2:1528 (1963).
403. J. D. Semel, G. M. Trenholm, A. A. Harris, J. E. Jupa, and S. Levin, *Am. J. Med.* 64:403 (1978).
404. C. F. Shaefer, R. C. Trincher, and J. P. Rissing, *Am. Rev. Respir. Dis.* 128:173 (1983).
405. B. Shaw and V. Baselski, *Clin. Microbiol. Newslett.* 4:87 (1982).
406. J. M. Shewan and T. A. McMeekin, *Annu. Rev. Microbiol.* 37:233 (1983).
407. S. Shigeta, K. Higa, M. Ikeda, and S. Endo, *Igaku No Ayumi* 88:336 (1974).
408. S. Shigeta, Y. Yasunaga, K. Honzumi, H. Okamura, R. Kumata, and S. Endo, *J. Clin. Pathol.* 31:156 (1978).
409. B. H. Shulman and M. S. Johnson, *J. Lab. Clin. Med.* 29:500 (1944).
410. K. Siboni, H. Olsen, E. Raon, P. Sogaard, A. Hjovth, K. N. Nielsen, K. Askgaard, B. Secher, J. Borghans, L. Khing-Ting, H. Joosten, W. Frederiksen, K. Jensen, N. Movtensen, and O. Sebbesen, *Scand. J. Infect. Dis.* 11:39 (1979).
411. O. F. Sieber and V. A. Fulginiti, *Acta Paediatr. Scand.* 65:519 (1976).
412. M. F. Sierra and S. J. Seligman, *Abstr. Annu. Meet. Am. Soc. Microbiol.* C3 (1978).
413. S. S. Sindhu, *J. Singapore Pediatr. Soc.* 13:31 (1971).
414. H. A. Sinsabaugh and G. W. Howard, *Int. J. Syst. Bacteriol.* 25:187 (1975).
415. D. C. G. Skegg, *N. Z. Med. J.* 83:117 (1976).
416. I. J. Slotnick, J. Hall, and H. Sacks, *Am J. Clin. Pathol.* 72:882 (1979).
417. D. L. Smalley, V. R. Hansen, and V. S. Baselski, *Antimicrob. Agents Chemother.* 23:161 (1983).
418. E. G. Smith, *Clin. Microbiol. Newslett.* 1/5:4 (1979).
419. S. Y. So, P. Y. Chau, Y. K. Leung, W. K. Lam, and D. Y. C. Yu, *Am. Rev. Respir. Dis.* 127:650 (1983).

420. J. D. Sobel, N. Hashman, G. Reinherz, and D. Merzbach, *Am. J. Med.* 73:183 (1982).
421. C. Solé-Vernin, C. M. Ulson, and M. Zuccolotto, *Rev. Inst. Med. Trop. Sao Paulo* 2:54 (1960).
422. A. C. Sonnenwirth, *Ann. N.Y. Acad. Sci.* 174:488 (1970).
423. W. B. Sorell and L. V. White, *Am. J. Clin. Pathol.* 23:134 (1953).
424. P. M. Southern and E. A. Kutscher, *J. Clin. Microbiol.* 13:1070 (1981).
425. P. M. Southern and M. L. Schneider, *Tex. Rep. Biol. Med.* 32:880 (1974).
426. M. S. Spaepen, H. A. Bodman, R. B. Kundsin, J. R. Berryman, and W. Fencl, *Health Lab. Sci.* 12:316 (1975).
427. J. D. C. Speller, *Br. Heart J.* 35:47 (1972).
428. J. D. C. Speller, M. E. Stephens, and A. C. Viant, *Lancet 1*:798 (1971).
429. M. Spotnitz, J. Rudnitzky, and J. J. Rambaud, *J. Am. Med. Assoc.* 202:950 (1967).
430. W. E. Stamm, J. J. Colella, R. L. Anderson, and R. E. Dixon, *N. Eng. J. Med.* 292:1099 (1975).
431. R. Y. Stanier, N. J. Palleroni and M. Doudoroff, *J. Gen. Microbiol.* 43:159 (1966).
432. A. T. Stanton and W. Fletcher, *Lancet 1*:10 (1925).
433. A. C. Steere, J. H. Tenney, D. C. Mackel, M. J. Snyder, S. Polakavetz, M. E. Dunne, and R. E. Dixon, *J. Infect. Dis.* 135:729 (1977).
434. M. Stephens, M. Potten, and A. J. Bint, *Infection* 7:109 (1979).
435. A. R. Stevens, J. S. Legg, B. S. Henry, J. M. Dille, W. M. M. Kirby, and C. A. Finch, *Ann. Intern. Med.* 39:1228 (1953).
436. R. M. Stirland and J. A. Tooth, *Br. Med. J.* 4:505 (1976).
437. D. B. Stoll, S. H. Murphey, and S. K. Ballas, *Postgrad. Med. J.* 57:723 (1981).
437a. G. B. Stover, D. R. Drake, and T. C. Montie, *Infect. Immun.* 41:1099 (1983).
438. D. A. Strandberg, J. H. Jorgensen, and D. J. Drutz, *Antimicrob. Agents Chemother.* 24:282 (1983).
439. I. Straus, *Arch. Med. Exp. Anat. Pathol.* 1:460 (1889).
440. J. M. Strauss, A. D. Alexander, G. Rapmund, E. Gan, and A. E. Dorsey, *Am. J. Trop. Med. Hyg.* 18:703 (1969).
441. J. M. Strauss, D. W. Ellison, E. Gan, S. Jason, J. L. Marcarelli, and G. Rapmund, *Med. J. Malays.* 24:94 (1969).
442. J. M. Strauss, S. Jason, H. Lee, and E. Gan, *J. Am. Vet. Med. Assoc.* 155:1169 (1969).
443. J. M. Strauss, S. Jason, and M. Mariappan, *Med. J. Malays.* 22:31 (1967).
444. J. M. Strauss, M. G. Groves, M. Mariappan, and D. W. Ellison, *Am. J. Trop. Med. Hyg.* 18:698 (1969).

445. M. Stutzer, *Zentralbl. Bakteriol. Parasitenkd. Infektionskr. Hyg. I. Abt., Orig. 91*:87 (1924).
445a. M. J. Stutzer, and A. Kwaschina, *Zentralbl. Bakteriol. I. Abt. Orig. 113*:219 (1929).
446. A. A. Sughatadasa, and S. N. Arseculeratne, *Br. Med. J. 1*: 37 (1963).
447. N. M. Sullivan, V. L. Sutter, M. M. Mims, V. H. Marsh, S. M. Finegold, *J. Infect. Dis. 127*:49 (1973).
448. V. Sutter, *Appl. Microbiol. 16*:1532 (1968).
449. V. Sutter, V. Hurst, and A. O. J. Landucci, *J. Dent. Res. 45*:1800 (1966).
450. W. P. Switzer, and E. D. Hubbard, *Am. J. Vet. Res. 58*:571 (1963).
451. R. J. S. Tan, E. W. Lim, and R. Sakazaki, *Jpn. J. Exp. Med. 47*:311 (1977).
452. D. Taplin, D. C. J. Bassett, and P. M. Mertz, *Lancet 2*:568 (1971).
453. D. Taplin and P. M. Mertz, *Lancet 2*:1279 (1973).
454. J. H. Tashijan, C. B. Coulam, and J. A. Washington, *Mayo Clin. Proc. 51*:557 (1976).
455. H. W. Tatum, W. H. Ewing, and R. E. Weaver, in *Manual of Clinical Microbiology*, 2nd ed. (E. H. Lennette, E. H. Spaulding, and J. P. Truant, eds.), American Society for Microbiology, Washington, D.C., 1974, p. 270.
456. D. Teres, *J. Am. Med. Assoc. 228*:732 (1974).
457. D. Thevenieau, L. Poli, P. Perzée, G. Le Gonidec, and J. Guélain, *Med. Mal. Infect. 11*:439 (1981).
458. P. Thibault, *Ann. Inst. Pasteur 100*:59 (1961).
459. R. N. T. Thin, *Lancet 1*:31 (1976).
460. A. D. Thomas, J. Forbes-Faulkner, and M. Parker, *Am. J. Epidemiol. 110*:515 (1979).
461. M. J. Thommassen, C. A. Demko, and S. Badger, *Abstr. ICAAC 508* (1980).
462. M. L. Thong, *Southeast Asian J. Trop. Med. Public Health 7*: 363 (1976).
463. M. L. Thong, S. D. Putchucheary, and E. L. Lee, *J. Clin. Pathol. 34*:429 (1981).
464. R. C. Tilton, O. Steingrimsson, and R. W. Ryan, *Am. J. Clin. Pathol. 69*:410 (1978).
465. M. J. Tong, *J. Am. Med. Assoc. 219*:1044 (1972).
466. R. Torronteras, C. Cintado, R. Garcia, P. Macias, and M. Menéndez, *Pédiatrie 33*:393 (1978).
467. N. Trowers, F. Stoller, E. Torres, and A. Gilman, *Abstr. Annu. Meet. Am. Soc. Microbiol.* 277 (1978).
468. K. Tsuchiya and M. Kondo, *Antimicrob. Agents Chemother. 13*:536 (1978).

469. S. Tumioka, Y. Kobayashi, and H. Uchida, *Curr. Chemother.* 1:440 (1977).
470. M. Uwaydah and A.-R. Taqi-Eddin, *J. Infect. Dis. Suppl. 134*: 28 (1976).
471. M. Valdivieso, B. Gil-Extremera, J. Zornoza, V. Rodriguez, and G. P. Bodey, *Medicine 56*:241 (1977).
472. P. Valenstein, G. H. Bardy, C. C. Cox, and P. Zwadyk, *Am. J. Clin. Pathol. 79*:245 (1983).
473. J. Vandepitte, G. Beeckmans, and R. Buttiaux, *Ann. Soc. Belge Med. Trop. 38*:563 (1958).
474. J. Vandepitte and J. Debois, *J. Clin. Microbiol.* 7:70 (1978).
475. J. Vandepitte, R. Eeckels, and V. Seynhaeve, *Ann. Soc. Belge Med. Trop. 45*:57 (1965).
476. M. Vaucel, *Bull. Soc. Pathol. Exot. 30*:10 (1937).
477. A. von Graevenitz, *Med. Welt 54*:177 (1965).
478. A. von Graevenitz, *Am. J. Clin. Pathol. 43*:537 (1965).
479. A. von Graevenitz, *Abstr. Annu. Meet. Am. Soc. Microbiol.* 123 (1972).
480. A. von Graevenitz, *Mt. Sinai J. Med. N.Y. 43*:727 (1976).
481. A. von Graevenitz, *Clin. Microbiol. Newslett.* 1:214 (1979).
482. A. von Graevenitz and C. Bucher, *Infection 10*:293 (1982).
483. A. von Graevenitz and C. Bucher, *Zentralbl. Bakteriol. I. Abt. Orig. A 254*:403 (1983).
484. A. von Graevenitz and G. O. Carrington, *Infection* 1:54 (1973).
485. A. von Graevenitz and M. Grehn, *Zentralbl. Bakteriol. I Abt. Orig. A 236*:513 (1976).
486. A. von Graevenitz and M. Grehn, *FEMS Lett.* 2:289 (1977).
487. A. von Graevenitz and J. J. Redys, *Health Lab. Sci.* 5:107 (1968).
488. A. von Graevenitz and G. Simon, *Appl. Microbiol. 19*:176 (1970).
489. A. von Graevenitz and J. Weinstein, *Yale J. Biol. Med. 43*:265 (1971).
490. J. A. Washington, *Ann. Intern Med. 86*:186 (1977).
491. K. C. Watson, J. G. Krogh, and D. T. Jones, *J. Clin. Pathol. 19*:79 (1966).
492. K. C. Watson, N. Wilson, and D. A. Glen, *J. Infect.* 7:278 (1983).
493. D. R. Weber, L. E. Douglas, W. G. Brundage, and F. C. Stallkamp, *Am. J. Med. 40*:234 (1969).
494. A. J. Weinstein, R. C. Moellering, Jr., C. C. Hopkins, and A. Goldblatt, *Am. J. Med. Sci. 265*:591 (1973).
495. L. Weinstein and E. Wasserman, *N. Engl. J. Med. 244*:662 (1957).
496. R. A. Weinstein, T. G. Emori, R. L. Anderson, and W. E. Stamm, *Chest 69*:336 (1976).
497. R. A. Weinstein, W. E. Stamm, L. Kramer, and L. Corey, *J. Am. Med. Assoc. 236*:936 (1976).

498. W. D. Welch, R. K. Porschen, and B. Luttrell, *Antimicrob. Agents Chemother.* 24:432 (1983).
499. S. Werthamer and M. Weiner, *Am. J. Clin. Pathol.* 57:410 (1972).
500. A. Whitmore and C. S. Krishnaswami, *Indian Med. Gaz.* 47:262 (1912).
501. A. Whitmore, *J. Hyg.* 13:1 (1913).
502. P. J. Whorwell, W. L. Beeken, I. W. Davidson, and R. Wright, *Lancet* 2:697 (1978).
503. M. M. Wishart and T. V. Riley, *Med. J. Aust.* 2:710 (1976).
504. N. G. Wright, H. Thompson, D. Taylor, and H. J. C. Cornwell, *Vet. Rec.* 93:486 (1973).
505. E. Yabuuchi, T. Ito, E. Tanimura, N. Yamamoto, and A. Ohyama, *Antimicrob. Agents Chemother.* 20:136 (1981).
506. E. Yabuuchi, N. Miyajima, H. Hotta, and A. Ohyama, *Med. J. Osaka Univ.* 21:1 (1970).
507. E. Yabuuchi and A. Ohyama, *Jpn. J. Microbiol.* 15:477 (1971).
508. E. Yabuuchi, A. Ohyama, H. Takeda, M. Sugiama, and S. Kono, *Jpn. J. Microbiol.* 14:241 (1970).
509. E. Yabuuchi, I. Yano, S. Goto, E. Tanimura, T. Ito, and A. Ohyama, *Int. J. Syst. Bacteriol.* 24:470 (1974).
510. M. Yamakado, H. Tagawa, and S. Tanaka, *J. Jpn. Soc. Intern. Med.* 64:816 (1975).
511. T. J. Yeh, I. N. Anabtawi, V. E. Cornett, A. White, W. H. Stern, and R. G. Ellison, *Ann. Thorac. Surg.* 3:29 (1967).
512. E. Yourassowsky, T. Prigogine, J. Geurts, and J. Hennecart, *Nouv. Pr. Med.* 8:2739 (1979).
513. V. L. Yu, T. L. Felegie, R. B. Yee, A. W. Pasculle, and F. H. Taylor, *J. Infect. Dis.* 142:602 (1980).
514. V. L. Yu, L. W. Rumans, E. J. Wing, R. McLeod, F. N. Sattler, R. M. Harvey, and S. C. Deresinski, *Arch. Intern. Med.* 138:1667 (1978).
515. F. Zappulla, M. Gatti, G. Rocchi, R. Tanas, O. Varoli, and A. Beccari, *Minerva Pediatr.* 31:679 (1979).
516. A. A. Zebral and E. Hofer, *Mem. Inst. Oswaldo Cruz* 71:171 (1973).
517. G. H. Zierdt and H. H. Marsh, *Am. J. Clin. Pathol.* 55:596 (1971).
518. J. J. Zuravleff and V. L. Yu, *Rev. Infect. Dis.* 4:1236 (1982).

7
Pathogenic Properties and Enzyme Profiles of *Pseudomonas* and *Acinetobacter*

J. MICHAEL JANDA and EDWARD J. BOTTONE   *The Mount Sinai Hospital and Mount Sinai School of Medicine, New York, New York*

INTRODUCTION

The past 10 years have witnessed a revolution in the field of bacterial pathogenesis during which a number of monumental contributions in infectious diseases and medical microbiology have been achieved owing in part to the recent application of molecular techniques to research of bacterial pathogenesis. These methodologies, including nucleic acid hybridization, plasmid analysis, restriction endonuclease mapping, RNA fingerprinting, Southern blot technique, gene sequencing, and genetic transformation studies, including gene cloning, have provided finite tools to investigate the molecular basis for the virulence of different microbial species and to determine the role of cell-associated factors and exoenzymes in the pathophysiology of host—microbe infections. Continuing studies will expand our practical knowledge of many of these processes and how they relate to infection, identification, and prophylaxis against species-related virulence factors.

That certain exotoxins play an important if not the major role in the virulence of certain gram-positive bacteria has long been known. Most of the knowledge referable to this subject derives from the pioneering work on the toxigenic clostridia, in particular *Clostridium botulinum*, *Clostridium tetani*, *Clostridium perfringens*, and with *Corynebacterium diphtheriae*. In the past few years, however, it has become increasingly apparent that factors such as exotoxins, responsible for the virulence of gram-positive bacteria, in large part, account for the virulence of many gram-negative bacteria as well. Studies of hemolysins and enterotoxins in *Vibrio* species and in *Escherichia coli* have clearly demonstrated their biological role in vivo as it relates to disease-associated syndromes. The advent of molecular technology as applied to *Pseudomonas aeruginosa*, the most commonly isolated gram-negative nonfermentative species recovered from clinical sources, has brought the realization that virulence of *P. aeruginosa* is a polygenic function, with many different bacterial factors playing and important role in pathogenesis in accordance with underlying host status and initial colonization site.

Regardless of the genera or species investigated, different bacterial strains within the same genus and species are not genetically identical and do not all produce the same exoenzymes, either on a qualitative or quantitative basis. Additionally, different exoenzymes may play important roles in pathogenesis at different body sites (e. g., pulmonary, gastrointestinal, wound). Thus we may envision that a pathogenic process initiated by a particular strain of bacterial species is not a random event, but one involving the selection of a particular isolate possessing the necessary virulence factors for infection at the site of introduction. The presence or absence of these virulence-associated factors may determine whether colonization [lack of virulence factor(s)] or infection [presence of factor(s)] will ensue. At the same time, equal importance must be accorded the overall physiological state of the patient, taking into consideration immune status, age, nutritional status, and presence of underlying disease. Interruption of intact and mucosal barriers (hyperalimentation, catheterization) and the administration of various forms of chemotherapy will also affect outcome. These host parameters taken in concert with those of the bacterium (innate virulence potential) will primarily dictate whether or not infection will ensue.

Although a primitive understanding of how a number of exoenzymes produced by *P. aeruginosa* relate to virulence, few studies to date have looked at the relative frequencies with which these enzymes are elaborated by clinical isolates. Indeed, excluding *P. aeruginosa*, there is a dearth of information concerning the exoenzymatic capability of other pseudomonads such as *Pseudomonas maltophilia* and *Pseudomonas cepacia*. Knowledge of the exoenzymatic potential of a given species is germane to a fuller definition of the species. For instance, the presence or absence of different exoenzymes may be species related and can aid in the identification of an unknown species. Secondly, the study of the exoenzyme profile of individual strains of a species may help to predict the invasive or disease-producing potential of a given isolate relative to source of recovery. Thirdly, enzyme profiling of nonfermentative bacteria may serve as an epidemiological tool for the fingerprinting of isolates associated with widespread infection, especially within the hospital setting. Enzyme profiling of nonfermenter strains of clinical and environmental (water, soil) origin may help to define and advance correlates of virulence with separating or enjoining strains of a particular species derived from these two ecological niches. These points together suggest that further studies on the exoenzymatic activities of nonfermentative bacteria should be undertaken to better assess the role of these products in identification, pathogenesis, and epidemiology. This chapter will advance the role of exoenzymes in the identification and pathogenic potential of P. aeruginosa mainly and, where available, present information relative to the exoenzymatic activity of other nonfermentative species.

PSEUDOMONAS AERUGINOSA

General Considerations

Of the nonfermentative bacteria comprising the microbial biosphere, P. aeruginosa stands out as the major cause of human infections usually of nosocomial origin. Notwithstanding the enormous diversity of pseudomonad species naturally inhabiting water and soil niches, P. aeruginosa apparently possesses or acquires a number of unique properties enabling it to become the preeminent human pathogen of this group. In fact, approximately 85% of all clinical isolations of nonfermentative bacteria irrespective of clinical significance are identified as P. aeruginosa.

Historically, one of the striking differences noted between P. aeruginosa and taxonomically related species was the remarkable number of extracellular enzymes produced by this species. The early studies of Liu and colleagues (1961) concerned the fractionation of specific enzymes of agar-grown P. aeruginosa strains. Presently more than 18 different extracellular enzymes have been reported for this species (Table 1), which far exceeds in number and extent (frequency) those produced by the closely related fluorescent pseudomonads *Pseudomonas fluorescens* and *Pseudomonas putida*. Prodigious exoenzyme production by P. aeruginosa is potentially correlated with its ability to incite human disease and to persist in the hospital environment.

Many of the exoenzyme activities reported for P. aeruginosa such as fibrinolysin, coagulase, and nucleases have only been qualitatively studied, whereas others, such as enterotoxin, are poorly described, or other enzymes (neuraminidase, exotoxin S) have only been found in one or a limited number of isolates. Other exoenzymatic activities, such as collagenase, while acting upon native collagen molecules, appear unable to degrade this substrate with the liberation of hydroxyproline analogous to more traditional collagenases such as those produced by *Clostridium haemolyticum*. There exists also a paucity of reports relative to the enzymatic capability of various P. aeruginosa morphotypes and those derived from different clinical sources and types of infection. Definitive studies determining both qualitative and quantitative levels of exoenzymes produced by a large number of P.

Table 1 Extracellular Enzymes Produced by *Pseudomonas aeruginosa*

Enzyme	References	Enzyme	References
Exotoxin A	27	Fibrinolysin	63
Exotoxin S	26	Coagulase	54
Enterotoxin	38	Gelatinase	52
Alkaline protease	50, 52	Staphylolytic enzyme	80
Elastase	52	Ribonuclease	80
Lecithinase (phospholipase)	15	Deoxyribonuclease	80
Collagenase	14, 66	Esterase	40, 68, 80
Hemolysin	31, 68	Alkaline phosphatase	25
Lipase	8, 40	Neuraminidase	42

aeruginosa isolates with respect to the above-mentioned properties are also lacking. While there is good clinical and experimental evidence supporting the role of exotoxin A in *P. aeruginosa* pathogenesis, little is known regarding the mechanism of its biological effect in vivo. The definitive role of lecithinase, protease, and elastase in the pathogenesis of *P. aeruginosa* infection still remains nebulous.

To date, only a limited number of studies have attempted to correlate exoenzyme formation (either qualitatively or quantitatively) with the disease-producing potential of *P. aeruginosa*. Despite these limitations, several general comments can be made. For most enzymes studied, a clear correlation between the presence or intensity of production and virulence of the strain or its source, biotype (pyocine), or serotype has not been established. Nevertheless, reports of the association of various exoenzymes with virulence have appeared. Muszynski (1973) found that 82 clinically derived *P. aeruginosa* strains exhibiting high virulence in mice quantitatively produced higher protease levels than less virulent strains. Janda et al (1980), using a protease plate assay, found that systemic isolates of *P. aeruginosa* produced substantially higher protease acitivty than those isolates originating from sputum. Janda and Bottone (1981) subsequently found that *P. aeruginosa* strains recovered from systemic sites (blood, bile, aspirates) more frequently (two to three times) produced elastase than those recovered from respiratory or urogenital sources. Earlier studies by Wretlind and collaborators (1973), however, comparing *P. aeruginosa* isolates from bacteremic and colonized individuals could not establish a similar association. Baltch et al. (1979), in a recent investigation of 100 isolates of P. aeruginosa recovered from patients, found no in vitro quantitative differences in lecithinase, protease, and elastase production among strains derived from patients dying from P. aeruginosa septicemia as contrasted to patients recovering from such infections. Baltch et al. did, however, find that the blood isolates were high elastase and protease producers when compared to those recovered from urine or sputum. Such reports in general seem to associate elevated protease(s) activity with systemic or blood isolates and hence strains of high virulence.

While most studies have utilized P. aeruginosa isolates from specific sources, less is known regarding the enzymatic properties of P. aeruginosa associated with specific clinical entities such as cystic fibrosis. Jagger et al. (1983) found that protease activity was restricted in mucoid isolates recovered from cystic fibrosis patients in contrast to nonmucoid strains derived from these individuals. In the initial steps of colonization of cystic fibrosis patients, increased proteolytic activity may aid establishment of the host mucosa, whereas in chronic residence the need for prodigious exoenzyme activity decreases. The association of certain exoenzymatic activities may be inverse in such individuals and may account for the more chronic and less insidious infections seen in cystic fibrosis patients than the fulminant type of pulmonary disease seen in immunocompromised individuals. Also of note, in none of the previously mentioned studies was the activity of exotoxin A investigated.

Frequency of Exoenzyme Production in Pseudomonas aeruginosa and Methods of Assay for Extracellular Enzymes

The frequency of production of 17 extracellular enzymes by P. aeruginosa is listed in Table 2. Most isolates tested have been shown to produce exotoxin A, esterase, hemolysin, lipase, and several proteolytic activities. Other enzymes such as amylase, chondroitinase, hyaluronidase, and lysozyme-like activity are uniformly absent. A number of hydrolytic enzymes, including those responsible for the degradation of nucleic acids, breakdown of elastin and lecithin, and the ability to clear the cell wall material of Staphylococcus aureus, have been reported to vary in frequency of production. The data listed in Table 2 do not take into account a special subgroup of P. aeruginosa, the apyocyanogenic (non-pyocyanin-producing) strains, which are more frequently recovered from environmental rather than clinical sources. Investigation of apyocyanogenic strains to date, principally by Gilardi (1976), suggest that a majority show markedly reduced enzymatic capabilities (lipase, protease, gelatinase) when compared to pyocyanin-producing strains. Little information is currently available concerning the enzymatic properties of P. aeruginosa (pyocyanogenic, apyocyanogenic) originating from the environment.

Table 2 Frequency of Production of Various Exoenzymes by Clinical Isolates of *Pseudomonas aeruginosa*

Enzyme	Phenotype[a]	Range (percent positive)
Amylase	−	0
Chondroitinase	−	0
Coagulase	V	0−76
Deoxyribonuclease	V	16−76
Elastase	V	41−86
Exotoxin A	+	83−89
Esterase	+	90
Fibrinolysin	+	82−100
Gelatinase	+	94
Hemolysin	+	77−98
Hyaluronidase	−	0
Lecithinase	V	40−73
Lipase	+	81−91
Lysozyme-like	−	0
Protease	+	94−98
Ribonuclease	V	20−87
Staphylolytic enzyme	V	34−80

[a] V, variable.
Source: Refs. 4, 21, 30, 55, 56, 59, and 80.

Some common qualitative and quantitative assays for determining eight major exoenzymes of *P. aeruginosa* are listed in Table 3. Many of the variable test results reported for clinical isolates of *P. aeruginosa* (Table 2) are a direct reflection of the method of assay, test conditions (length of assay, incubation temperature, etc.), and type of substrate utilized for the growth of *P. aeruginosa* prior to testing. The age of *P. aeruginosa* cultures may be an additional critical factor in the expression of certain exoenzymatic properties. For instance, Morihara (1964) and Morihara and Tsuzuki (1977) have found that elastase activity of *P. aeruginosa* isolates is dependent upon both the

Table 3 Some Assay Methods for Detection of Exoenzymes Produced by *Pseudomonas aeruginosa*

Enzyme	Qualitative	Quantitative
Exotoxin A	Agar well assay (58) Modified Elek plate (4) Skin test (guinea pig) (59) CIE (59)	Reverse passive hemagglutination assay (58) ADP ribosylation assay (10, 75) Radioimmunoassay (12)
Protease	Dialyzed brain—heart infusion—skim milk agar (70) Skim milk agar (7)	Assays using hide powder azure (9, 49, 64) or azocasein (33)
Alkaline protease	Immunoprecipitation (28)	Radioimmunoassay (13)
Elastase	*Pseudomonas* isolation agar containing elastin-Congo red (28) Solubilization of elastin (22) Elastin agar plates (82)	Release of sulubilized elastin (36)
Collagenase	Solubilization of native collagen (22)	^{14}C-Acetylated acid soluble collagen (32)
Hemolysin	Supernatant titration against rabbit red blood cells (34) Trypticase soy agar base containing erythrocytes of various species (55)	Measurement of hemoglobin release (34)
Lecithinase	Lecithin agar (55) TSA with egg yolk enrichment (16)	Assays using p-nitrophenylphosphorylcholine (3) or egg yolk phosphatidylcholine (3)
Lipase	Tributyrin agar (82) TSA with polyethylene sorbitan monooleate (Tween 80) (16)	p-Nitrophenylcaprylate (82)

age of the culture and the chemical composition of the medium used for assay. Stock cultures or strains retained for long periods of time through preservative measures may potentially lose their ability to degrade elastin. Recent cultures of P. aeruginosa may in addition possess distinct morphotypes which differ in their ability to degrade various substrates. To what extent this phenomenon occurs and its significance remain to be determined.

Extracellular Enzymes or Products Produced by Pseudomonas aeruginosa

Exotoxin A

By far the most extensively studied exoenzyme produced by P. aeruginosa is exotoxin A. Almost 100 years ago it was speculated that the clinical disease manifestation of P. aeruginosa infection was principally due to the elaboration of a toxin. Decades later, Liu et al. (1973) undertook the initial purification of exotoxin A, which provided the impetus for subsequent studies on its chemical, biological, and pathophysiological properties.

Exotoxin A is one of few bacterial toxins whose mode of action at the molecular level is known. The native molecule, a proenzyme, has an approximate molecular weight of 71,550 and consists of two chain positions: an NH_2 terminal (fragment A of molecular weight 26,000), which confers enzymatic activity to the molecule, and a COOH terminus (fragment B, of molecular weight 45,000), accounting for the binding of the toxin to the appropriate cell membrane receptor. Iglewski and Kabat (1975) have determined that the molecular mode of action of exotoxin A is to transfer adenosine diphosphate ribose from nicotinamide adenine dinucleotide (NAD) to elongation factor 2 (EF-2). Transfer of the adenosine diphosphate ribose moiety to EF-2 by covalent linkage irreversibly inactivates EF-2 and thus terminates peptide elongation and protein synthesis within the affected cell. Although the toxin molecules of C. diphtheriae and P. aeruginosa are structurally dissimilar and have different binding sites at the cell surface, their activities with respect to inactivation of EF-2 appear analogous. Conversion of the proenzyme into an active form (fragment A) can be accomplished through either denaturation and reduction or cleavage by freeze thawing or incubation of the reduced molecule in the presence of proteolytic enzymes (such as chymotrypsin) and NAD+ or through chemical cleavage (Vasil et al., 1977; Lory and Collier, 1980).

Exotoxin A appears to have a fairly wide range of activities and pathophysiological effects on laboratory animals and on eukaryotic cells in vitro, although not all species show equal sensitivities to the toxin (Table 4). Injection of purified exotoxin A has been shown to produce death in a variety of animals, including rats, rabbits, dogs, and monkeys (Pollack, 1980; Bodey et al., 1983). Physiological effects noted from exposure to exotoxin A differ depending upon the species of animal challenged, route, and amount administered, but when elicited include decreased cardiac output, hypotension and shock, disruption of clotting function, hepatic dysfunction, and leukopenia. Inhibition of protein synthesis in human polymorphonuclear leukocytes and mouse macrophages exposed to toxin A has been reported (Weber et al., 1982), although phagocytic function (relative to *P. aeruginosa*) did not appear to be altered. In mouse corneal studies traumatized cornea exposed to exotoxin A resulted in a number of pathological changes discerned through electron and scanning electron microscopy (Hazlett et al., 1981). Histologically, these changes included stromal swelling, dispersion of collagen (undamaged) fibrils, death of epithelial and endothelial cells, and loss of proteoglycan ground substance. In the rat pulmonary model, toxin-deficient mutants were less virulent than their wild-type counterparts in which toxin-positive strains produced parenchymal changes in the lung and caused monomuclear cell infiltrates (Woods et al., 1982). Dermatonecrotic lesions produced in guinea pigs, similar to ecthyma gangrenosum lesions observed in human bacteremic *P. aeruginosa* infections, were found to be produced by subcutaneous inoculation and local elaboration of toxin A and proteases by the bacterium (Young and Pollack, 1980). Inoculation of *P. aeruginosa* strains into the skin of guinea pigs which were deficient in exotoxin A production resulted in lesions with focal necrosis without erythyema, suggesting that the later dermatological change noted in humans and animals is the direct result of exotoxin A production. Toxin A has been also shown to inhibit the proliferative activity of precursor granulocyte and macrophage cells arising from bone marrow (Stuart and Pollack, 1982). Picogram amounts of toxin elicit this response, which may account for the accompanying leukopenia often seen in human *Pseudomonas* infections.

Data accumulated over the past several years suggest that exotoxin A is one of the most important exoenzymes produced by *P. aeruginosa*, being 10,000–20,000 times more lethal than the *P. aeruginosa* lipopolysaccharide (Liu, 1974; Young, 1980) and possessing the same magnitude of activity as the alpha toxin of *C. perfringens* (Pollack, 1983). Normal, colonized, infected, and bacteremic patients show an increasing immunological response to exotoxin A, with the highest titers occurring in patients with bacteremia with exotoxin A+ strains of *P. aeruginosa* (Pollack and Young, 1979; Cross et al., 1980). Higher antibody levels in this latter group were correlated with a better prognosis (survival) than in individuals in whom a sparse or no immunological response was forthcoming. In one study (Cross et al., 1980), of 11 *P. aeruginosa* bacteremic patients, 7 of 8 patients infected with toxin A+ strains succumbed, while 3 infected with toxin A− isolates survived. This finding along with in vitro and animal studies suggests a major role for this exoenzyme in the pathogenesis of *P. aeruginosa* infections.

Table 4 Mode of Action and Potential Cellular Effect of Some Exoenzymes Produced by *Pseudomonas aeruginosa*

Enzyme	Mode of action	Pathophysiological effect	Target	
			Cells	Organ/system
Exotoxin A	Interruption of protein synthesis inhibition of EF-2	Edema, hemorrhage, necrosis	Bone marrow Blood mononuclear Alveolar septal Cornea epithelial and endothelial Stromal	Eye Pulmonary Vascular Liver
Pyocyanin	Alteration in electron transport through respiratory chain	Enhanced colonization potential	Other bacteria	Unknown
Proteases	Destruction of proteinaceous tissues containing elastin and collagen fibrils; conversion of	Hemorrhagic lesions; necrosis	Alveolar septal Corneal, stromal Alveolar type I,	Wounds (burn) Eye Pulmonary

			epithelial Capillary endothelial
	protein molecules into peptides digestion of structural components of tissue		
Lecithinase	Liberates phosphorylcholine from lecithin	Central abscess; erythema and induration; necrosis, edema	Stromal Surfactant (compound)
			Lung Liver (?)
Hemolysin	Solubilizes phospholipids	Unknown	(?)
			Eye (corneal) Pulmonary
Enterotoxin	Unknown	Release of fluid and electrolytes into lumen	Unknown
			Gastrointestinal
Neuraminidase	Hydrolysis of α-2-3-glycosidic linkages of N-acetylneuraminlactose	Unknown	Mucins
			Pulmonary

Source: Refs. 2, 5, 9, 11, 18, 19, 20, 22, 23, 27, 33, 34, 37, 38, 42–44, 46, 53, 61, 62, 71, 73, 77–79, 83, and 84.

Although there is a serological response to exotoxin A in bacteremic patients, the exact role or function of exotoxin A in man is still poorly understood, as no active toxin has ever been isolated from the serum or tissue of infected individuals (Pollack, 1980). In fact, some bacteremic patients recover from *P. aeruginosa* infection without ever mounting a neutralizing antibody response to the toxin. Additionally, 10–15% of infected individuals harbor *P. aeruginosa* isolates which are exotoxin A—, suggesting that the exoenzyme is not related to virulence in a susceptible host. Besides these observations, clinical studies indicate that other cellular components such as the lipopolysaccharide of *P. aeruginosa* produce an immunological response which is somewhat protective. In the burned rat model, neutralizing antibodies to exotoxin A administered to animals is not protective to subsequent challenges with *P. aeruginosa* (Walker et al., 1979). Another disturbing fact is that many of the genetic studies concerning exotoxin A were derived from the study of a single *P. aeruginosa* isolate (PA-103), which, although a very excellent exotoxin producer, is almost completely avirulent in most systems studied (Young, 1980). Finally, an intriguing aspect concerning the role of exotoxin A in virulence involves the discrepancy between the types of clinical diseases produced by *C. diphtheriae* and *P. aeruginosa*, both of which produce an exotoxin whose mode of action is identical. With *C. diphtheriae*, elaboration of the toxin from a localized infection site (pharynx) produces systemic manifestations associated with toxemia. However, in the case of *P. aeruginosa* the widespread effects of exotoxin A are not seen during colonization, but only after invasion and local proliferation in which toxin is present concomitant with the microorganism. These facts suggest that the virulence potential of *P. aeruginosa* is related to many factors and that the role of exotoxin A in pathogenesis is probably infection site associated, that is, minimal in burn wounds, and is probably only fully manifested upon systemic invasion.

Proteases

Pseudomonas aeruginosa, when compared to other species of the genus, or members of the family Enterobacteriaceae, displays marked proteolytic activity manifested by 95% of clinical isolates. Although the exact number and substrate specificities of the proteases produced by *P. aeruginosa* are not entirely defined, these enzymes have been shown to be active on a wide range of biological substrates, including casein, milk (Fig. 1), elastin, and fibrin (Liu, 1974). Various investigators have proposed that from three to five distinct proteases are produced by different *P. aeruginosa* isolates. Three, however, a general protease, an alkaline protease, and an elastase, are recognized based upon pH otpima, substrate specificity, and physical properties.

Protease activity has been implicated as an important virulence factor in the pathogenesis of *P. aeruginosa* through animal and genetic

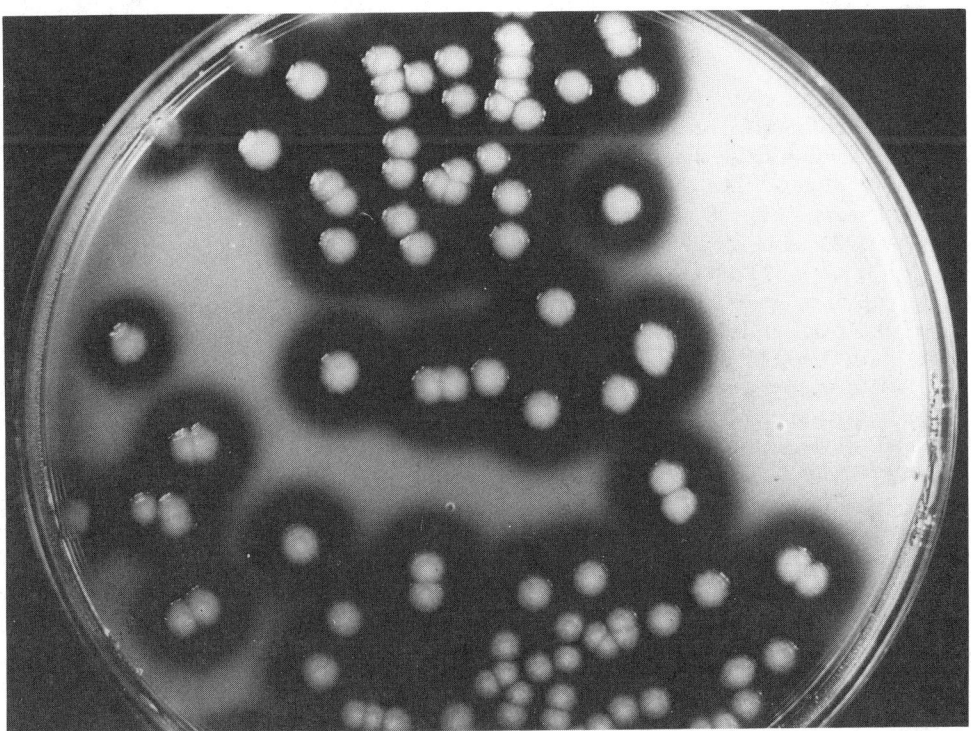

Fig. 1 Protease activity of isolated colonies of *P. aeruginosa* on dialyzed brain—heart infusion—skim milk agar.

studies. The major pathological processes elicited from protease expression are tissue necrosis and destruction of the natural architecture of viable tissue (Young 1980) (Table 4). In animal studies, injection of purified proteases results in a tissue response that mimicks the pathological processes often seen in clinical *P. aeruginosa* infections. Gray and Kreger (1979) found that intratracheal administration of proteases into rabbit lung resulted in intra-alveolar hemorrhage and necrosis of alveolar and septal cells, a response similar to that often observed in the human lung in necrotizing *Pseudomonas* pneumonia. Similar studies (Kreger and Gray, 1978) also implicated proteases in the destructive process in *P. aeruginosa* corneal infections. Additionally, high-dose intraperitoneal injection of distinct fractionated proteases has been shown to produce dermatonecrosis and subcutaneous hemorrhage in newborn mice (Wretlind and Wadstrom, 1977). The most elegant sutdies to date concerning the potential role of proteases in human infections come from the work of Holder (1983) using the burned mouse model. Weakly proteolytic strains were shown to have generation times twice that of highly proteolytic isolates when grown in the presence of burned skin extract (Cicmanec and Holder, 1979). When exogenous proteases were added, the generation time of the weak producing strain was found to be similar to that of the high-protease-producing strain. In addition, for expression of maximum virulence as manifested by systemic invasion, both proteases and elastase were required (Holder and Haidaris, 1979). Addition of purified proteases along with a weakly proteolytic and avirulent strain (PA-103) to the burned skin surface enhanced the pathogenicity of this isolate. These studies suggest that proteases are important in burns and wound infections by providing release of and degradation of essential nutrients for the multiplication and subsequent destruction of these tissues.

The *P. aeruginosa* elastase, a zinc metalloendopeptidase with an approximate molecular weight of 35,000, is found in relatively few bacerial species and is a potent protease capable of degrading native elastin molecules (Fig. 2) (Kessler et al., 1982). Besides possessing many of the properties described for pseudomonal proteases, elastase has been shown to degrade both elastin and cartilage proteoglycans (Kessler et al., 1982) and inactivate complement components C1, C3, C5, C8, and C9 in either the fluid or cell bound form of both (Schultz and Miller, 1974). Mutants temperature sensitive in the elastase gene demonstrate reduced virulence in the rat lung model when compared to the wild-type isolate, suggesting that elastase is required for maximal virulence (Woods et al., 1982). Because of the rarity of this enzyme among microbial species and the paucity of clinical and experimental data regarding elastase, it has been postulated that this enzyme may be an important virulence factor in the invasion of *P. aeruginosa* from local sites (Mull and Callahan, 1965; Holder and Haidaris, 1979; Janda and Bottone, 1981). Indeed, *Pseudomonas* vasculitis may reflect the action of elastase in this invasive process as attested to in many instances by erosion of the vessel wall and parenchymal spread (Fig. 3). Although the exact role of proteases in the infectious process is still evolving, studies clearly implicate these enzymes as important contributors in pulmonary, wound (burn) and corneal infections.

Hemolysin and Phospholipase C (Lecithinase)

Most strains of *P. aeruginosa* isolated in the clinical laboratory produce detectable hemolysins when grown on a trypticase soy agar containing susceptible red blood cells. At least two separate and distinct molecules are reportedly responsible for this activity: a heat-stabile glycolipid and a heat-labile phospholipase (Liu, 1966) (Table 4). Phospholipase C, the better characterized of the two molecules, is able to hydrolyze phosphatidylcholine in erythrocyte membranes and release phosphorylcholine and diacylglycerol (Kurioka and Liu, 1967). It has been suggested to act synergistically with the

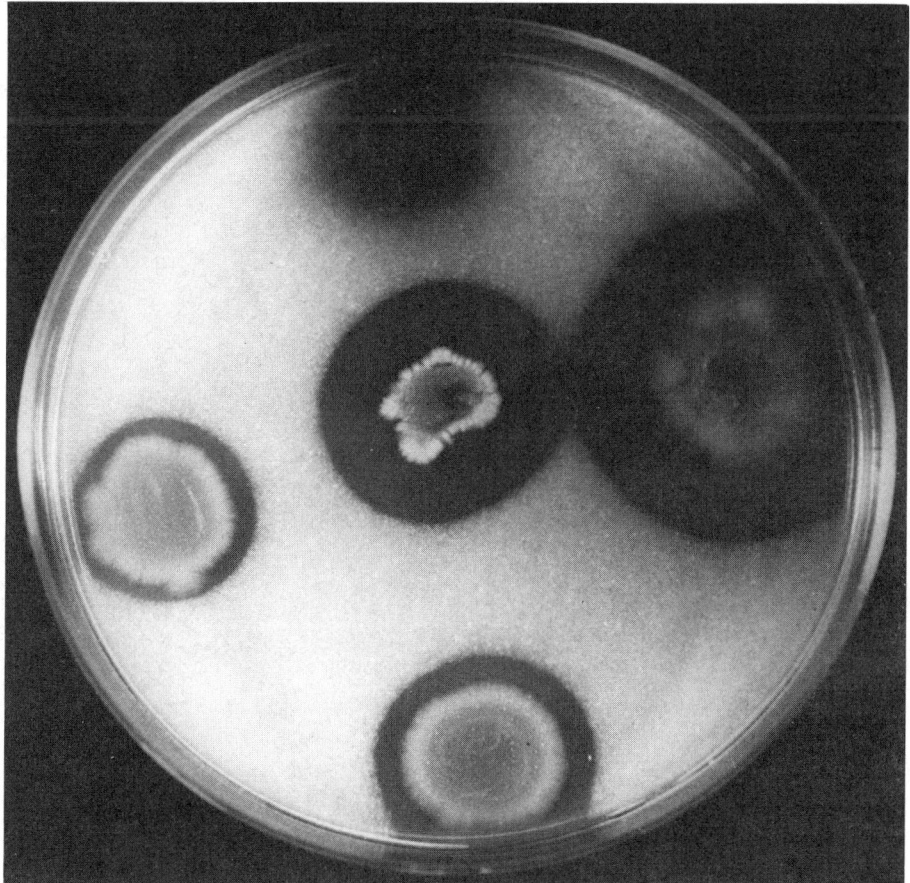

Fig. 2 Elastase activity of five P. aeruginosa isolates assessed by spot inoculation of each strain onto bovine neck ligament elastin agar.

P. aeruginosa alkaline phosphatase to release inorganic phosphate from phospholipid (Esselman and Liu, 1961). Under conditions of high levels of inorganic phosphate, phospholipase C is repressed (Stinson and Hayden, 1979). Berka and colleagues (1981) analyzed 249 strains of P. aeruginosa from different sources, comparing their overall lecithinase acitivity to a standard (PA01) strain. Although 100% of the isolates produce phospholipase C, their relative activity varied with the clinical source, being highest in strains recovered from the urogenital system. Among cystic fibrosis isolates tested, 80% of the nonmucoid isolates produced higher levels of phospholipase C than PA01, whereas only 20% of their mucoid counterparts exhibited similar synthetic capability.

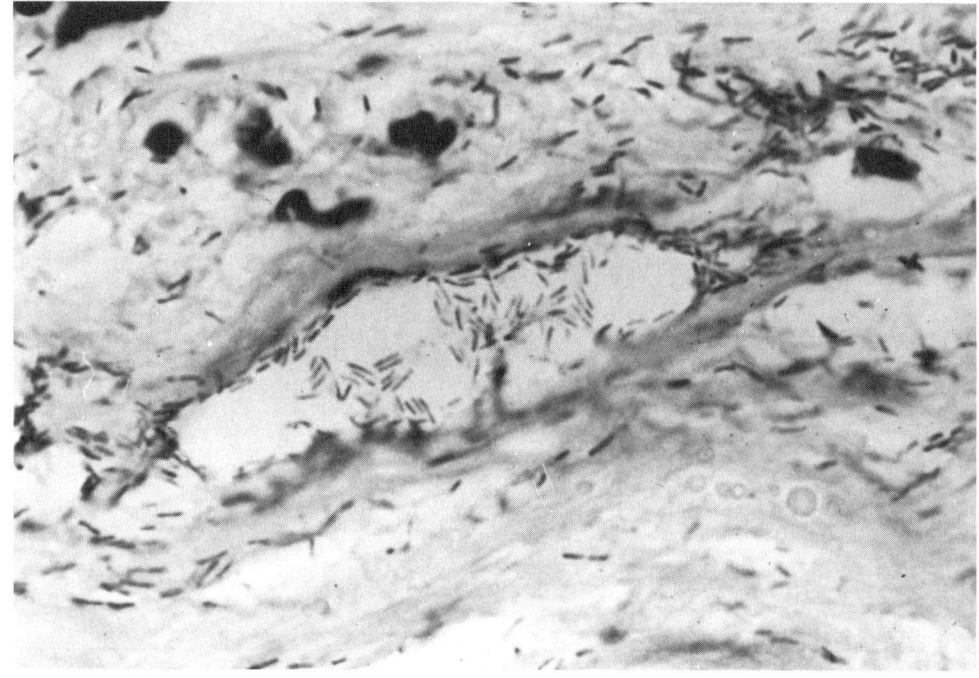

Fig. 3 Erosion of *P. aeruginosa* through the vessel wall of a patient with metastatic *Pseudomonas* infection.

Several studies have implicated the hemolysins produced by *P. aeruginosa* in the pathogenesis of this organism. Johnson and Allen (1978) found that high-virulence isolates (those recovered from corneal ulcers or postoperative endophthalmitis) were more hemolytic (100% with titers of at least 1:8) than less virulent strains (65% with titers under 1:8) recovered from mild conjunctivitis or of unknown clinical significance. Injection of purified hemolysin into the eye of a rabbit resulted in corneal opacification with an enormous polymorphonuclear infiltrate. Phospholipase has also been suggested to play a role in pulmonary infections as an antisurfactant (Liu, 1974), perhaps aiding colonization of respiratory structures (Southern et al., 1970).

Pyocyanin

Pyocyanin is a 1-hydroxy-5-methyl phenazine pigment which is both water and chloroform soluble. This pigment is expressed by greater than 95% of clinical isolates and serves as a major diagnostic criterion for this bacterium, since it is unique to *P. aeruginosa*.

The interaction of ecological and host adaptation factors concerning the formation of this compound (environmental versus clinical) and the physiological function of this pigment has been the topic of much speculation over the past years. Recently reports by Hassan and Fridovich (1980) and Baron and Rowe (1981) have shed some light on its potential significance in vivo. Pyocyanin has been found to oxidize nicotinamide adenine dinucleotide and produce oxygen-free radicals in the absence of enzymatic catalysis through the electron transport system. The pigment has also been shown to be bactericidal to a large number of gram-positive bacteria as well as inhibitory to the growth of some gram-negative rods such as *E. coli, Proteus vulgaris,* and *Acinetobacter* sp. (Baron and Rowe 1981). These studies allow one to hypothesize that pyocyanin, when released in vivo under aerobic conditions, might inhibit many gram-positive and gram-negative organisms through alterations in their electron transport system. This effect, therefore, could suppress the normal host microflora and facilitate *P. aeruginosa* colonization (Table 4).

Exoenzymes Produced by other Pseudomonas Species and Members of the Genus Acinetobacter

Introduction

From the above it is apparent that many of the exoenzymes produced by *P. aeruginosa* are relevant to its virulence and inherent ability to incite disease from diverse focal sites. Unfortunately, while a plethora of reports have appeared with respect to *P. aeruginosa*, little information is currently available concerning the exoenzymatic activity of other members of this genus. This is undoubtably due to the reduced frequency of isolates of non-*aeruginosa* species from clinical material, and, additionally, studies of the exoenzymatic activity of non-*aeruginosa* species have not been especially correlated with virulence potential as assessed through animal pathogenicity studies or clinical infections. Presented here, therefore, is a brief résumé of the reported exoenzymatic activity of selected pseudomonads.

Pseudomonas maltophilia

The taxonomic status of *P. maltophilia* is currently under investigation and a proposal has been made to transfer this microorganism to the genus *Xanthomonas* (Swings et al., 1983). Despite these taxonomic controversies, *P. maltophilia* is the second most commonly isolated pseudomonad from human clinical sources and has been associated with hospital-acquired infections, particularly in patients with underlying disease or receiving immunosuppressive therapy. The major exoenzymes produced by *P. maltophilia* are listed in Table 5. Of interest are enzymatic activities directed against chitin (O'Brien

Table 5 Extracellular Enzymes Produced by *Pseudomonas maltophilia*

Generally positive (>85%)	Variable (15–18%)	Usually negative (<15%)
Deoxyribonuclease	Urease	Amylase
Ribonuclease	Alginase	Chondroitinase
Gelatinase	Elastase	Collagenase
Esterase	Albuminase	Keratinase
Lipase	Staphylolytic enzyme	Lecithinase
Chitinase		
Hyaluronidase		
Hemolysin[a]		
Mucinase		
Fibrinolysin		
Protease		

[a]Figures on the overall hemolytic activity of *P. maltophilia* vary from author to author.
Sources: Refs. 24, 55–57, and 76.

and Davis, 1982) and algin (von Riesen, 1980), substrates which are usually degraded only by marine-associated bacteria. Additonal enzymes of potential diagnostic significance include deoxyribonuclease, mucinase, and elastase (Nord et al., 1975; Holmes et al., 1979; O'Brien and Davis, 1982). Presently no studies have been reported linking any of these activities with virulence, biotype, or ecological selection.

Pseudomonas cepacia

Pseudomonas cepacia is usually resistant to a large number of antimicrobial agents and has been increasingly isolated from patients with cystic fibrosis (McKevitt and Woods, 1984). Recent ecological studies utilizing a variety of phenotypic properties suggest that at least two separate biovars exist, namely, strains isolated from plants and those of clinical origin (Gonzales and Vidaver, 1979). *Pseudomonas cepacia* isolates of plant origin were found to be more pectinolytic, produced a bacteriocin active against two indicator strains, and were capable of producing a postive response in the onion maceration test when inoculated in low numbers. Clinical isolates lacked these properties. Other studies (Nord et al., 1975; O'Brien and Davis, 1981; McKevitt and Woods, 1984) have reported variable frequencies of production of the following enzymes: hemolysin, gelatinase, protease, lipase, lecithinase, DNase, RNase, and fibrinolysin. Only the proteolytic, lipolytic, and gelatinolytic activities have been reported to be produced by a significant majority (≥75%) of the strains tested to date. Cytotoxins and exotoxins similar to those produced by P. aeruginosa apparently are not produced (McKevitt and Woods, 1984). As yet no clear-cut relation exists between the exoenzymes produced by P. cepacia and physiological function.

Pseudomonas paucimobilis

Recently, P. paucimobilis (group IIk, biotype 1) has been subjected to enzyme profiling by Smalley (1982) using the API ZYM system (Analytab Products, Plainview, New York). Of the 19 enzymes tested for in chromogenic-dependent substrate reactions, 16 were positive for P. paucimobilis. The author postulated that some of these enzymes might be potentially related to P. paucimobilis virulence.

Acinetobacter

The genus Acinetobacter was recently reviewed by Juni (1978) and little more than two paragraphs dealt with the hydrolytic enzymes elaborated by species of this genus. In general, Acinetobacter species have been found to be lipolytic but do not produce amylase or deoxyribonuclease. The production of hemolysin and gelatinase activities within the species Acinetobacter calcoaceticus is biotype dependent (Rubin et al., 1980), being produced by the varieties *haemolyticus* and *alcaligenes*. Hemolysis has been shown to be related to phospholipase (lecithinase) activity and at least two types have been described (Lehmann, 1973).

One psychrophilic Acinetobacter species has been found to produce both lipase and esterase activity (Breuil and Kushner, 1975). Lipase but not esterase activity was optimalized at lower (20°C) and not higher (30°C) incubation temperatures.

REFERENCES

1. Baltch, A. L., Griffin, P. E., and Hammer, M. (1979). *J. Lab. Clin. Med.* 93:600.
2. Baron, S. S., and Rowe, J. J. (1981). *Antimicrob. Agents Chemother.* 20:814.
3. Berka, R. M., Gray, G. L., and Vasil, M. L. (1981). *Infect. Immun.* 34:1071.
4. Bjorn, M. J., Vasil, M. L., Sadoff, J. C., and Iglewski, B. H. (1977). *Infect. Immun.* 16:362.

5. Bodey, G. D., Bolivar, R., Fainstein, V., and Jadeja, L. (1983). *Rev. Infect. Dis.* 5:279.
6. Breuil, C., and Kushner, D. J. (1975). *Can. J. Microbiol.* 21:423.
7. Brown, M., and Foster, J. (1970). *J. Clin. Pathol.* 23:172.
8. Chakrabarty, A. N., Adhya, S., and Pramanik, M. K. (1970). *J. Appl. Bacteriol.* 33:397.
9. Cicmanec, J. F., and Holder, I. A. (1979). *Infect. Immun.* 25:477.
10. Collier, J., and Kandel, J. (1971). *J. Biol. Chem.* 246:1496.
11. Cross, A. S., Sadoff, J. C., Iglewski, B. H., and Sokol, P. A. (1980). *J. Infect Dis.* 142:538.
12. Crowe, K. E., Bass, J. A., Young, V. M., and Straus, D. C. (1982). *J. Clin. Microbiol.* 15:115.
13. Cryz, S. J., and Iglewski, B. H. (1980). *J. Clin. Microbiol.* 12:131.
14. Diener, B., Carrick, L., and Berk, R. S. (1973). *Infect. Immun.* 7:212.
15. Esselman, M., and Liu, P. V. (1961). *J. Bacteriol.* 81:939.
16. Gilardi, G. L. (1976). *Mt. Sinai J. Med.* 43:710.
17. Gonzales, C. F., and Vidaver, A. K. (1979). *J. Gen. Microbiol.* 110:161.
18. Gray, L., and Kreger, A. (1979). *Infect. Immun.* 23:150.
19. Hassan, H. M., and Fridovich, I. (1980). *J. Bacteriol.* 141:156.
20. Hazlett, L. D., Berk, R. S., and Iglewski, B. H. (1981). *Infect. Immun.* 34:1025.
21. Hedberg, M., Miller, J. K., and Tompkins, V. N. (1969). *Am. J. Clin. Pathol.* 39:631.
22. Holder, I. A., and Haidaris, C. G. (1979). *Can. J. Microbiol.* 25:593.
23. Holder, I. A. (1983). *Rev. Infect. Dis.* 4:S914.
24. Holmes, B., Lapage, S. P., Easterling, B. G. (1979). *J. Clin. Pathol.* 32:66.
25. Hou, C. I., Gronlund, A. F., and Campbell, J. J. R. (1966). *J. Bacteriol.* 92:851.
26. Iglewski, B. H., Sadoff, J., Bjorn, M. J., and Maxwell, E. S. (1978). *Proc. Nat. Acad. Sci. U.S.A.* 75:3211.
27. Iglewski, B. H., and Kabat, D. (1975). *Proc. Nat. Acad. Sci. U.S.A.* 72:2284.
28. Jagger, K. S., Bahner, D. R., and Warren, R. L. (1983). *J. Clin. Microbiol.* 17:55.
29. Janda, J. M., Atang-Namo, S. A., Bottone, E. J., and Desmond, E. P. (1980). *J. Clin. Microbiol.* 12:626.
30. Janda, J. M., and Bottone, E. J. (1981). *J. Clin. Microbiol.* 14:55.
31. Jarvis, F. G., and Johnson, M. J. (1949). *J. Am. Chem. Soc.* 71:4124.

32. Jensen, S. E., Phillippe, L., Tseng, J. T., Stemke, G. W., and Campbell, J. N. (1980). *Can. J. Microbiol.* 26:77.
33. Johnson, M. K., and Allen, J. H. (1978). *Invest. Ophthal. Visual Sci.* 17:480.
34. Johnson, M. K., and Boese-Marrazzo, D. (1980). *Infect. Immun.* 29:1028.
35. Juni, E. (1978). *Annu. Rev. Microbiol.* 32:349.
36. Kessler, E., Isreal, M., Landshman, N., Chechick, A., and Blumberg, S. (1982). *Infect. Immun.* 38:716.
37. Kreger, A. S., and Gray, L. D. (1978). *Infect. Immun.* 19:630.
38. Kubota, Y., and Liu, P. V. (1971). *J. Infect. Dis.* 123:97.
39. Kurioka, S., and Liu, P. V. (1967). *J. Bacteriol.* 93:670.
40. Lawrence, R. C. (1967). *Dairy Sci. Abstr.* 29:1.
41. Lehmann, V. (1973). *Acta Pathol. Microbiol. Scand. Sec. B* 81:427.
42. Leprat, R., and Michel-Briand, Y. (1980). *Ann. Microbiol. Paris* 131B:209.
43. Liu, P. V., Abe, Y., and Bates, J. L. (1961). *J. Infect. Dis.* 108:218.
44. Liu, P. V. (1966). *J. Infect. Dis.* 116:481.
45. Liu, P. V. Yoshii, S., and Hsieh, H. (1973). *J. Infect. Dis.* 128:514.
46. Liu, P. V. (1974). *J. Infect. Dis.* 130:594.
47. Lory, S., and Collier, R. J. (1980). *Infect. Immun.* 28:494.
48. McKevitt, A. I., and Woods, D. E. (1984). *J. Clin. Microbiol.* 19:291.
49. McManus, A. T., Moody, E. E., and Mason, A. D. (1979). *Burns* 6:235.
50. Morihara, K. (1963). *Biochim. Biophys. Acta* 73:113.
51. Morihara, K. (1964). *J. Bacteriol.* 88:745.
52. Morihara, K., and Tsuzuki, H. (1977). *Infect. Immun.* 15:679.
53. Mull, J. D., and Callahan, W. S. (1965). *Exp. Mol. Pathol.* 4:567.
54. Muszynski, A. (1973). *Path. Microbiol.* 39:135.
55. Nord, C. E., Sjoberg, L., Wadstrom, T., and Wretling, B. (1975). *Med. Microbiol. Immunol.* 161:79.
56. O'Brien, M., and Davis, G. H. G. (1981). *Microbios Lett.* 16:41.
57. O'Brien, M., and Davis, G. H. G. (1982). *J. Clin. Microbiol.* 16:417.
58. Ohman, D. E., Sadoff, J. C. and Iglewski, B. H. (1980). *Infect. Immun.* 28:899.
59. Pollack, M., Taylor, N. S., and Callahan, L. T. (1977). *Infect. Immun.* 15:776.
60. Pollack, M., and Young, L. S. (1979). *J. Clin. Invest.* 63:276.

61. Pollack, M., (1980). *N. Engl. J. Med.* 302:1360.
62. Pollack, M. (1983). *Rev. Infect. Dis.* 5:S979.
63. Rangam, C. M., Gupta, J. C., Bhagwat, R. R., and Bhagwatt, A. G. (1961). *Indian J. Med. Res.* 49:232.
64. Rinderknecht, H., Geokas, M. C., Silverman, P., and Haverback, B. J. (1968). *Clin. Chim. Acta* 21:197.
65. Rubin, S. J., Granato, P. A., and Wasilauskas, B. L. (1980). In *Manual of Clinical Microbiology* (E. H. Lennette, A. Balows, W. J. Hausler, Jr., and J. P. Truant, eds.), American Society for Microbiology, Washington, D.C., p. 263.
66. Schoellman, G., and Fisher, E. (1966). *Biochim. Biophys. Acta* 122:557.
67. Schultz, D. R., and Miller, K. D. (1974). *Infect. Immun.* 10:128.
68. Sierra, G. (1960). *Antonie van Leeuwenhoek J. Microbiol. Serol.* 26:189.
69. Smalley, D. L. (1982). *J. Clin. Microbiol.* 16:564.
70. Sokol, P. A., Ohman, D. E., and Iglewski, B. H. (1979). *J. Clin. Microbiol.* 9:538.
71. Southern, P. M., Mays, B. B., Pierce, A. K., and Sanford, J. P. (1970). *J. Lab. Clin. Med.* 76:548.
72. Stinson, M. W., and Hayden, C. (1979). *Infect. Immun.* 25:558.
73. Stuart, R. K., and Pollack, M. (1982). *Infect. Immun.* 38:206.
74. Swings, J., De Vos, P. Van den Mooter, M., and De Ley, J. (1983). *Int. J. Syst. Bacteriol.* 33:409.
75. Vasil, M. L., Kabat, D., and Iglewski, B. H. (1977). *Infect. Immun.* 16:353.
76. Von Riesen, V. L. (1980). *Appl. Environ. Microbiol.* 39:92.
77. Walker, H. L., McLeod, C. G., Leppla, S. H., and Mason, Jr. A. D., (1979). *Infect. Immun.* 25:828.
78. Weber, B., Nickol, M. M., Jagger, K. S., and Saelinger, C. B. (1982). *Can. J. Microbiol.* 28:679.
79. Woods, D. E., Cryz, S. J., Friedman, R. L., and Iglewski, B. H. (1982). *Infect. Immun.* 36:1223.
80. Wretlind, B., Heden, L., Sjoberg, L., and Wadstrom, T. (1973). *J. Med. Microbiol.* 6:91.
81. Wretlind, B., and Wadstrom, T. (1977). *J. Gen. Microbiol.* 103:319.
82. Wretlind, B., Sjoberg, L., and Wadstrom, T. (1977). *J. Gen. Microbiol.* 103:329.
83. Young, L. S., and Pollack, M. (1980). In *Pseudomonas aeruginosa: The Organism, Diseases It Causes, and Their Treatment* (L. D. Sabath, ed.), Hans Huber Publisher, p. 119.
84. Young, L. S. (1980). *J. Infect. Dis.* 142:626.

8
In Vitro and In Vivo Tests to Determine Virulence of *Pseudomonas aeruginosa*

IAN ALAN HOLDER *University of Cincinnati College of Medicine and Shriners Burns Institute, Cincinnati, Ohio*

INTRODUCTION

Since 1882, when *Pseudomonas aeruginosa* was first isolated, numerous studies on the pathogenicity of this microorganism have appeared in the world's literature. It is not within the scope of this chapter to present a historical review, especially since a brief survey of the older literature on toxins has been published recently as part of a book on *P. aeruginosa* (Liu, 1979). Furthermore, a very large number of factors have been described as playing some role in the virulence of *P. aeruginosa* or in the pathogenicity of infections due to this microorganism. A partial list of these factors is presented in Table 1.

Exotoxin A and the proteolytic enzymes have received the most attention in recent years. Therefore this review will focus on both the in vitro and in vivo methods that have been used to assess the elaboration and activity of these factors and the role(s) they may play in the pathogenesis of *Pseudomonas* infections.

IN VITRO METHODS

Proteolytic Enzymes

Agar Plate Methods

A variety of agar plate assays, utilizing various protein substrates, have been described to screen *P. aeruginosa* isolates for proteolytic activity.

Table 1 Factors Associated with Virulence and/or Pathogenicity of *Pseudomonas aeruginosa*

Factor	References
Polymorphonuclear leukocyte inhibitor	Nonoyama et al. (1979)
Vascular permeability factor	Kusama and Suss (1972)
Exoenzyme S	Iglewski et al. (1978)
Leucocidin	Scharmann (1967a;b)
Lecithinase	Liu (1966a)
Pigments	Armstrong et al. (1971)
Slime	Sensakovic and Bartell (1974)
Hemolysins	Esselmann and Liu (1961), Sierra (1961)
Motility and chemotaxis	McManus et al. (1980), Craven and Montie (1981)
Exotoxin A	Liu (1966b), Liu et al. (1973), Callahan (1974)
Protease (alkaline protease and elastase)	Morihara (1964), Morihara et al. (1965)

Since both alkaline protease and elastase are capable of degrading casein, a variety of agar plates containing skim milk as the protein substrate have been used. Proteolytic activity can be determined by either culturing the test organism directly on the agar surface or, if activity in supernatant fluid is to be determined, these fluids can be placed into wells cut into the surface of the skim milk agar. Plates are read after incubation at 37°C for 24–48 hr. Zones of clearing, indicating hydrolysis of the protein substrate, around the growing colony or the test well containing the supernatant fluid indicate positive protease activity. The results of a skim milk agar proteolytic plate assay for a protease-producing and a non-protease producing strain of *P. aeruginosa* are presented in Figure 1. Proteolytic activity in the supernatant fluids of the respective strains are also included. The basal medium and percentage of skim milk used in such plate assays has been varied by different investigators. Brown and Foster (1970), for example, used a basal medium consisting of 25% nutrient broth plus 2% agar and 10% skim milk granules, whereas

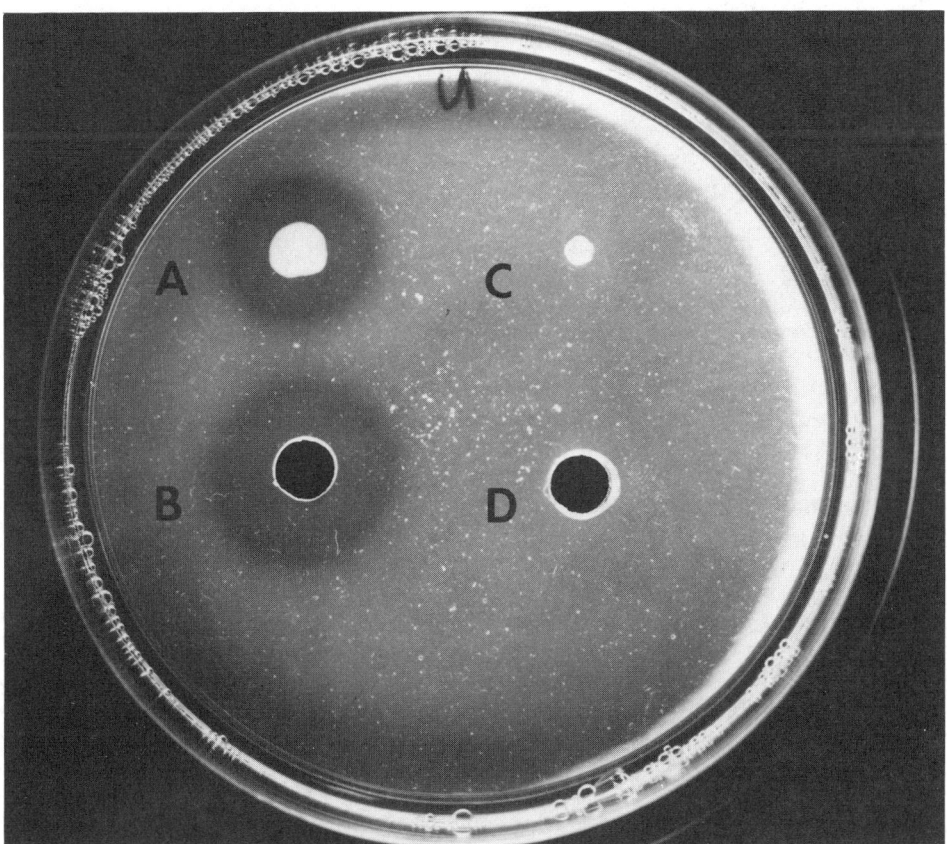

Fig. 1 A 3% skim milk-trypticase soy agar proteolytic assay plate: (A and C) overnight growth of a protease-producing and a non-protease producing strain of *P. aeruginosa* and (B and C) the well assay in which 10 times concentrated, sterile, cell-free supernatant fluids from the respective *Pseudomonas* strains were placed in the wells. The clear zones around the microbial growth in A and the well in B represent proteolytic activity.

Wretlind et al. (1977) employed brain—heart infusion containing 15% skim milk. The composition of the basal medium and the percentage of skim milk used can affect the sensitivity of skim milk test agar plates.

Compared to plates prepared as previously described, $2\frac{1}{2}$–10 times larger zones of hydroloysis were found when P. aeruginosa strains were tested using dialyzed brain—heart infusion (BHI) milk medium (Sokol et al., 1979). Dialyzed BHI milk medium is prepared by dissolving 18.5 g of BHI broth in 50 ml of water and dialyzing against 1 liter of water for 18 hr at 4°C. Agar is added to the dialysate to a concentration of 3%. A 3% (wt/vol) solution of skim milk is prepared, and the solutions are autoclaved separately. Equal volumes of the two sterile solutions are mixed at 60°C. The lower concentration of skim milk in the medium apparently results in a more sensitive assay.

Protease assays using plates consisting of BHI broth supplemented with 1.5% agar and 3% skim milk (Janda et al., 1980) or trypticase soy agar supplemented with 3% skim milk have also been used with success (Holder and Haidaris, 1979; Cicmanec and Holder, 1979).

Agar plate assays using protein substrates other than casein have been used also. Holder (1983) used plates containing hide powder blue as a protein substrate to determine proteolytic activity of supernatant fluids from *Pseudomonas* cultures. This method utilizes 200 mg of hide powder blue azure as a protein substrate, 2 g of agar, and 100 ml of 0.05 M tris buffer (pH at 37°C) containing 10^4 M $ZnCl_2$ and 10^{-3} M $CaCl_2$. The hide powder blue azure is pulverized with a mortar and pestle before addition to the molten agar—buffer mixture. After thorough mixing, plates are poured. The zinc and calcium are added to ensure optimal enzyme activity. To test more specifically for elastolytic activity, elastin—congo red can be substituted for hide powder blue azure. Elastin—containing plates must be incubated for at least 72 hr before being read. Since these plates contain no nutrient materials to support the growth of organisms, this method can only be used to test solutions thought to contain enzymic activity. Wretlind et al (1977) used BHI agar plates containing elastin to test the ability of *Pseudomonas* strains to produce elastase.

These plate assays are both simple and cheap and can be used to determine not only proteolytic activity associated with strains of P. aeruginosa, but also the effect of protease inhibitors on the enzyme activity contained culture filtrates or liquid enzyme solutions (Holder and Haidaris, 1979; Holder, 1983).

Spectrophotometric Methods

A number of investigators have used spectrophotometric methods to determine protease activity in culture filtrates from strains of P. aeruginosa. A common procedure empolying casein as the protein substrate is that of Kunitz (1946/1947). Other workers have used this method as originally described or with slight modification (Morihara, 1964; Johnson et al., 1967; Carrick and Berk, 1975; Wretlind and Wadstrom, 1977). The procedure spectrophotometrically measures (280 µm) the concentration of split products of proteolysis after precipitation of the remaining native protein in the protease—substrate mixture after a fixed period of incubation. A photometric method for the determination of elastase activity measures the amount of dye released from dyed elastin by interaction with elastolytic enzymes (Sachar et al., 1955). Either orcein- or congo red-dyed elastin has been used to measure the elastolytic activity of *Pseudomonas* enzymes (Johnson et al., 1967; Pavlovskis and Wretlind, 1979). A dye relase procedure, originally described by Rinderknecht et al. (1969), was employed by Cicmanec and Holder (1979) to measure total protease activity in *Pseudomonas* culture filtrates. In this case hide powder blue served as the dyed protein substrate.

Immunological Methods

Recently a highly sensitive and specific radioimmunoassay for alkaline protease has been described (Cryz and Iglewski, 1980). This assay takes advantage of the fact that highly purified *Pseudomonas* enzymes are available and can be used to prepare specific antibody. Briefly, a fixed amount of ^{125}I-labeled alkaline protease plus culture supernatant fluid under investigation is placed in a reaction mixture after which anti-alkaline protease serum is added. After incubation an IgG adsorbant is added, the alkaline protease:anti-alkaline protease— IgG adsorbant complexes are sedimented by centrifugation and the pellets counted in a gamma counter. The amount of alkaline protease in the unknown sample can be estimated, relative to a control sample without unknown added, from the amount of radioactivity displaced. In this assay purified elastase showed no cross-reactivity.

Since specific antibodies can be made against *Pseudomonas* proteases, this also allows standard immunological assays such as gel diffusion procedures to be used for the detection of these antigens.

Enzyme Profiling

A system of enzyme profiling of clinical isolates of *P. aeruginosa* as an assessment of their invasive potential has been described (Janda and Bottone, 1981). The profiling, which measures the ability of the strains to produce a variety of enzymes, uses skim milk and elastin-containing plates. Since enzyme profiling of *P. aeruginosa* is the subject of another chapter in this book, it will not be described further here.

Exotoxin A

Inhibition of Protein Synthesis

The fact that toxic substances are produced by *P. aeruginosa* has been known for years (Liu et al., 1961). In 1966 Liu described a proteinaceous lethal toxin produced by *P. aeruginosa* strains grown on a variety of media (Liu, 1966a,b). Intraperitoneal injections of *Pseudomonas* culture filtrates from these media were lethal for mice. Liu also determined that a lethal toxin was elaborated by *P. aeruginosa* strains growing in the skin lesions of rabbits after intracutaneous infections were produced. Homogenization and centrifugation of infected skin separated a supernatant fluid which was lethal for mice upon intraperitoneal injection. Bartell et al. (1968) demonstrated that in spite of significant reduction in the number of infecting *Pseudomonas* by treatment with specific *Pseudomonas* bacteriophage, *Pseudomonas*-infected mice died early in the course of infection. They speculated that a "lethal event" had occurred rapidly in the course of infection and the consequences of this event could not be reversed by significant reduction of bacterial number. This observation was consistent with the interpretation that a lethal factor was produced early in the infection process. In a similar study burned mice were infected in the burn site with small numbers of *P. aeruginosa*. When organisms in the burned skin tissue reached a critical concentration, there was generalized toxemia with subsequent mortality; the process was not reversible at this stage, even by reducing substantially the numbers of infecting organisms (Stieritz and Holder, 1979). These authors concluded that a "toxin" was produced early in the infection process, and at a critical concentration the infected animals were "committed to die" even though dramatic reductions in the number of infecting organisms could be brought about by local and systemic antibiotic treatment.

The purification of exotoxin A (Liu et al., 1973; Callahan, 1974, 1976; Leppla, 1976) allowed study of the action of pure toxin in experimental animals (Pavlovskis and Shackelford, 1974; Pavlovskis et al., 1976) and in cells in culture (Pavlovskis and Gordon, 1972; Middlebrook and Dorland, 1977a–c; Michael and Saelinger, 1979; Weber et al., 1982; Moehring and Moehring, 1983).

A major breakthrough in the study of exotoxin A from *P. aeruginosa* was the determination of its mechanism of action (Iglewski and Kabat, 1975). Using a rabbit reticulocyte cell-free lysate system for the incorporation of radioactive [^{14}C]amino acid into acid-precipitable protein, Iglewski and Kabat found that exotoxin A was a potent inhibitor of protein synthesis. In addition, they compared the aminoacyl transferase activity of exotoxin A to that of diphtheria toxin fragment A. Both bacterial toxins were found to inhibit protein synthesis by catalyzing the transfer of the ADP ribosyl moiety of nicotinamide adenine dinucleotide (NAD) to mammalian elongation factor (EF-2), a protein essential to protein synthesis. The ADP ribose moiety is covalently bound to mammalian cell EF-2; the resulting EF-2–ADP ribose complex is nonfunctional, and protein synthesis ceases as functional EF-2 is depleted. The effects of toxin on the incorporation of radioactivity into protein and the NAD-dependent nature of the reaction are shown in Figure 2.

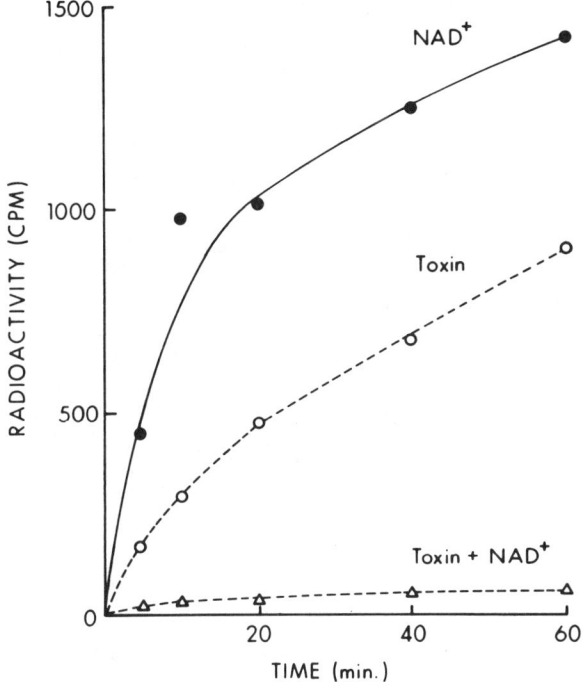

Fig. 2 Effect of P. aeruginosa (PA) toxin and NAD on [^{14}C]leucine incorporation into protein in a reticulocyte lysate. Aliquots of the cell-free system were incubated in either the presence or absence of NAD and of PA toxin. Aliquots were sampled at intervals.

Direct measurement of ADP ribosylation activity is one method to determine the activity of exotoxin A in culture filtrates, tissue extracts, and serum of infected animals. Also, determination of functional EF-2 levels in the organs of burned animals following infection with various strains of P. aeruginosa can be used as an assay of in vivo exotoxin A production (Saelinger et al., 1977; Iglewski et al., 1977; Pavlovskis et al., 1978; Snell et al., 1978; Holder, 1983). All of these studies use the procedure of Gill and Dinius (1973) with or without modification. This method determines levels of active EF-2 in homogenates of tissue added to a reaction mixture containing [^{14}C]NAD in the presence or absence of diphtheria toxin. After incubation the reaction is stopped by the addition of trichloracetic acid. Radioactive counts are done on the trichloroacetic acid precipitates and the difference between radioactive counts incorporated with and without diphtheria toxin (or fragment A) represents the ADP−ribose:EF-2 complex formed.

Tissue Culture

A wide variety of cell lines are susceptible to the action of exotoxin A. Middlebrook and Dorland (1977a) used a cytotoxic assay to examine the sensitivity of 21 mammalian cell lines to exotoxin A. The range of sensitivities of the various cell lines ranged over four orders of magnitude, with tissue culture LD_{50} values falling between 0.10 and 3×10^2 ng/ml. The type of serum used in the culture medium significantly influenced the response of the cells to toxin (Middlebrook and Dorland, 1977b); fetal calf serum, which did not have demonstrable antibody to toxin, did not protect cells and is the serum of choice for toxicity assays. Other investigators also have shown various cell lines to be susceptible to the action of exotoxin A (Vasil et al., 1976; Middlebrook and Dorland, 1977c; Vasil and Iglewski, 1978). Michael and Saelinger (1979) demonstrated a reduction in functional EF-2 in cultured mouse cells after incubation with *Pseudomonas* toxin. Clearly, tissue culture provides an additional method for determing exotoxin A activity contained in various fluids and also can be adapted to assay the toxin neutralization capacity of antiserum.

IN VIVO METHODS

Intact Animal Models

Intact animals, mainly mice, have been used frequently to test the effects of *Pseudomonas* culture filtrates suspected of containing either lethal or toxic exoproducts.

While Liu et al. (1961) demonstrated that substances produced in vitro were toxic for HeLa cells grown in cluture, it was not until 1966 that these substances were shown to be lethal when injected into animals (Liu, 1966b). Carney and Jones (1968) fractionated, by gel filtration, culture filtrates from two mouse-virulent and one avirulent strain of *P. aeruginosa*. Of four fractions prepared in a similar manner from each culture filtrate, one was highly lethal when inoculated intraperitoneally into mice; the fractions prepared from the culture filtrate in which the virulent cells were grown were more toxic than were the same fractions prepared from the avirulent strain filtrate. The more lethal filtrates contained higher activity of the four enzymes measured, that is, lecithinase, hemolysin, protease and elastase.

When the nature of the toxic substances contained in *Pseudomonas* culture supernatant fluids was identified, these factors were purified and similar animal experiments were carried out using these purified materials. For example, Atik et al. (1968) injected dogs intravenously with partially purified exotoxin A. Immediate anaphylactoid reactions and circulatory changes followed by death were demonstrated. Using mice and a similar protocol, Pavlovskis and Shackelford (1974) and

Pavlovskis et al (1976) demonstrated a rapid decrease in protein synthesis in the liver with characteristic histopathological lesions upon autopsy. These experiments defined the liver as the major target organ for toxin in vivo.

Various investigators have used intact animals to determine the effects of crude or purified solutions of *Pseudomonas* alkaline protease and elastase (Johnson et al., 1967; Meinke et al., 1970; Diener et al., 1973; Kawaharajo et al., 1975a,b; Wretlind and Wadstrom, 1977). These studies showed these proteolytic enzymes to be lethal when injected into animals. Both the LD_{50} and the organs in which pathology occurred varied according to the route of injection. Intraperitoneal injection caused hemorrhagic lesions in the gastrointestinal tract, whereas pulmonary lesions were caused by both intravenous and intranasal inoculations.

While the experiments described above provided information about the lethality and pathological effects of *Pseudomonas* exoproducts in live animals, thus supporting suggestions that these factors might be virulence associated, they provided no information as to whether these substances were produced during the course of natural or experimental infection. Because of this, these models do not serve to show the causal relationship between elaboration of an exoproduct and the virulence of *P. aeruginosa*. A more appropriate model would require infection by the microorganism, thus permitting the contribution of the various exoproducts to the infectious process to be assessed.

Various studies with intact animals have shown that *P. aeruginosa* is not highly virulent. Bartell et al. (1968) demonstrated that it took 2.5×10^8 of a *P. aeruginosa* strain, injected intraperitoneally with 5% mucin into mice, to kill 80–100% of the animals in 48 hr. No animals died when the dose was reduced by 1 log. Kobayashi (1971) tested the virulence of 22 clinical *P. aeruginosa* isolates by intraperitoneal injection in mice. The LD_{50} values ranged between 10^5 and $10^{7.5}$. Similar results were obtained for mice infected via intravenous route (Klyhn and Gorrill, 1967). Therefore it was necessary to develop animal models in which the host is rendered more susceptible to infection in order to sudy factors associated with the virulence of *P. aeruginosa*.

Burn-Compromised Models

In the late 1950s clinicians treating burn victims became aware that severe infections caused by *P. aeruginosa* were becoming more prevalent, and by the 1960s *Pseudomonas* infections proved to be the most common and most devastating infections in burn patients.

In response to this, many investigators of thermal trauma turned their attention to the development of animal models in which a burn was imposed prior to *Pseudomonas* infection. Burned animal models were developed, therefore, to study the pathogenesis of *Pseudomonas* infections and to provide a means of evaluating the efficacy of new treatment modalities. Such models, it was hoped, would provide data relavent to the patient population of concern. In fact, some interesting host-parasite interactions occurred when burned, rather than intact, animals were subjected to *P. aeruginosa* infections.

In 1961, Liedberg described a guinea pig model in which a brass cylinder, heated to 97°C, was used to inflict the burn. The heated cylinder (5.7-mm base) was applied to the shaved trunks of anesthesized animals. The extent of burn could be adjusted by using multiple sites for burning. The animals were infected intraperitoneally 2 hr postburn, using 1 ml of various dilutions of a *P. aeruginosa* culture containing approximately 10^9 CFU/ml in the undiluted sample. He found that 11 of 12 guinea pigs died when injected with a 1:64 dilution of the culture; no unburned animals died after identical injection. Another model used anesthesized mice dipped into hot water (70°C for 5 sec) and sustaining burns over 33% of their total body surface (Millican et al., 1966). While the burn itself caused little mortality, 90% of the animals were dead 48 hr after intraperitoneal injection of 10^6 *P. aeruginosa*; only 10% of control mice died under similar experimental conditions. Burns to the rumps of mice (animals immersed for 3—5 sec in 70°C water to a level slightly above their genitalia) were used by Markley and Smallman (1968). When mice burned in this manner had their rumps dipped into a culture containing 10^7 *P. aeruginosa*, a 90% mortality occurred in 3 days.

An alternative to body burns was described by Rosenthal (1966). In this model the tails of anesthesized mice were dipped into 70°C water for 5 sec. The time of dipping and the water temperature could be varied to change the extent of injury. When the burned tails were dipped into a 1:1 dilution of an overnight *Pseudomonas* culture, adjusted to an optical density of 0.7 at 660 μm, most animals died by day 7 postinfection.

Walker and Mason (1968) described a standard burn produced in rats. This model used a hollow, cylindrical burning device fashioned with an aperture of precise dimensions which allowed a burn of defined size to be made on the shaven back of anesthesized rats strapped into a chamber. The burn was made by dipping the exposed back of the rat into boiling water for various periods of time.

While most of these models were developed to test the efficacy of various treatment, some have been used to assess the role of virulence factors in infections due to *P. aeruginosa*. Kusama and Suss (1972) used the Rosenthal model (1966) in their studies on the vascular permeability factor of *P. aeruginosa*, and this same model was used to determine the protective effects of immunization with various *Pseudomonas* exoproducts on mortality in burned mice (Kawaharajo and Homma, 1977). The Walker—Mason (1968) scalded rat model has been used to evaluate the role of exotoxin A in burn wound sepsis (Walker et al., 1979), and role of motility in the virulence of P. aeruginosa was first described using this system (McManus et al., 1980).

Whereas any of these burned animal models can be used to determine the role of various Pseudomonas virulence-associated products in the pathogenesis of these infections, the large inocula required to infect the animals are of concern to some. It is felt that burn patients are initially colonized with low number of bacteria, and only after these initial inocula have increased to very large numbers does burn wound sepsis occur. Indeed, the elegant study of Teplitz et al. (1964) showed no systemic invasion to occur until the numbers of P. aeruginosa infecting a burn wound was more than 10^5 per gram wet weight skin. This relation between Pseudomonas numbers in skin and burn wound sepsis has been confirmed repeatedly in Pseudomonas-infected patients. Therefore models in which the size of the challenging inoculum far exceeds 10^5 lose some of their clinical relevance.

McRipley and Garrison (1964, 1965) showed that there was a highly increased susceptibility of burned rats to low-dose Pseudomonas challenge. Using a metal cylinder, with one side molded in the general shape of a rat back and the other end attached to a steam generator to cause a scald burn, they demonstrated that rats thus burned became exquisitely susceptible immediately postburn to lethal infection with small numbers of P. aeruginosa. This was true whether the animals were challenged subcutaneously or intraperitoneally. This enhanced susceptibility to infection was transient, with LD_{50} values returning to those in unburned rats by 1—5 days, depending on the route of challenge (Table 2).

Stieritz and Holder (1975) demonstrated a similar enhanced susceptibility to low-dose Pseudomonas challenge in burned mice, using a burning procedure previously described (Holder and Jogan, 1971). This procedure involves pressing an asbestos template with a 1 X 1.5 in. window against the shaved back of an anesthetized mouse. Ethanol is spread evenly over the area of the back outlined by the window, ignited, and allowed to burn for 10 sec. This causes a partial-thickness burn of approximately 30% of the total body surface area which, with 0.5-ml saline given intraperitoneally as fluid resuscitation, is a nonlethal burn. Mice burned in this manner are highly susceptible to fatal infections by subcutaneous challenge of P. aeruginosa in the burn site (Table 3). In contrast to the data of McRipley and Garrison (1964) using rats, burned mice infected by routes other than subcutaneously in the burn site require large challenge doses to be lethal. While these LD_{50} values are higher than those obtained by subcutaneous challenge, they are lower than LD_{50} values in unburned mice for the same route of infection. Of interest is the fact that the extremely enhanced subcutaneous susceptibility to infection of mice burned in this manner is restricted to challenge with P. aeruginosa (Table 4). No difference was found in LD_{50} values between burned

Table 2 Effect of Burns on Susceptibility of Rats to *P. aeruginosa*

Route of injection	Burned rats		Normal rats LD_{50}[a]
	Time of challenge (post burn)	LD_{50}[a]	
Subcutaneous (normal area)	2 min	<10 (100)	2.0×10^9 (45)
	1 day	1.5×10^7 (60)	
Subcutaneous (burned area)	2 min	<10 (200)	
	1 day	<10 (200)	
	5 days	1.3×10^7 (60)	
	10 days	2.0×10^8 (50)	
	15 days	1.5×10^9 (40)	
Intraperitoneal	2 min	<10 (100)	5.6×10^7
	1 day	1.9×10^5 (90)	
	3 days	3.0×10^7 (60)	

[a]Figures in parentheses indicate the numbers of rats used for determination of the LD_{50}.

Table 3 LD_{50} Values of *Pseudomonas aeruginosa* (Strain M-2) in Normal Mice and Burned Mice Challenged Immediately After the Burn by Various Routes

Route of injection	Normal mice	Burned mice
Subcutaneous	—	<10
Burned skin	1.3×10^6	4.0×10^4
Unburned skin	6.5×10^6	1.5×10^3
Intraperitoneal	6.5×10^7	1.5×10^3
Intravenous	1.3×10^7	1.9×10^3

and unburned mice challenged with a variety of other organisms. Quantitiative bacterial counts in burned skin, blood, and organs taken at various times after injection of 100 *P. aeruginosa* into the burned site indicated that organisms were not found in the blood or major organs until counts in the skin exceeded 10^5/per gram wet weight tissue. These data show that the progression of infection from burned skin to systemic invasion follows the same course defined by Teplitz et al. (1964) for burn wound sepsis. Thus this burned

Table 4 The LD_{50} by Subcutaneous Challenge in Normal Mice and Burned Mice Immediately After the Burns

Organism, strain	Normal mice	Burned mice
Pseudomonas aeruginosa		
M-2	1.3×10^6	<10
U.M.	1.1×10^6	<10
M-6	—	<10
PAO	—	<10
Escherichia coli	$>1.5 \times 10^8$	1.5×10^7
Klebsiella species	9.1×10^7	1.2×10^8
Staphylococcus aureus	$>1.8 \times 10^8$	$>1.8 \times 10^8$
Candida albicans	$>4.7 \times 10^6$	$>4.7 \times 10^6$

mouse model appears to be both clinically relevant and unique to the study of P. aeruginosa pathogenesis in burns. The model has been used to provide both indirect (Stieritz and Holder, 1979) and direct (Saelinger et al., 1977; Pavlovskis et al., 1978) evidence for exotoxin A production in the course of P. aeruginosa infections. Moreover, it has been used to assess protection against infection in mice immunized with exotoxin A (Pavlovskis et al., 1981).

Elastase and alkaline protease also were shown to be virulence-associated factors in Pseudomonas infections using this model. Snell et al. (1978) and Holder and Haidaris (1979) demonstrated enhanced mortality and increased depletion of liver EF-2 in mice infected with a protease-deficient strain supplemented with sublethal injections of proteases, as compared to groups of burned mice infected with the same strain without protease supplementation. Enhanced survival and protection of liver EF-2 was seen when the proteolytic activity of the Pseudomonas enzymes, produced in vivo, was neutralized by local treatment (subcutaneous in the burned site) with the protease inhibitor α 2-macroglobulin. The LD_{50} values of wild-type, protease-producing strains of P. aeruginosa were significantly lower than those of elastase-deficient mutants when tested in this model (Pavlovskis and Wretlind, 1979). The effects on mortality of both active immunization with exoproduct toxoids or passive immunization using anti-exoproduct serum were tested using this model, also (Okada et al., 1980; Cryz et al., 1983a). Studies testing antibody against P. aeruginosa contained in immune globulins prepared for intravenous use in humans were done in this model (Pollack, 1983; Holder and Naglich, 1984). Furthermore, protection against mortality by immunization with high molecular weight Pseudomonas polysaccharides was tested in this model (Pollack et al., 1984).

The in vivo elaboration of exoenzyme S was shown to occur in this model (Iglewski et al., 1978) and studies confirming the relation between virulence and motility and testing the efficacy of flagella "vaccine" were done in this model (Holder et al., 1982; Montie et al., 1982). The model has been used to evaluate the efficacy of topical antimicrobial therapy in the treatment of Pseudomonas-infected burn wounds as well (Stieritz et al., 1982).

Recently another burned mouse procedure was described by Collins and Roby (1983) in studies testing the anti-P. aeruginosa activity of an intravenous human immunoglobulin preparation. In this model an asbestos cloth with a 2 X 4 cm oval "window" cut in it is placed over the shaven dorsum of an anesthesized mouse. The burn is made by subjecting the area of the mouse's back exposed by the window to a 5-sec burn using a Fisher gas burner. A procedure similar to this had been described previously for rats (Howerton and Kolmen, 1972).

Any of the burned rodent models described above can be used to determine the role of various exoproducts in the pathogenesis of Pseudomonas infections.

OTHER MODELS

Pseudomonas aeruginosa infections cause mortality and morbidity in a wide variety of populations other than burn patients. *Pseudomonas* pneumonia is common in patients debilitated because of some underlying disease process, and such infections continue to be the principal cause of severe, ultimately life-threatening chronic pulmonary infections in patients with cystic fibrosis. *Pseudomonas* infections are one of the leading bacterial causes of human corneal ulcers and are common nosocomial infections in granulocytopenic patients. Because of this, models have been devised to study the roles of *Pseudomonas* virulence factors in these different infection processes. The next section outlines some models that have been described to try to mimic the pathogenesis of *Pseudomonas* infections in these various patient groups.

Acute Lung Infection Models

In 1968 Southern et al. described a complex apparatus in which a large group of mice could receive an aerosol challenge of *P. aeruginosa* at the same time. The apparatus consists of 66 tapered cylindrical tubes where the small end of the cylinders enter into a common aerosolizing chamber. The mice are placed in the cylinders with their heads toward the narrow end of the tube. Stoppers inserted into the tubes, behind the mice, ensure an airtight chamber and forces the noses of the mice toward the exposure chamber. Mice are then exposed to 30 min of *P. aeruginosa* aerosol. Quantitation of the number of CFU in the lung tissue of mice sacrificed immediately after aerosolization showed that "although there was a variation in count of $1.5 \times \log_{10}$, the vast majority of counts are within $\log_{10}$ deviation." Therefore this apparatus could be used to provide a standardized lung challenge to many mice at the same time. Using this model, the same group (Southern et al., 1970) tested the pulmonary clearance of 14 strains of *P. aeruginosa*. Three patterns of clearance were seen: (a) multiplication of the organisms in lungs for 1 hr after nebulization, followed by progressive clearance; (b) relatively little change in bacterial clearance in lungs up to 4 hr after nebulization; and (c) progressive clearance. There was no correlation between clearance patterns and the production of various exoproducts or intraperitoneal LD_{50} values. Organisms producing type 1 clearance, however, produced significantly greater amounts of extracellular lecithinase, and this suggested that alterations in pulmonary surfactant by lecithinase may enable some *Pseudomonas* strains to multiply in the lungs, predisposing to pneumonia.

A reliable method to produce acute, bilateral, hemmorrhagic pneumonia in guinea pigs was described by Pennington and Ehrie (1978).

The tracheas of anesthesized guinea pigs are surgically exposed, and a *P. aeruginosa* inoculum, in saline, is injected intratracheally using a syringe fitted with a 25-gauge needle. The trachea is then briefly occluded proximally to prevent coughing, after which the neck is sutured closed. This model was used to study the influence of prior immunosuppression, challenge strain variation, and inoculum size on survival and *Pseudomonas* killing in the lung (Pennington, 1979; Pennington and Miler, 1979). The majority of *P. aeruginosa* isolates from cystic fibrosis patients have a mucoid coat. The contribution of this coat to virulence in lung infections was assessed using this model (Blackwood and Pennington, 1981). No differences in pulmonary killing were found when mucoid and nonmucoid strains were used as challenge organisms. *Pseudomonas aeruginosa* strains that produced exotoxin A or elastase and strains deficient in the production of these products also were tested in this model (Blackwood et al., 1983). Strains deficient in elastase production were more easily cleared from the lung and appeared more virulent than strains that did not produce exotoxin A. These data suggested that elastase was an important virulence factor during acute pneumonia due to *P. aeruginosa*. Several studies had shown previously that purified *Pseudomonas* proteases or *Pseudomonas* culture filtrates containing protease caused hemorrhagic lung lesions when instilled intranasally or into the tracheas of animals (Meinke et al., 1970; Kawaharajo et al., 1975a; Gray and Kreger, 1979).

Shimizu et al. (1976) reported a hemorrhagic pneumonia model using mink as the experimental animal. In this model a vinyl tube, 1.5 mm in diameter, is inserted into the animal's nostril, and the inoculum, in 0.5 ml of saline, is given via the tube syringe. Homma et al. (1978) used this model to test the effectiveness of a multifactorial vaccine in preventing hemorrhagic pneumonia. The multifactor vaccine containing toxoids of elastase and alkaline protease, as well as original endotoxin protein of *P. aeruginosa*, was compared with a vaccine containing only original endotoxin protein. The mink vaccinated with the multifactor vaccine were resistant to 100 times the amount of bacterial challenge compared to the original endotoxin protein-immunized groups. In 1979 Aoi et al. used the multifactor vaccine, shown to be protective in the mink hemorrhagic pneumonia model, to immunize animals on mink farms experiencing epidemics of hemorrhagic pneumonia. Significant decreases in mortality rates in vaccinated mink were observed compared to unvaccinated groups, and considerable economic loss to the mink farmers was prevented.

Chronic Lung Infection Model

A rat model of chronic *P. aeruginosa* respiratory infection was reported by Cash et al. (1979). In this model, *P. aeruginosa*, at a final concentration of 10^6 organisms per ml, is prepared in melted agar maintained at 50°C. Then 10 ml of the agar-bacteria mixture are added, with stirring, to heavy mineral oil at 50°C, after which the oil—agar mixture is rapidly cooled by placing ice around the vessel. During the cooling procedure the agar droplets solidify into beads. After a series of washes to remove the oil, the loosely packed bacteria-containing agar beads are suspended in phosphate-buffered saline. For infection, the trachea of anesthesized rats are surgically exposed, the trachea incised, and 0.05 ml of the bead suspension placed in a distal bronchus using a bead-tipped curved needle. After instillations of 10^4 organisms in agar beads, quantitative cultures of lung tissue showed that the numbers per lung had reached 10^6 by day 3 and remained at this level for 35 days. While there was some colonization of the kidney when mucoid strains were used for the agar bead inoculum, there was no gross kidney pathology. None of the animals died when this type of respiratory infection was created. Histological examinations of the infected lungs showed lesions characteristic of those seen in humans with acute or chronic P. aeruginosa pneumonia. Using mutant strains of P. aeruginosa deficient in the production of single virulence factors, Woods et al. (1982) demonstrated that both exotoxin A and elastase production are required for maximum virulence of Pseudomonas in the chronic lung infection model. This is in sharp contrast to the fact that elastase, but not exotoxin A, serves as a virulence factor in acute P. aeruginosa pneumonia (Blackwood et al., 1983). The "agar bead" model was used by Pennington et al. (1981) to show that, while immunization with lipopolysaccharide antigens might confer protection against subsequent Pseudomonas lung infection, it was without benefit if the immunization was carried out during established infection. Immunization with a "cell surface" vaccine was shown to prevent histological changes in the lung of rats challenged with the bacteria—agar bead when compared to the pathology observed in nonimmunized counterparts (Klinger et al., 1983). However, bacterial counts in the lungs of immunized and nonimmunized infected rats were equivalent, causing the authors to speculate that the vaccine may have contained not only components of cell surface proteins, but virulence exoproducts as well.

Corneal Ulcer Models

Pseudomonas aeruginosa is one of the leading bacterial causes of central ulcers in the human cornea. After trauma these infections proceed on a rapid and destructive course, usually resulting in permanent corneal damage or loss of vision. Because of the antimicrobial resistance patterns of Pseudomonas, these infections are difficult to treat. In order to understand the mechanisms by which P. aeruginosa exerts its pathogenicity for eye tissue and to try to discover alternate methods of treatment, a variety of animal models of corneal infection have appeared in the literature. Models have been described where either purified exoproducts and/or exoproduct-containing fluids are injected into the eye or where actual P. aeruginosa eye infections are established.

Models Which Use Purified Exoproducts

Fisher and Allen (1958) showed degradation of rabbit cornea homogenates when incubated with a partially purified fraction of a P. aeruginosa culture filtrate having proteolytic activity. Furthermore, intracorneal inoculation of this enzyme-containing fluid into the eyes of rabbits caused extensive tissue damage, resulting in perforation of the cornea. Kessler et al. (1977) also used rabbit corneal homogentates to test the effects of Pseudomonas protease. They showed that, while purified protease had little collagenolytic acitivty, there was extensive degradation of the protein backbone of the corneal proteoglycans.

The effect of purified alkaline protease and elastase were tested in a mouse eye model in which the central area of the cornea of the anesthesized mouse was incised three times with a needle (Kawaharajo et al., 1974). One drop (0.01 ml) of enzyme-containing fluids was then deposited on the traumatized eye. Kawaharajo et al. demonstrated that small amounts (0.8–2 µg) of either enzyme caused corneal opacity, with ulcer formation occurring when higher doses were used. Similar histological changes in the cornea were observed when either enzyme was used.

Kreger and Gray (1978) injected purified Pseudomonas protease preparations intracorneally into rabbits and followed corneal pathology by histological and electron-microscopic methods. In some cases the epithelial layer of the corneas was surgically removed just before the intracorneal injections were given. A series of pathological events, some starting as early as 4–6 hr after the intracorneal injection of submicrogram amounts of protease, were shown to occur. The results of this study supported other data which showed that proteases elicited corneal damage by causing the destruction of the proteoglycan ground substance.

Ocular damage produced by exotoxin A was studied in adult mice in which the corneas were damaged by making three nonpenetrating wounds on the corneal surface (Hazlett et al., 1982a,b). Some of the damage observed microscopically was similar to that seen in Pseudomonas eye infections, but there were differences: for example, infection induced a more purulent exudate with higher and more rapid polymorphonuclear leukocyte infiltration; exotoxin A produced cataract of the ocular lens, whereas infection did not. A relation between exotoxin A ocular challenge and the age of mice has been established

(Berk et al., 1981a,b). Mice up to 10 days old showed some lethal effects when small amounts of toxin (up to 0.49 µg) were injected under the fused eyelids. In mice that were 16 days or older and whose eyelids were open, no lethality was observed when up to 15 µg of toxin were applied topically; however, cataract formation was common in eyes that had received toxin. Higher amounts of toxin were necessary to cause cataracts as the age of the mice increased to 30 days. Hazlett et al. (1978) demonstrated that some mouse strains are naturally resistant to intracorneal infections by *P. aeruginosa*. They used the model in which exotoxin A was injected beneath the fused eyelids of mouse pups, following ocular damage histologically (Hazlett et al., 1981). In contrast to infection where lysis of the ocular epithelium occurs within minutes, it took toxin several hours to induce lytic changes. Furthermore, the phagocytic cellular response observed in eye infections was absent in the toxin-treated eyes. However, Kreger and Griffen (1974) had shown previously that injections of sterile culture filtrates of a toxin-producing but nonprotease-producing strain of *P. aeruginosa* into rabbit corneas caused no damage at all. These findings suggest that the animal species employed affects the results.

Eye Infection Models

Because earlier studies suggested that corneal destruction in *P. aeruginosa* eye infections might have resulted from the proteolytic enzymes produced in vivo by this organism, Wilson (1970) used an eye infection model to test the therapeutic effects of the protease inhibitor disodium ethylenediamine tetracetic acid (Na_2EDTA). In this model, rabbit corneas are simultaneously damaged and inoculated by having a silk suture, dipped in an overnight culture of *P. aeruginosa*, run through the superficial stroma. Antibiotics were used to treat all infected eyes and, in some, Na_2EDTA was also applied. All animals inoculated by this method showed evidence of corneal infections between 18 and 20 hr postchallenge. By the seventh day, 15 of the 24 corneas infected had perforated in the group treated with antibiotics alone. This occured in spite of the fact that all eye cultures were negative by day 3. Eyes treated with Na_2EDTA were resistant to the necrosis observed when only antibiotics were used as treatment. The continuation of eye damage, in spite of negative eye cultures, supported a role for extracellular products in the pathological events seen in this infectious process. Gerke and Magliocco (1971) described an eye infection model in which the cornea is incised by the use of a 26-gauge needle prior to inoculation with bacterial culture. Alternatively, they also inoculated live cultures into the corneal stroma of the unwounded eye. Various parameters were assessed as the infection progressed. The series of pathological events observed in this model were similar to those seen in humans and other animals. These authors established a system of grading the pathology of the eye as the infection progressed which should be useful to others studying this type of infection.

A similar mouse incised cornea model was used by Kawaharajo and Homma (1975) to test the effects of inoculation of the eye with protease-producing and non-protease-producing strains of *P. aeruginosa*. The histological destruction, enlargement, cellular infiltration, and ulceration seen when protease-producing strains were used were absent in corneas infected with protease-negative organisms. Toxin A mutants and an elastase mutant of *P. aeruginosa* were tested in the mouse incised eye model and the damage caused by these strains were compared to damage caused by their wild-type parent strains (Ohman et al., 1980). The results of this study indicated that while exotoxin A was not involved in the establishment of corneal infections, it did appear necessary for maintenance of the infection; toxigenic strains persisted in the eye for considerably longer periods of time than did the nontoxin mutants. Corneal damage occurred if the challenge strain produced alkaline protease; elastase production was not necessary. The influence of iron concentration in the medium in which strains are grown before being inoculated into the eye was assessed using this model (Woods et al., 1982). Strains cultured in high-iron containing media were less virulent than strains grown in low-iron media. This same model was used to test the therapeutic affects of active immunization using a multifactor vaccine or passively transferred gammaglobulin from rabbits immunized with this vaccine (Kawaharajo and Homma, 1976; Hirao and Homma, 1978).

The ability of *P. aeruginosa* to adhere to injured corneas in the mouse incised cornea model was compared to the adherence observed in uninjured corneas. The results of the study showed that *P. aeruginosa* preferentially adheres to the injured cornea and that adherence increases with time of contact with the cornea (Ramphal et al., 1981). The role of complement in mouse corneal infections caused by *P. aeruginosa* was investigated using this model (Cleveland et al., 1983). In this case, however, rather than using strains of bacteria with different capacities for exoproduct production, mouse strains that were normocomplementeric or complement deficient (C'5) were used. The study showed that C'5 plays no role in susceptibility or resistance to *Pseudomonas* eye infection. On the other hand, normocomplementeric mice, depleted of C'3 by cobra venom factor, had a diminished ability to restore a clear cornea after *Pseudomonas* infection.

Neutropenic Animal Models

Since patients rendered neutropenic as the result of either underlying malignancy or treatment with immunosuppressive agents are highly susceptible to *P. aeruginosa* infections, animal models have been designed to mimic these infections. Rats, made neutropenic by intraperitoneal injection with cyclophosphamide (100 mg/kg, day 0; 75 mg/kg, day 4) were challenged intraperitoneally, on day 5 with a *Pseudomonas* culture in 5% mucin (Lumbush and Norden, 1976). The effects of antibiotic treatments on the infections thus created, were tested in this model. The mortality rate for untreated neutropenic rats was 90—98%, depending on the size of the challenge dose. Another model in which cyclophosphamide was used to cause leukopenia employed mice as the experimental animal (Cryz et al., 1983b). In this model 150 µg of cyclophosphamide per gram mouse weight was injected intraperitoneally on days 0, 2, 4. Anesthesized mice were challenged on day 4 by a 5-µl inoculum placed directly into a 0.5 cm surgical incision made on the shaved back of the mouse. When two strains of *P. aeruginosa* were tested for the ability to cause mortality in mice rendered leukopenic, a challenge dose of 10^1 CFU caused 83—100% mortality; no nonleukopenic mice died using a challenge dose of 10^5 CFU. The enhanced susceptibility to infection of the leukopenic mice was restricted to *P. aeruginosa* challenge; *Staphylococcus aureus* and *Klebsiella pneumoniae* could not substitute for the *Pseudomonas*.

Ziegler and Douglas (1979) made rabbits neutropenic by the injection of nitrogen mustard mechlorethamine (3.0 mg/kg) into the marginal ear vein. A total of 72 hr after nitrogen mustard treatment the rabbits were challenged by various dilutions of *P. aeruginosa* instilled into the conjunctival sac, gentle rubbing of the lower eyelid being used to facilitate absorbtion. Of 28 rabbits inoculated (10^8 CFU), 22 developed severe conjunctivitis and hemorrhagic necrosis of the lower eyelid and 82% died of bacteremia. No nonneutropenic control animal treated in a similar manner developed conjunctivitis, and there was no mortality in this group. Of 26 patient isolates tested, 12 were virulent, causing death in 50—100% of the experimental groups. There appeared to be a correlation between *P. aeruginosa* strains isolated from patients' blood and virulence, but proteolytic enzyme production alone did not account for the virulence of these strains. *Escherichia coli* or *K. pneumoniae* could not substitute for *P. aeruginosa* in this model.

EPILOGUE

Pseudomonas aeruginosa causes severe infections in a variety of patients. Because of this, investigators have devised models in which to study the role(s) of *Pseudomonas* virulence-associated factors in the pathogenesis of these infections. The preceding pages have described some, but by no means all, of the models used to study the various infectious processes.

Taken as a whole, the results cited above, in which animal models are used to study *Pseudomonas* virulence factors, indicate that the pathogenesis of these infections is exceedingly complex and that the role played by any one virulence factor may not be simple. The use of models can greatly facilitate an understanding of the processes involved; however, investigators who use animal models must use care. We have seen that the species of animal, the strain used, and the age of the animal can affect the results obtained. Therefore broad generalizations based on the results obtained from experiments using any one animal model must be viewed with caution. This caveat applies to anyone using animal models for whatever purpose.

ACKNOWLEDGMENTS

The efforts of Catharine B. Saelinger and Herman C. Lichstein in reviewing this manuscript are acknowledged with gratitude.

REFERENCES

Aoi, Y., Noda, H., Yanagawa, R., Homma, J., Abe, C., Morihara, K., Goda, A., Takeuchi, S., and Ishihara, T. (1979). Jpn. J. Exp. Med. 49:199.

Armstrong, A. V., Stewart-Tull, D. E. S., and Roberts, J. S. (1971). J. Med. Microbiol. 4:249.

Atik, M., Liu, P. V., Hanson, B. A., Amini, S., and Rosenberg, C. F. (1968). J. Am. Med. Assoc. 205:84.

Bartell, P. V., Orr, T. E., and Garcia, M. (1968). J. Infect. Dis. 118:165.

Berk, R. S., Beisel, K., and Hazlett, L. D. (1981a). Infect. Immun. 34:1.

Berk, R. S., Iglewski, B., and Hazlett, L. (1981b). Infect. Immun. 33:90.

Blackwood, L. ., and Pennington, J. (1981). Infect. Immun. 32:443.

Blackwood, L. L., Stone, R., Iglewski, B., and Pennington, J. (1983). Infect. Immun. 39:198.

Brown, M., and Foster, J. (1970). J. Clin. Pathol. 23:172.

Callahan, L. T. (1974). Infect. Immun. 9:113.

Callahan, III, L. T. (1976). Infect. Immun. 14:55.

Carney, S. A., and Jones, R. J. (1968). Br. J. Exp. Pathol. 49:395.

Carrick, L., and Berk, R. S. (1975). Biochim. Biophys. Acta. 391:422.

Cash, H. A., Woods, D., McCullough, B., Johanson, Jr., W., and Bass, J. (1979). Am. Rev. Respir. Dis. 119:453.

Cicmanec, J. F., and Holder, I. A. (1979). Infect. Immun. 25:477.

Cleveland, R. P., Hazlett, L., Leon, M., and Berk, R. (1983). *Invest. Ophthalmol. Vis. Sci.* 24:237.
Collins, M. S., and Roby, R. (1983). *J. Trauma* 23:530.
Craven, R. C., and Montie, T. C. (1981). *Can. J. Microbiol.* 25:458.
Cryz, S., and Iglewski, B. H. (1980). *J. Clin. Microbiol.* 12:131.
Cryz, Jr., S., Furer, E., and Germanier, R. (1983a). *Am. J. Microbiol.* 39:1072.
Cryz, Jr., S., Furer, E., and Germanier, R. (1983b). *Infect. Immun.* 39:1067.
Diener, B., Carrick, Jr., L., and Berk, R. (1973). *Infect. Immun.* 7:212.
Esselmann, M., and Liu, P. V. (1961). *J. Bacteriol.* 81:939.
Fisher, Jr., E., and Allen, J. H. (1958). *Am. J. Ophthalmol.* 46:249.
Gerke, J. R., and Magliocco, M. (1971). *Infect. Immun.* 3:209.
Gill, D. M., and Dinius, L. L. (1973). *J. Biol. Chem.* 248:654.
Gray, L., and Kreger, A., (1979). *Infect. Immun.* 23:150.
Hazlett, L. D., Rosen, D. D., and Berk, R. S., (1978). *Infect. Immun.*, 20:25.
Hazlett, L. D., Berk, R., and Iglewski, B. (1981). *Infect. Immun.* 34:1025.
Hazlett, L. D., Iglewski, B., and Berk, R. (1982a). *Ophthalmic Res.* 14:401.
Hazlett, L. D., Wells, P., and Berk, R. (1982b). The Mouse Cornea as a Model for Pseudomonas Infection: Further Studies on Penetration and Lysis of the Unwounded Eye, in *The Structure of the Eye* (J. G. Holly field, ed.), Elsevier—North Holland, New York, pp. 279—296.
Hirao, Y. and Homma, J. (1978). *Jpn. J. Exp. Med.* 48:41.
Holder, I. A. (1983). *Rev. Infect. Dis.* 5:S914.
Holder, I. A., and Haidaris, C. G. (1979). *Can. J. Microbiol.* 25:593.
Holder, I. A., and Jogan, M. (1971). *J. Trauma* 11:1041.
Holder, I. A., and Naglich, J. (1984). *Am. J. Med.* 76(3A):161.
Holder, I. A., Wheeler, R., and Montie, T. (1982). *Infect. Immun.* 35:276.
Homma, J. Y., Abe, C., Tanamoto, K., Hirao, T., Morihara, K., Tsuzuki, H., Yanagawa, R., Honda, E., Aoi, Y., Fujimoto, Y., Goryo, M., Imazeki, N., Noda, H., Goda, A., Takeuchi, S., and Ishihara, T. (1978). *Jpn. J. Exp. Med.* 48:111.
Howerton, E. E., and Kolmen, S. (1972). *J. Trauma* 12:335.
Iglweski, B. H., and Kabat, D. (1975). *Proc. Nat. Acad. Sci. U.S.A.* 72:2284.
Iglewski, B. H., Liu, P. V., and Kabat, D. (1977). *Infect. Immun.* 15:138.
Iglewski, B. H., Sadoff, J., Bjorn, M. J., and Maxwell, E. S. (1978). *Proc. Nat. Acad. Sci. U.S.A.* 75:3211.

Janda, J. M., and Bottone, E. J. (1981). *J. Clin. Microbiol.* 14: 55.
Janda, J. M., Atang-Nomo, S., Bottone, E. J., and Desmond, E. P. (1980). *J. Clin. Microbiol.* 12:626.
Johnson, G. G., Morris, J. M., and Berk. R. S. (1967). *Can. J. Microbiol.* 13:711.
Kawaharajo, K., and Homma, J. (1975). *Jpn. J. Exp. Med.* 45:515.
Kawaharajo, K., and Homma, J. Y. (1977). *Jpn. J. Exp. Med.* 47: 495.
Kawaharajo, K., Abe, C., Homma, J., Kawano, M., Gotoh, E., Tanaka, N., and Morihara, D. (1974). *Jpn. J. Exp. Med.* 44: 435.
Kawaharajo, K., and Homma, J. (1976). *Jpn. J. Exp. Med.* 46:155.
Kawaharajo, K., Homma, J. Y., Aoyama, Y., Okada, K., and Morihara, K. (1975a). *Jpn. J. Exp. Med.* 45:79.
Kawaharajo, K., Homma, J. Y., Aoyama, Y., and Morihara, K., (1975b). *Jpn. J. Exp. Med.* 45:89.
Kessler, E. H., Kenneth, E., and Brown, S. (1977). *Invest. Ophthalmol. Vis. Sci.* 16:488.
Klinger, J. D., Cash, H., Wood, R., and Miler, J. (1983). *Infect. Immun.* 39:1137.
Klyhn, K. M., and Gorrill, R. H. (1967). *J. Gen. Microbiol.* 47: 227.
Kobayashi, F. (1971). *Jpn. J. Microbiol.* 15:295.
Kreger, A. S., and Gray, L. (1978). *Infect. Immun.* 19:630.
Kreger, A. S., and Griffin, A. (1974). *Infect. Immun.* 9:828.
Kunitz, M. (1946/1947). *J. Gen. Physiol.* 30:291.
Kusama, H., and Suss, R. H. (1972). *Infect. Immun.* 5:363.
Leppla, S. H. (1976). *Infect. Immun.* 14:1077.
Liedberg, C. F. (1961). *Acta Chir. Scand.* 120:88.
Liu, P. V. (1966a). *J. Infect. Dis.* 116:112.
Liu, P. V. (1966b). *J. Infect. Dis.* 116:481.
Liu, P. V. (1979). Toxins of *Pseudomonas aeruginosa*, in *Pseudomonas aeruginosa: Clinical Manifestations of Infection and Current Therapy* (R. G. Doggett, ed.), Academic, New York, p. 63–85.
Liu, P. V., Abe, Y., and Bates, J. L. (1961). *J. Infect. Dis.* 108: 218.
Liu, P. V., Yoshii, S., and Hsieh, H. (1973). *J. Infect. Dis.* 128: 514.
Lumbush, R. M., and Norden, C. (1976). *J. Infect. Dis.* 133:538.
McManus, T. T., Moody, E. E., and Mason, A. D. (1980). *Burns* 6:235.
McRipley, R. J., and Garrison, D. W. (1964). *Proc. Sco. Exp. Biol. Med.* 115:336.
McRipley, R. J., and Garrison, D. W. (1965). *J. Infect. Dis.* 115: 159.

Markley, K., and Smallman, E. (1968). *J. Bacteriol.* 96:867.
Meinke, G., Barum, J., Rosenberg, B., and Berk, R. (1970). *Infect. Immun.* 2:583.
Michael, M., and Saelinger, C. B. (1979). *Curr. Microbiol.* 2:103.
Middlebrook, J. L., and Dorland, R. B. (1977a). *Can. J. Microbiol.* 23:175.
Middlebrook, J. L., and Dorland, R. B. (1977b). *Can. J. Microbiol.* 23:175.
Middlebrook, J. L., and Dorland, R. B. (1977c). *Can. J. Microbiol.* 23:183.
Millican, R. C., Evans, G., and Markley, K. (1966). *Ann. Surg.* 163:603.
Moehring, J. M., and Moehring, T. J. (1983). *Infect. Immun.* 4:998.
Montie, T. C., Doyle-Huntzinger, D., Craven, R., and Holder, I. A. (1982). *Infect. Immun.* 38:1296.
Morihara, K. (1964). *J. Bacteriol.* 88:745.
Morihara, K., Tsuzuki, H. T., Oka, T., Inoue, H., and Ebata, M. (1965). *J. Biol. Chem.* 240:3295.
Nonoyama, S. H., Kojo, H., Mine, Y., Nishida, M., Goto, S., and Kuwahara, S. (1979). *Infect. Immun.* 24:399.
Ohman, D. E., Burns, R., and Iglewski, B. (1980). *J. Infect. Dis.* 142:547.
Okada, K., Kawaharajo, K., Kasai, T., and Homma, J. (1980). *Jpn. J. Exp. Med.* 50:63.
Pavlovskis, O. R., and Gordon, F. B. (1972). *J. Infect. Dis.* 125:631.
Pavlovskis, O. R., and Shackelford, A. H. (1974). *Infect. Immun.* 9:540.
Pavlovskis, O. R., and Wretlind, B. (1979). *Infect. Immun.* 24:181.
Pavlovskis, O. R., Voelcker, F. A., and Shackelford, A. H. (1967). *J. Infect. Dis.* 133:253.
Pavlovskis, O. R., Iglewski, B., and Pollack, M. (1978). *Infect. Immun.* 19:29.
Pavlovskis, O. R., Erdman, D., Leppla, S., Wretlind, B., Lewis, L., and Martin, K. (1981). *Infect. Immun.* 32:681.
Pennington, J. E. (1979). *J. Infect. Dis.* 140:73.
Pennington, J. E. and Ehrie, M. G. (1978). *J. Infect. Dis.* 137:764.
Pennington, J. E., and Miller, J. J. (1979). *Infect. Immun.* 25:1029.
Pennington, J. E., Hickey, W., Blackwood, L., and Arnaut, M. (1981). *J. Clin. Invest.* 68:1140.
Pollack, M. (1983). *J. Infect. Dis.* 147:1090.
Pollack, M., Pier, G. B., and Prescott, R. K. (1984). *Infect. Immun.* 43:759.
Ramphal, R., McNiece, M., and Pollack, F. (1981). *Ann. Ophthamal.* 13:421.

Rinderknecht, H., Goekas, M. C., Silverman, P., and Haverback, B. J. (1969). *Clin. Chim. Acta.* 21:197.
Rosenthal, S. M. (1966). *Proc. Soc. Exp. Biol. Med.* 123:347.
Sachar, L. A., Winter, K. K., Secher, N., and Frankel. S. (1955). *Proc. Soc. Exp. Biol. Med.* 90:323.
Saelinger, C. B., Snell, K., and Holder, I. A. (1977). *J. Infect. Dis.* 136:555.
Scharmann, W. (1976a). *J. Gen. Microbiol.* 93:283.
Scharmann, W. (1976b). *J. Gen. Microbiol.* 93:292.
Sensakovic, J. W., and Bartell, P. F. (1974). *J. Infect. Dis.* 129:101.
Shimizu, T., Homma, J., Abe, C., Tanamoto, K., Aoyama, Y., Okada, K., Yanagawa, R., Fujimoto, Y., Noda, H., and Takashima, I. (1976). *J. Vet. Res.* 37:1441.
Sierra, G. (1961). *Antonie van Leeuwenhoek J. Microbiol. Serol.* 26:189.
Snell, K., Holder, I. A., Leppla, S. A., and Saelinger, C. B. (1978). *Infect. Immun.* 19:839.
Sokol, P. A., Ohman, D. E., and Iglewski, B. H. (1979). *J. Clin. Microbiol.* 9:538.
Southern, Jr., P. M., Pierce, A., and Sanford, J. (1968). *Appl. Microbiol.* 16:540.
Southern, Jr., P. M., Mays, B., Pierce, A., and Sanford, J. (1970). *J. Lab. Clin. Med.* 76:548.
Stieritz, D. D., and Holder, I. A. (1975). *J. Infect. Dis.* 131:688.
Stieritz, D. D., and Holder, I. A. (1979). *J. Med. Microbiol.* 11:101.
Stieritz, D. D., Bondi, A., McDermott, D., and Michaels, E. (1982). *J. Antimicrob. Chemother.* 9:133.
Teplitz, C., Davis, D., Mason, Jr., A. D., and Moncrief, J. A. (1964). *J. Surg. Res.* 5:200.
Vasil, M. L., and Iglewski, B. H. (1978). *J. Gen. Microbiol.* 108:333.
Vasil, M. L., Liu, P. V., and Iglewski, B. H. (1976). *Infect. Immun.* 12:1467.
Walker, H. L., and Mason, Jr., A. D. (1968). *J. Trauma* 8:1049.
Walker, H. L., McLeod, Jr., C. G., Leppla, S. H., and Mason, Jr., A. D. (1979). *Infect. Immun.* 25:828.
Weber, B., Nickol, M. M., Jagger, K. S., and Saelinger, C. B. (1982). *Can. J. Microbiol.* 28:679.
Wilson, L. A. (1970). *Br. J. Ophthalmol.* 54:587.
Woods, D. E., Sokol, P., and Iglewski, B. (1982). *Infect. Immun.* 35:461.
Woods, E. E., Cyrz, S., Friedman, R., and Iglewski, B. (1982). *Infect. Immun.* 36:1223.
Wretlind, B., and Wadstrom, T. (1977). *J. Gen. Microbiol.* 103:319.

Wretlind, B., Sjöberg, L., and Wadström, T. (1977). *J. Gen Microbiol.* 103:329.
Ziegler, E. J., and Douglas, H. (1979). *J. Infect. Dis.* 133:288.

9
Pseudomonas aeruginosa:
Serology, Phage, and Pyocin

CHARLES H. ZIERDT *Clinical Center, National Institutes of Health, Bethesda, Maryland*

INTRODUCTION

This chapter is intended to be more than simple instruction in typing techniques. The development of major categories of typing is followed from earliest inception to the current state of the art. Thorough references are provided, so that further reading is facilitated for research and development in the various subjects. By gleaning the reference sections of key papers a comprehensive literature is available to research workers and to those with special needs in a major area.

Need for Pseudomonas Typing

It is a truism for *Pseudomonas aeruginosa,* as for most pathogens, that some strains cause more infections than others. Cystic fibrosis is an example (Zierdt and Williams, 1975) where a particular strain of mucoid *P. aeruginosa* causes most of the infections. The purpose of typing is to identify strains within the species. The typing description of a strain serves as the fingerprint for that strain, by which it can be followed in its activities and travels between humans and their environment.

Typing of *P. aeruginosa* is desirable for best hospital care and will increase in importance as this adaptive and versatile bacterium continues to enlarge its sphere of influence through increasing and more diverse infections in patients.

It has been often asserted that typing should be invoked only when there is a need in the form of a very definite infectious outbreak within a hospital or community. This view is most often expressed when the speaker's laboratory lacks the expert staff to carry out typing. The flaw in this thinking is that the first indication of an infectious outbreak is more often than not learned from surveillance typing data. This has long been a belief of the author and is a conclusion of Cross et al. (1983) reporting a very serious outbreak of infections in a general military hospital. The outbreak was not realized or acted upon until strain typing was done, demonstrating the presence and transfer of epidemic strains of the bacterium. Without the confidence gained by knowledge of the epidemic strains and identification of employee and patient carriers, the epidemiology team did not convince the directors to initiate a stringent program of patient isolation, treatment of carriers, restriction of catheterization, glove policy, and enforced handwashing to slow the infections rate. With proof of epidemic bacterial strains linked to specific infections and fatal infections the need for action became compelling.

Feasibility of Typing Systems

Although all of the known typing systems are described in this chapter, it is not expected that all will be put to use in a single laboratory. Different laboratories have different needs. Technical help, if available, can be put to use on time-consuming bacteriophage typing, which for *P. aeruginosa* is by far the most sensitive typing system in delineating strain differences but at the same time has the highest requirement for technical expertise in performance and interpretation. Phage typing on paper is straightforward, but in practice responds only to the best talent in the laboratory. Research applications are most likely to benefit from phage typing. For example, when phage and serotyping are performed together, it becomes evident that within the confines of a ward outbreak of infections caused by strains of a single serotype those strains are likely to be a single phage type. However, *P. aeruginosa* strains of the same serotype from widespread outbreaks might well have different phage types from one outbreak to another, and even within the same outbreak.

It can easily be understood that strains separated in geography and in time may belong to the same serogroup, but have little else in common that is subject to mutation. The first variable quality that comes to mind might be pathogenicity. Serotype O:11, however, may be exceptional in the sense that it has in a few years spread rapidly around the world and remains a distinct and stable strain. This is due to its extraordinary virulence, which thus far shows no signs of diminishing. Serotype II strains are not seen commensurately in the environment unless seeded there by humans.

Hospital infections and hospital epidemiology committees are charged with surveillance of *P. aeruginosa*. A current file of typing reactions on the resident and visiting *P. aeruginosa* strains is a valuable tool to these committees. Typing confers to the committee the ability to trace a particular strain in the environment, the patients, and the staff and to determine whether a patient came into the hospital with a certain strain or if the strain was acquired nosocomially (within the hospital). It is important to know whether a patient's infection is of endogenous or exogenous origin, that is, from a strain that he carries on his body or a strain from another source. Most *P. aeruginosa* infections are exogenous.

Hospital epidemiology services and infection control committees may not initially recognize the great help that fingerprinting bacterial strains can be. Once familiar with a typing procedure and the information that it can supply, these groups regard typing results as essential to infections surveillance.

There are ancillary benefits to typing *P. aeruginosa*. In this research hospital a computer file of all typing results is maintained. Analysis of data routinely collected for many years has formed the basis for research reports, particularly for strains from cystic fibrosis patients and the flux of these strains in the wards. Such computer analysis has uncovered basic new epidemiological data for *P. aeruginosa* (Zierdt and Williams, 1975; Farmer et al., 1982) and *Staphylococcus aureus* (Zierdt et al. 1980; Zierdt 1982).

The epidemiology team is thus weeks or months ahead of the game when typing surveillance of stringently selected cultures is done routinely, rather than playing catch-up with an outbreak in full swing.

Background for Typing and Infections Control

Pseudomonas aeruginosa and *S. aureus* compete for the dubious honor of being the leading cause of hospital-acquired infections. The balance may change on the basis of what services the hospital provides, the relative size of those services within the hospital, the changing antibiotic resistance of resident bacterial strains, and, very important, the relative virulence of resident strains. During 1975 the balance of infections on the pediatric oncology wards in the Clinical Center went from *P. aeruginosa* to *S. aureus*, and the single reason for the reversal was the introduction into the hospital at that time of an extraordinarily virulent *S. aureus* strain (phage pattern 94/96/292). This strain was in the process of moving into U.S. hospitals at that time.

Pseudomonas aeruginosa has access to hospitals and to patients by routes unavailable to its chief competitor. It flourishes wherever there may be moisture and high humidity, even without the benefit of an obvious nutrient source (Holder, 1977). Chrysanthemums and other flowering plants in the patients' room harbor the organism, from where it has been traced directly to patients (Schroth et al., 1977). Salads, uncooked vegetables, and other foods served to patients frequently contain high numbers of *P. aeruginosa* (Kominos et al., 1977), and infections have been traced to these sources. Ensuing recommendations were that no live plants should be in the hospital rooms of patients at risk, and only cooked food should be served. Reservoirs of *P. aeruginosa* have been pinpointed throughout the hospital and traced directly to infections in patients. Disinfectant solutions of all types have been cited. Their killing power when freshly diluted and unused may be just adequate, except for quaternary ammonium disinfectants, which sometimes permit growth of *P. aeruginosa* even in freshly made solutions. Diminishing the disinfectant solution's power by volatization, combination with protein and other organic material, or degradation on standing often results in a culture medium for *P. aeruginosa* and subsequent challenge of the patient with very large numbers of the organism contaminating catheters, suction tubes, surgical and other instruments, oxygen tents, dressings, counter and table surfaces, respirometers, delivery room resuscitators, hydrotherapy tanks, ointments, and wet mopping equipment (Fierer et al., 1967; Holder, 1977; Moody, 1977; Schroth et al., 1977). Infections have resulted from growth in hospital water supplies, in sinks, and on faucets (Favero et al., 1971). The importance of sinks in spread of *P. aeruginosa* to patients was discounted by Chadwick (1976a,b), who considered patient-to-patient transmission more important. The relative importance of environmental, instrumental, and human-to-human spread depends heavily on individual hospital and ward problems, and particularly on the type of patient. Chadwick's view would certainly be corroborated on burn wards. Fecal transmission of *P. aeruginosa* was demonstrated by Shooter et al. (1966).

Burn patients provide a continued challenge in the control of *P. aeruginosa* growth and pathology on the burned area (Edmonds et al., 1972; MacMillan et al., 1973). Immunization, topical ointments, systemic antibacterials, debridement, and *P. aeruginosa* quantitation in eschar are the standard safeguards employed (Kominos et al., 1972; Lowbury et al., 1970). The great majority or, in some hospital units, all of the burn and cystic fibrosis patients become infected. Organ transplant patients (particularly kidney transplants) are at severe risk because of the use of immunosuppressive drugs (Leigh, 1971). *Pseudomonas aeruginosa* is the primary pathogen among acute leukemia patients (Reynolds et al., 1975) and solid tumor patients (Moody, 1977; Moody et al., 1971). Epidemics of *P. aeruginosa* infections are common in nurseries (Bobo et al., 1973; Fierer et al., 1967; Falcoa et al., 1972). Primary skin and ear infections, pneumonias, and urinary tract infections are becoming more frequently reported (Hoadley, 1977), particularly, but not exclusively, among swimmers. Two extensive *P. aeruginosa*-caused outbreaks in hospitals were reported by Ayliffe et al. (1965, 1966).

Earlier on, most reports on *P. aeruginosa* infections included two to three typing methods. These might be serotyping followed by pyocin typing, phage typing, or both (Kominos et al., 1977, Asheshov, 1974). The type of a strain is usually the only marker separating a strain that is identical in other measurable characteristics to other strains of the species. Another practical use of typing is that it provides species identification in a single test. Although the longer diagnostic process usually identifies the species, followed by typing, there are examples where typing provides single-test speciation, or serves as the final arbiter when other methods leave room for doubt. This is certainly an ancillary use of typing systems.

SEROLOGY

History

The first report on serotyping of *P. aeruginosa* was that of Klieneberger (1907). He used unheated antigens and the resulting sera were not absorbed. As a consequence he found a plethora of serotypes. Klieneberger used patient's sera to show elevated titers as proof of *P. aeruginosa* infection and to separate strains recovered from infection. Aoki (1926), using techniques similar to those of Klieneberger, reported 37 serotypes, with many cross-reactions, with unheated antigens. Subsequent attempts by Jacobsthal (1912) and Trommsdorff (1916) also used unheated antigens. It is interesting that Jacobsthal, even in 1912, recognized the importance of typing strains just after isolation rather than later after extensive subculture. In 1924 Brutsaert reported the use of boiled versus unheated cell suspensions and described increased specificity of boiled antigens, but Kanzaki (1934) returned to unheated antigens, with predictable results.

Pseudomonas aeruginosa differs from *Salmonella* in its response to formalin, alcohol, and heat. It became evident from a number of reports that unheated whole-cell antigens yielded sera with excessive cross-reactions that were not amenable to improvement by absorption with cross-reacting strains. Grün et al. (1967), Kleinmaier et al. (1958), Köhler (1957), Mayr-Harting (1948), Van Den Ende (1952), Ansorg (1978), and Ansorg et al. (1978), even in 1978, were using formalin intended to destroy H antigen. Earlier Munoz et al. (1949) and Gaby (1946) employed dilute formaldehyde in the belief that its actions would increase specificity and leave only O antigens. Gaby reported that sera from these formalized strains agglutinated all of the test strains, leading him to conclude that *P. aeruginosa* strains were serologically homogeneous. Unlike the situation in *Salmonella*, where H antigens are destroyed in dilute formaldehyde, leaving only somatic (O) antigens, the *P. aeruginosa* antigen mosaic is little affected. Indeed, an unheated but formalinized suspension of *P. aeruginosa* injected into a rabbit results in flagellar (H) antibody in high titer. Flagellar antibodies are strain specific and may be used for epidemiological study (Pitt, 1980). They comprise only four or five serotypes and therefore are most useful in extending strain-separating sensitivity of O-antigen typing (Pitt, 1980; Lanyi, 1970).

Boiled or autoclaved cells finally became standard in developing sets of typing antisera (Christie, 1948; Kleinmaier, 1957; Kleinmaier and Muller, 1958; Kleinmaier et al., 1958; Mayr-Harting, 1948). Habs' (1957) classic paper described 12 O-antigen types in P. aeruginosa. The Habs immunogen strains have since been widely distributed and have become the nucleus for most of the typing sets now in use, although there are those who continue to develop their own sets.

Veron (1961) added to Habs' set types O:2b, O:2ab, O:5cd; Sandvik (1960) added type O:13; Wahba (1965a) type O:14. Wokatsch (1964) added "thirteen newly listed O-groups." Smirnova (1976) recently described a new set of 20 O antisera from an immunogen strain set that he developed.

Recent Work

An ever-broadening front of attack on the complex P. aeruginosa cell has somewhat recovered time lost in decades of neglect while species of Enterobacteriaceae were the prime research subjects. Pseudomonas aeruginosa has proven to be "different" in many areas of molecular biology, including basic mechanisms of genetic control (Holloway and Krishnapillai, 1975). Three abundant antigens have been described from water-washed whole P. aeruginosa cells (Tunstall and Gowland, 1975; Hobbs et al., 1964). The first is an abundant, strain-specific, high molecular weight protein comprising the loose outer layer of the organism. The second is common to all strains, a polysaccharide or lipopolysaccharide, lying beneath the first layer. The third is the mucopeptide of the cell wall, also common to all strains.

New approaches have shed much light on the antigenic complexity of P. aeruginosa, taking advantage of modern techniques combining precipitation reactions with electrophoresis. Lányi (1966–1967) and Lányi et al. (1975) used extracts of boiled cell suspensions prepared from his O-antigen typing series to immunize rabbits. The soluble antigens and antisera were then combined in gel electrophoresis. The resulting precipitation lines (limited to one or two) proved to be type specific, with one exception.

Hoiby (1975a,b,c; 1977) and Hoiby and Axelsen (1973) prepared a standard unheated antigen (StAg) by ultrasound homogenization of four strains from cystic fibrosis patients. The four strains chosen were from Mikkelsen's (1970) O antigen types 3, 5, 6, and 11. The antibody developed from StAg was called standard antibody (StAb). StAg and sera from cystic fibrosis patients were combined in crossed immunoelectrophoresis with intermediate gel. Stained gel slabs revealed a complex pattern of precipitation curves. The titer of cystic fibrosis antibody was estimated by measuring the area encompassed by the precipitation lines with the area of the lines obtained with the StAb—StAg reaction. By this technique, 51 of the 55 antigens demonstrated in the reference system were present in all of Mikkelson's 13 O group strains. The corresponding antibodies in the reference system (StAb) were completely absorbed by each O group strain. Of the four remaining antigens of the reference system (StAg), three were distributed among the O group strains, and the last was a separate O group antigen.

Hoiby and Axelsen (1973) found no antibody to the mucopolysaccharide slime characteristic of cystic fibrosis strains of *P. aeruginosa*. It is unlikely that there was no cystic fibrosis slime antibody in their StAb. One or more of their 50-plus antigens is indeed from the mucous component, but it is present in all strains of *P. aeruginosa*. In other words, the overt slime of cystic fibrosis strains represents an increased slime synthetic capacity that is also present to a lesser degree in nonmucoid strains. Hoiby's explanation that the washing procedure might have removed the slime seems unlikely. The mild washing employed would not remove all of the mucopolysaccharide in the cell cover.

The difficulties encountered in all attempts to use unheated or formalin-prepared antigenic preparations in typing systems become clearer in this work. Of 55 antigens present in StAg, 51 were destroyed by heat, and these 51 were common to all strains! In Hoiby's studies, strain-specific O antibody in patient's sera were rarely detected using StAg. It probably is necessary to use typing-set strains or the patient's own strain to adequately detect O-group strain-specific precipitins by the crossed immunoelectrophoresis with technique of intermediate gel.

It has already been mentioned that heat-stable serotyping differences between *P. aeruginosa* strains depend on differences in the outer membrane lipopolysaccharide. Concerning Hoiby's discovery by crossed immunoelectrophoresis of over 50 proteins of *P. aeruginosa*, all but 4 or 5 of them common to all serogroups, Mutharia et al. (1982) described major outer membrane proteins F, H2, and I. These proteins were antigenically related in all of the 17 international (International Antigenic Typing System, IATS) serotype strains.

Using the techniques of Lüderitz et al. (1966, 1968, 1971) for lipopolysaccharide extraction and purification, Chester and Meadow (1973) extracted lipopolysaccharide from Habs' strains. All of Habs' strains contained the same lipid moiety (lipid A) composed of the same 10- and 12-carbon, 2- or 3-hydroxy fatty acids. The polysaccharide portion had two components, one of which was a low molecular weight core consisting of glucose, heptose, rhamnose, alanine, and galactosamine. This core was common to all strains. The other component consisted of higher molecular weight side chains that were characteristic of each serotype and conferred its serological specificity. These side chains contained little neutral sugar, but much amino sugar of known composition (particularly glucosamine) and up to five unidentified amino sugars. There were essentially specific precipitin reactions between lipopolysaccharide from the different type strains and homologous antisera prepared with boiled whole-cell suspensions. It was concluded that the strain-characteristic lipopolysaccharide fraction was the lone determinant of O-antigenic specificity in *P. aeruginosa*.

Numerous techniques have been reported concerning precipitin and hemagglutination reactions using lipopolysaccharide extracts (Adám et al., 1971; Chester and Meadow, 1973; Fisher et al., 1969; Hanessian et al., 1971; Hoiby, 1975a; Hoiby et al., 1975; Hoiby and Axelsen, 1973; Schiotz and Hoiby, 1975). Unusual 2-amino sugars from lipopolysaccharide were found to confer antigenic specificity to 13 of Habs' *P. aeruginosa* strains (Suzuki et al., 1977). These 2-amino sugars were D-fucosamine, DL-fucosamine, and quinovosamine. These were in addition to glucosamine, galactosamine, muramic acid, and two unidentified amino sugars. Specificity of pyocin killing as well as serological reactions may be conferred by characteristic amino sugars in the lipopolysaccharide (Koval and Meadow, 1975).

Flagellar antibodies have been relatively little studied. Much of the published work is suspect because of technical deficiencies. Only a few investigators used purified flagella and the use of controls such as nonmotile mutants of the research strains.

Lányi (1970) studied flagellar antigens by treating highly motile strains with 0.5% formalin and immunizing rabbits; O antibodies were removed by absorption with the boiled homologous strains. Of 541 *P. aeruginosa* strains, 288 belonged to 1 strong H type (type 1) and 246 were divided into 6 subtypes of type 2. By combining H and O typing results, Lányi devised a system of 53 serotypes. It is quite possible that pilus antibodies were present in the H antisera, since nothing was done that would separate one from the other.

Ansorg (1978) and Ansorg et al. (1978) described O-H (somatic—flagellar) antibodies from formalinized suspensions of five selected strains. Cross-absorption between these strains was successful in achieving five high-titered (1:6400) non-cross-reactive sera designated a_0, a_1, a_2, a_3, and a_4. These were used for indirect fluorescence on air-dried smears of unknown strains, with microscopic examination, to find 17 H types (as combinations of reactions using the 5 H antisera). These H types, matched with Habs' O serotypes, determined 99 serovars among clinical isolates.

Pitt and Bradley (1975) earlier made use of mutants with and without flagella and fimbriae to make high-titer antisera further purified by appropriate absorption. Use of these mutants obviated the need to attempt to remove the particular antigen by heat or physical or chemical action, and greatly simplified absorption steps. Unlike

Ansorg, Pitt and Bradley found the anti-O titer in rabbits to fall sharply by the 3rd week and to almost disappear at 4 weeks. The anti-H titer reached peak titer at 20 days and only decreased slightly thereafter. Anti-H antibodies typically inhibited motility at 1:25,000 to 1:100,000 dilutions in deep tubes of soft nitrate agar. The test is specific for anti-H antibodies in the presence of anti-O and anti-pilus antibodies. Four H types were reported.

The usual method of removing excess cross-reacting antibody is absorption with selected strains. Once the cross-reactions are known within the set of immunogen strains and their antisera, the absorption process is routine. Practices varies widely. A single absorbing P. aeruginosa strain was used by Duncan et al. (1976) to remove cross-reacting antibodies from many of the typing sera. Another practice to increase specificity is absorption with small numbers of the homologous immunogen strains (Shionoya et al., 1977).

Almost all P. aeruginosa strains have pili, the tubelike protein extensions of bacterial walls that function variously in attachment to specific cell sites in animals and plants (Hohmann and Wilson, 1975; Swanson, 1975), in phage attachment and entry (Bradley, 1972), in agglutination reactions (Bradley and Pitt, 1975), and in mechanisms of genetic exchange. Bradley and Pitt selected strains with nonretractile pili since these were better antibody producers. The pili could be prevented from retracting with osmium treatment. Bradley and Pitt reported four immunotypes of pili and found these useful in epidemiological typing of P. aeruginosa when combined with flagellar and somatic typing.

Established Systems

Homma and the Japanese Study Group have done extensive investigation into the technique and clinical practice of immunotyping P. aeruginosa (Homma et al., 1970; Homma, 1974, 1976). Correlation of the different O-antigen typing schemes has been done by Homma et al. (1977), Matsumoto and Tazaki (1969), Muraschi et al. (1966), and Kodama and Ishimoto (1976). Each system has a different numbering order for serotypes, but most use Arabic numerals (Table 1). Verder and Evans (1961) used Roman numerals and Homma in his most recent version used capital letters. In the author's practice, using the Homma, IATS (Liu or Difco), and Fisher systems, correlations between the three systems has been surprisingly good. When a strain during repetitive typing changes type in one system, it changes to the corresponding type in the other systems. When the correlation fails, it is usually when a strain becomes nontypable in one or two systems but remains typable in the remaining system(s).

The very thorough study of seven worldwide immunotype systems by Homma et al. (1971, 1977), compared the systems of Homma, Liu et al. (1983), Habs, Verder and Evans, Meitert et al. (1976), Lanvi, and Fisher and delineated the typing performance and expected cross-reactions.

Table 1 *Pseudomonas aeruginosa* Agglutination Reactions: Correspondence of O-Antigen Types of the Major Typing Antisera Schemata

Habs	Vernon	IATS (Difco)	Pasteur Institute	Homma (old)	Homma (new)	Lányi	Meitert	Sandvik	Fisher (Parke-Davis)	Verder and Evans
1[a]	1	1	1	10	I	6	13	VII	4	IV
2[a]	2	2	2	7		3	2		3	I
3[a]	3	3	3	1	A	1	5	III		VI
4[a]	4	4	4	6	F	11	8	IV		
5[a]	5	5	5	2	B (2,7,13,16)		6		7	X
6[a]	6	6	6	8	G	4	4	I	1	II
7[a]	7	7	7				1			
8[a]	8	8	8	3	C	5	3	VIII	6	VIII
9[a]	9	9	9	4	D	10	14	V	5	IX
10[a]	10	10	10	9	H	2	11			
11[a]	11	11	11	5	E	7	15	VI	2	III
12[a]	12	12	12	14	L	13	7			VII

13	15	11	J	12[a]	
	13	12	K		II[a]
	14	13[a]			
	16	15	M (15, 17)	9	16
		16			
		17			
		18		8	
					9
	17				10[a]
					12
					17
					V[a]

[a] Include in the IATS (Difco) schema. All *P. aeruginosa* strains included in the IATS are available from the American Type Culture Collection (ATCC).

Shionova et al. (1977) painstakingly elaborated the properties of Homma's immunogenic types in rabbits. These included details of typing serum production such as days to peak titer, cross-agglutinin titers, absorption techniques, and correlation with the Verder—Evans and Habs schemas.

Liu's types 7, 8, 13, and 14 were cross-related to types 3 and 12 of Homma (Terada et al., 1977). Common antigens were found among Homma types 2, 7, 13, and 16 and also 15 and 17. On the basis of this work, Homma (1976) established a new 13-type system using capital letters A to U. Type B comprised previous Homma types 2, 7, 13, and 16 and type U comprised previous Homma types 15 and 17. This is the set manufactured by the Toshiba Kagaku Kogyo Company. The original long set in the author's laboratory did not incur excessive multitype reactions and furthermore had the sensitivity to identify the most common cystic fibrosis-related strain (serotype 8-Homma) which had a multination distribution. The immunogen strain for type 8 was then changed. The cystic fibrosis-associated strain was also identified with the author's phage set. None of the currently distributed serotyping sets have this unique sensitivity.

Fisher et al. (1969) at Parke-Davis selected seven P. aeruginosa strains, all exotoxin A producers (Pollack et al., 1977) whose purified lipopolysaccharide was combined to make a vaccine for human use. Antisera to these seven lipopolysaccharides proved to be useful in epidemiological work and were widely used. The company was unsuccessful in its bid to market a vaccine, even though immune efficacy was established in many studies, particularly among patients in burn units. The project was dropped, and until the antisera are produced again, the typing set is not available.

The Habs schema for serotyping was combined with active pyocin typing (Wahba, 1965a) using 10 pyocin indicator strains. Members of pyocin types G, K, L, O, and P belonged to serogroups O:14, O:1, O:3, O:9, and O:6, respectively. Serogroup O:1 was divided between pyocin types C and K, and the nontypable strains. Serogroup O:6 strains were divided among pyocin types C, D, and P. Wahba believed the combined systems to be useful and complementary in epidemiological studies. Many examples were given indicating that some patients were infected by two strains of P. aeruginosa when typed by both systems, but only one strain by serotyping alone. On the other hand, it is quite possible that this enhanced sensitivity through adding pyocin to serotyping may serve no useful prupose in some outbreaks of infection. If two or more pyocin types merely subdivide the one serotype that is the epidemic strain, the result may be confusing rather than enlightening. The subdivision by pyocin typing may be valid but serve no essential prupose. In an epidemic it is a requisite to identify the P. aeruginosa strain or strains involved. Further subtyping may be unnecessary, unless of course one embraces the belief that infecting and noninfecting strains of the same serotype are involved. This situation may occur rarely.

There has been much investigation of variations in typing techniques. A study of serotyping by slide agglutination and tube agglutination using heat-killed as well as live P. *aeruginosa* suspensions found that all combinations gave equal results (Kono and Sei, 1977). Kono and Sei reported the slide test with heat-killed suspensions to be more practical and recommended a limit of 60 sec for reading the agglutinations. This time limit is now generally recommended to avoid weak reactions involving minor antigens. A further point was made that negative tests using live antigen should be repeated using autoclaved antigen, sometimes leading to a positive test.

Agglutination of strains by more that one antiserum occurs frequently, but some laboratories have a much higher percentage of these multitypes. A multitype of as many as five serotypes is not uncommon. One suggestion often given to correct this is to further dilute the antisera from the usual 1:10 dilution. This procedure may occasionally work very well on a specific problem strain, but diluted sera should not be used routinely. Antiserum dilution that eliminates cross-reactions for "problem" strains may be the dilution of extinction for many agglutination reactions for unknown strains.

Multitypes are repeatable and stable patterns in our laboratory. They are just as useful as monotypes in tracing strains during an infections outbreak. Those who report multitypes as nontypable are handicapping themselves in their investigations. A purist adherence to monotypes has its place, but not during an investigation of an epidemic that involves human lives. The multitype obviously is more of a fingerprint than the monotype, and in that sense is more useful.

When possible, a single epidemiological study should be completed without changing lot numbers of typing antisera, whether those antisera are made in-house or are obtained from commercial sources. Confidence that titer and specificity remain unchanged from lot to lot is misplaced confidence. Not only does each rabbit respond somewhat differently to the same antigen, but for a number of reasons no two separately prepared antigenic preparations are out identical. The pooling of sera from, say, 10 rabbits for each immunogen strain is an excellent procedure to counteract differences in individual response.

The problem of changes in serotype is handled differently by investigators. While everyone recognizes the problem in its broadest terms, it is often deemphasized in reports, not with intent to mislead, but to maintain the chief thrust of an epidemiological report. There are reports addressed specifically to spontaneous serotype, phage type, and pyocin type changes in P. *aeruginosa*, which, of the major pathogens, is most notorious for these changes. Changes in serotype, pyocin, and phage types are probably based on cell wall changes mediated by a refluxing of the cell's extremely active multiple lysogenic and autolytic state.

Homma et al. (1972) found startling serotype switches on infrequent subculture at room temperature, reporting that 4 of 11 strains had changed serotype. A serotype strain 1 that had changed to serotype 9 was studied more closely by selecting 35 colonies from one plate and typing growth from each of these. Homma et al. reported that "out of the 35 substrains 20 were identified to belong to serotype 9, seven to serotype 8, two to serotype 2, one to serotype 3 and one to serotype 7. Only two substrains were identified as the original serotype O:1, and two strains were nonagglutinable." The same phenomenon was noted for one other of the 11 strains subjected to this type of analysis. Thus 2 of 11 strains showed marked serotype changes. This corresponds roughly to the experience with retyping at 6-month intervals in the author's laboratory (unpublished data). It is fortunate that the rate of change is not higher. If this is a general phenomenon, then we would do well to follow Homma's strictures to type newly isolated strains immediately, and type the 1-a colony form where possible. Again, this phenomenon might explain capricious changes in the intensity of the agglutination reactions through admixture of mutants on the plate during preparation of the agglutinogen suspension.

Serotype changes were reported in 431 P. aeruginosa strains that were subcultured for 85 days, with storage also at 5°C (Kawaharajo, 1973). Brokopp et al. (1977) reported successful use of the seroset from Difco (Liu's Internation Subcommittee seroset) in epidemiological investigations. Bébéar and Dulong de Rosnay (1973) used serotyping to follow infections strains in their hospital. They found exogenous spread of infections usual when only a few serotypes were found, such as in their burns unit, and in the general services more varied serotypes were found, indicating infections of endogenous origin. Loiseau-Marolleau (1973) also used serotyping to trace hospital infections. Pitt and Erdman (1977) reported on specificity of serotyping.

Kurup and Sheth (1976) found that patients retained their individual serotypes for long periods. They favored immunotyping as a single procedure, but reported that typability goes to 98 from 83% when pyocin typing is done jointly. Al-Dujaili and Harris (1974) combined a 13-serum immunotyping set (essentially Habs' set, made at the Pasteur Institute) and the 10-pyocin indicator set of P. aeruginosa strains from Gillies and Govan (1966). They favored immunotyping over pyocin typing if a single system was to be used. In an interesting continuation they subdivided pyocin type 1 (active typing as above) by passive pyocin typing (Govan and Gillies, 1969) and reported that the second pyocin typing was useful.

The short 7-serum typing set of Fisher (Parke-Davis) was preferred over a 27-pyocin indicator strain set in a 3-year study of P. aeruginosa isolates in a military hospital (Baltimore et al., 1974). In this study most long-term patients again retained the same strain for extended periods of time and did not change that strain for another during or after antibiotic therapy. This argues against conclusions reached by some studies using pyocin, and particularly phage typing, that *P. aeruginosa* strain changes in patients were frequent. The frequency of strain change in the patient in many of these studies owes to misinterpretation of pyocin and phage pattern differences. Young and Moody (1974) also successfully used the Fisher system.

At one time Homma (1974) fractionated rabbit antiserum to *P. aeruginosa* typing-set strains and found most agglutinating activity to be IgM. Immunoglobulin was of lesser titer, but its activity was increased when anti-rabbit serum goat antibody was added to the system. Homma used the adjuvant action of the goat anti-rabbit antibody to enhance detection of antibodies in sera of patients chronically infected with *P. aeruginosa*. Formalin-killed cells were used in place of autoclaved cells, as sensitivity was greatly decreased. Since strain specificity usually is not as important in detecting the level of antibody in patients's sera, the compromise is a reasonable one.

Pollack et al. (1983) reported that serum response to exotoxin A was a far better indication of *Pseudomonas* disease than was response to specific lipopolysaccharides of the infecting strains.

Using serotyping, Yabuuchi et al. (1971) showed that 21 of 31 apyocyanogenic strains of *P. aeruginosa* were typable, but only 7 of 40 melanogenic (pyorubrin producing) strains were typable.

Standardization of Serotyping

Liu has led a drive toward a standardized worldwide serotyping schema, initiated by the Subcommittee on Pseudomonadaceae at the 1970 meeting of the International Committee on Systematic Bacteriology. Habs' system of 12 "serogroups" was chosen for a base of heat-stable immunogen types. One of Sandvik's strains was chosen as serogroup 13, the Verder—Evans group V became serogroup 14, Lányi's serogroups 12 and 3 became serogroups 15 and 16, and serogroup 10 of Meitert became serogroup 17. This came about after a great deal of work involving four laboratories. Difco Laboratories (Detroit, Michigan) was asked to prepare antisera to the 17 immunotypes. These unabsorbed antisera, diluted 1:10 for use, proved surprisingly free of cross-reactions when tested on unknown strains. This typing set has been marketed for a number of years and has proven to be a practical, easy-to-use tool in epidemiological work. Suspensions of the 17 immunogen strains are available for controls and cross-agglutination checks. Adoption of this schema as the International Antigenic Typing System (IATS) was partially based on the fact that it was the best known in Europe, America, and Asia. Also, it was felt that it would form a solid base of well-documented serogroups which would remain unchanged while leaving open the possibility of including new and additonal serogroups as they were described.

The 13-serotype set and three pools developed by Homma and the Japanese Study Group is also widely used. It is manufactured by Toshiba Kagaku Kogyo Co., Ltd., and is marketed by Nichimen Co., Ltd. (KHX Section, 1185 Avenue of the Americas, New York, N.Y. 10036).

The production section of the Pasteur Institute (25 rue Dr. Roux Paris, 75115, France) markets a set of 13 antisera also based on Habs' schema, but modified and expanded by Veron. Three serotype pools are available: A (types 1, 3, 5, 6, and 10), and B (types 2, 5, 7, and 8), and C (types 9, 11, 12, and 13). This seroset is also available from API Products, Ltd. (Plainview, New York).

Recommended procedure for each of these shcemas calls for slide agglutination reactions using live cell antigens. *Pseudomonas aeruginosa* growth taken from overnight growth on an appropriate nutrient agar may be suspended with a transfer loop into drops of antisera on a glass plate or the squared-off bottom of a disposable Petri dish (such as Falcon No. 1013 150 X 25 mm with 20-mm grid). Reading is done after hand rotation for 1 min.

Testing is much faster, and a more uniform bacterial suspension is available by suspending *P. aeruginosa* growth to the No. 10 MacFarland turbidity in saline, rather than suspend growth with a loop. The suspension can sit at room temperature for up to 4 hr without deterioration or may be kept overnight at 4°C if necessary. This invokes a 1:2 dilution of the sera as one drop of *P. aeruginosa* suspension is added to one drop of IATS antiserum (already diluted 1:10). This has not changed typing reactions in our hands, as predicted that it would not. We have found unheated cell suspensions to have fewer autoagglutination problems, and heating of itself causes some suspensions to clump. Treatment with an anionic detergent (Tween 20) and a protease mixture (Rhozyme 41) experimentally has reduced the percentage of nontypable strains (to be published), but we have not utilized it routinely as yet.

Dominance of Type O:11

The increasing dominance of type O:11 as the agent of infectious outbreaks is a worldwide phenomenon. A hospital in Cairo reported *P. aeruginosa* serotype O:11 to be the predominant infectious strain (Hassan, 1979).

Of typable strains, serotype O:11 was predominant among ear and skin infections reported from whirlpools in a very large recreational park in the Netherlands (Havelaar et al., 1983).

Pseudomonas aeruginosa serotype O:11 was identified among endocarditis cases in Cook County, Illinois (Rajashekaraiah et al., 1981). The drug addicts afflicted were using barbiturate and pyribenzamine tablets (T's and blues) triturated in water in bottle caps, filtered through cigarette filters into a syringe, and injected intravenously.

Heroin was the drug used in some cases. The chicago addicts were environmentally isolated from one another, so that the emergence of type O:11 as the strain growing in trituration vessels and syringes was probably based on contamination of syringes from infected individuals. This selection was in turn based on the extraordinary pathogenicity of type O:11 itself. A series of 10 endocarditis cases in Detroit addicts were infected exclusively with type O:11 (Reyes and Lerner, 1983).

Respiratory, bloodstream, and urinary tract infections in hospitals have not become so predominantly caused by serotype O:11 as water-associated disease and disease in drug addicts, but in this area also it is a major infectious agent (Sherertz and Sarubbi, 1983).

Occurrence of P. aeruginosa type O:11 in endocarditis is regional, confined first to Detroit, and now Detroit and Chicago. *Serratia marcescens* has been reported in such cases in San Francisco, and *Pseudomonas cepacia* in New York (Reyes and Lerner, 1983). Movement of people will undoubtly extend serotype O:11 to these areas.

McCausland and Cox (1975) provided the earliest report of whirlpool-caused dermatitis. This first outbreak was caused by P. *aeruginosa* type O:1. Jacobson et al. (1976) reported swimming pool-associated infections caused by serotype O:11.

A large outbreak of dermatitis derived from the whirlpool of an Atlanta hotel (Khabbaz et al., 1983). This recent outbreak was unique because it involved P. *aeruginosa* serotype O:9 and not O:11. This was a newly opened hotel and type O:11 may not have been carried to this new environment as yet.

An outbreak of type O:11-caused dermatitis at a West Yellowstone, Montana, motel (Hopkins et al., 1981) involved not a whirlpool but a dry sauna and swimming pool. A small New Zealand spa pool, essentially a hot tub, badly maintained and containing 1×10^7 P. *aeruginosa* organisms per milliliter in the cloudy water, resulted in a series of type O:11 infections, some with lymphadenitis and axillary abscesses (Stafford et al., 1982).

It seems probable that if P. *aeruginosa* type O:11 did not exist, there would be fewer reported infections, and these would be of lesser severity.

Vogt et al. (1982) reported a small outbreak of dermatitis among users of a Vermont ski resort whirlpool. The P. *aeruginosa* strains isolated were unusual in that they were serotype O:1.

Pseudomonas aeruginosa type O:11 contamination of an olympic-sized swimming pool in Scotland (Reid and Porter, 1981) resulted in a large outbreak of otitis externa. It is interesting that only otitis externa is found in some outbreaks, while many types of dermatitis occur exclusively in others. The degree of superhydration of different skin areas is an important if not overriding factor (Hojyo-Tomoka et al., 1973).

Outbreaks of modular folliculitis from whirlpool exposure were reported by Burkhart and Shapiro (1980), Rasmussen and Graves (1982), and Sansker et al. (1978). Kush and Hoadley (1980) reported type O:11 infections from whirlpool baths at eight Atlanta locations. A generalized pruritic pustular rash occurred in 32 of 61 persons (53%) using the whirlpool in a Minnesota motel (Washburn et al. 1976).

Two separate outbreaks of urinary tract infection in two Atlanta hospitals were caused by type O:11 strains that contaminated surgical instruments such as resectoscopes. Insured sterilization of these instruments stopped the outbreaks (Strand et al. 1982).

A massive aerosol challenge during 90 min in a home whirlpool heavily contaminated with *P. aeruginosa* (serotype O:1) resulted in a life-threatening pneumonia (Rose et al., 1983; Rinke, 1983).

Urinary tract infections without dermatitis caused by type O:11 and other serotypes have been reported (Salmen et al., 1983). These infections were acquired in home and spa whirlpool baths.

Commercial whirlpool baths at eight locations (all of those studied) in Atlanta were found to be contaminated with *P. aeruginosa*. Serotype O:11 was isolated from more samples than any other serotype.

Noone (1983) reported nosocomial spread of *P. aeruginosa* type O:11 through the intensive care unit of a Middlesex hospital. Most cross-infections were mediated by personnel during periods of frantic activity within the unit.

Swimming pool-associated skin rash caused by *P. aeruginosa* type O:11 was reported from Georgia by Jacobson et al. (1976).

BACTERIOPHAGE TYPING

History

The development of phage systems for *P. aeruginosa* followed by a decade the initiation by the American bacteriologist Fisk (1942a) of a phage system for typing *S. aureus*. There is no international set of *P. aeruginosa* typing phages. Jacob (1952) was the first to induce bacteriophage lysis in *P. aeruginosa*, using ultraviolet light. Dickinson (1948) at Boots Pure Drug Company used the model to test viricidal compounds and found proflavine effective against *P. aeruginosa* bacteriophage. Dickinson and Codd (1952) attempted to relate phage to the semilytic phenomenon in *P. aeruginosa* called variously metallic lysis, iridescent lysis, and autoplaque phenomenon. Warner (1950) had investigated this same phenomenon and concluded correctly that phage was unrelated to iridescent lysis, as had D'Herrelle (1926), but the phenomenon continued to be rediscovered, with many reports unfortunately and incorrectly linking the two. There was no evident thought of using these phages for typing. At this time Hadley (1924), Jadin (1932), Rabinowitz (1934), and Fastier (1945) isolated bacteriophages against *P. aeruginosa* from sewage, feces, and pus.

Don and van den Ende (1950) recovered phages from lysogenic strains during a study of the iridescent phenomenon. This has turned out to be the best source. All P. aeruginosa strains are lysogenic for at least one phage and most are lysogenic for more than one phage. Shionoya et al. (1967) more recently recovered 10 distinct phages from one strain (P-1-111). Yamamoto and Chow (1968) used mitomycin C for phage induction, but most of these phages do not have to be induced via ultraviolet light or mitomycin C. Amazingly, they are present as free phage in broth cultures in titers as high as 1:10,000 confluent lysis or 1×10^6 to 1×10^8 plaque-forming units/per milliliter. There is no visual evidence of lysis in broth culture and P. aeruginosa strains seldom spontaneously lyse to a clear state in broth cultures. Thus the phage-free, cryptic lysogenic state is abrogated in sufficient cells in broth cultures of all P. aeruginosa strains to provide lytic cycles and a high titer of phage without lysing the mass of the cells to extinction. This provides an available reservoir of candidate phages for a typing set. Broth supernatant fluid (it may be called lysate) from these simple cultures may then be diluted and plated with candidate host strains. There may result more than one plaque morphology representing distinct phages. Single plaques showing clear lysis with sharply demarcated edges are picked and plated on a selected host strain until a usable volume of lysate is achieved. A fervent hope is maintained that during successive propagations the phage will not have lost its desirable lysing characteristics. Mutation during propagation occurs rarely among S. aureus phages but nevertheless is a problem faced by every worker in this area. *Pseudomonas aeruginosa* phages are more subject to mutation because of the multiply lysogenic state of the host strains. There is no reliable method of curing P. aeruginosa strains of the lysogenic state.

The phages selected for the final typing set will be those that are restricted in lytic range on a large, representative indicator set of distinctive P. aeruginosa strains, acquired from a number of hospitals. When the phages are put together as a set, phage patterns derived from the indicator set are compared. If more than one phage produces identical reactions, one is retained and the others are discarded or stored. The same applies to testing phage sets acquired from other laboratories. Finally a new set is chosen that is most useful to the hospital and the geographical area where the set will be used.

Acquiring or Establishing a Phage Typing Set

A decision must be made whether to isolate phages from P. aeruginosa strains and from sewage to create a typing set or to request an existing set or sets from other laboratories. Requests for phages may not always be rewarded. One reason for this is that old sets may have deteriorated. Requests to national centers and to those creating the most recent phage sets may be most fruitful.

The final phage set in the author's laboratory consisted of 26 phages. Phages Pa and Pb of Dickinson and Codd (1952) were acquired from the American Type Culture Collection (ATCC). These are ATCC 12055 (B-1 and B-2). The host is P. aeruginosa ATCC 12055. Of the 16 phages received from Hoff and Drake (Hoff and Drake, 1961; Drake, 1966), 9 were selected for this set. The remaining 15 phages were from a very large set isolated by the author from broth culture supernatant fluids. Postic and Finland (1961) reported good reproducibility of patterns using 13 phages, including 3 Dickinson phages and 4 that were isolated by Asheshov (ATCC 12175 B-1, B-2, B-3 and B-4). This author found the Dickinson and Asheshov phages so similar in their action on the indicator set that only phages B-1 (Pa) and B-2 (Pb) of Dickinson were included in his typing set. Mutations had probably occurred during many propagations as these phages were sent around the world. Feary et al. (1963) also recorded reactions on their indicator set indicating the sameness of the Dickinson and Asheshov phages.

Because of the very good possibility of mutation during serial propagation of these phages, large lots of broth lysate should be made, enough to last for an entire study. These lysates are mixed with equal volumes of double-strength skim milk, divided into small volumes, and preserved at −80°C or lyophilized (Zierdt, 1959).

Details of P. aeruginosa phage propagation techniques in broth, in soft agar layers, or on agar are available in most of the reports cited in this section. Although there are differences in media, for example, any one of the reported techniques can be copied successfully, if certain pitfalls are avoided, which will be described. Adams (1959) describes widely used, more or less standard techniques for phage isolation, propagation, and titration. Bergan (1972) assembled a final phage typing set from sets that he collected around the world and provides details of all aspects of phage typing. Other detailed reports are those of Lindberg and Latta (1974), Postic and Finland (1961), Sjoberg and Lindberg (1968), Smith (1972), Sutter et al. (1965), and Vieu (1969).

It is most important that rapid sparkling lysis be achieved during propagation of the typing phage, as this best ensures a uniform population of mature and infectious phage particles. Cloudy lysates are apt to contain immature, noninfectious phages, pyocins, and many other kinds of interfering particles. Importantly, they also will contain more of the mature phage from the lysogenic reservoir of the host strain. At this point some lysate should be held back for stock, so that a propagation failure does not leave one without good lysate for another try.

Lysates used only for typing should never be filtered, as absorption of phage particles to membrane filters is particularly marked (Zierdt, 1979). A 10- to 100-fold loss of phage titer is usual by filtration. It is better to get rid of residual host organism with ether.

Ether (saturated) treatment incurs no loss of phage viability and 1 hr of treatment at 37°C is adequate. The ether is then allowed to dissipate by placing the lysate in plastic dishes for about 1 hr.

Use of the Phage Typing Set

A particular typing medium is difficult to recommend, since there are almost as many nutrient agars used as there are phage typing reports. A semisynthetic medium (Sutter et al., 1963) was reported to lessen many confusing lawn growth phenomena, such as iridescence, that contribute to confusing phage lysis reactions. Secondary growth in phage-lysed areas was reduced. The author's laboratory uses a similar synthetic medium (Shionoya et al., 1967) with the pH reduced to 6.0. The important advantage at pH 6.0 is complete elimination of the iridescent lysis phenomenon (Zierdt, 1971), which is the most confusing lawn change accompanying phage typing reactions. It is an autolytic action presenting as partial lysis with a strong metallic sheen, either diffuse but more often distinctly like phage plaques. Unlike phage plaques, they become larger on continued incubation. They are also distinguishable by their metallic sheen, which phage plaques never have, and by failure to lyse the lawn enough to bare the agar. However, the iridescent phenomenon is frequently phage induced in the area about the phage plaque. This presents a picture difficult to decipher when complicated with an admixture of phage plaques, secondary growth over the plaques, resistant colonies in the phage plaque area, and frequent slime production. Thus the use of pH 6.0 agar is a worthwhile step in cutting down on confusing reactions.

Many workers have added 0.01 m $CaCl_2$ to the agar. The value of this cation for *P. aeruginosa* phages has not been documented, but is done with the assumption that it may be beneficial, as for *Escherichia coli* phages. The bacterial lawn is best made from suspending agar growth (less than 24 hr old). It forms a more even suspension than growth in broth, which is likely to have a heavy pellicle and flocculent growth.

The surface of the uninoculated typing plates is dried by raising the dish lids in the 37°C incubator. Usually 1 hr of drying time suffices, but this depends on humidity and air movement and must be controlled so that overdrying does not occur. If the drying step is omitted, as may perhaps be done successfully if older stored plates are used, the inoculum, the phages, or both may not fully absorb. The dropped phages then may spread over the agar and mix with neighboring phages.

The usual concentration of agar in the medium is 1.5%, but some workers have used 2.5% agar poured to only a depth of 2 mm, which they believe increases lysis and makes the reactions easier to read.

The bacterial inoculum may be applied by flooding the plate with a faintly turbid suspension of the test organism, tilting the plate, and siphoning off the excess, as is done in antibiotic sensitivity testing. Another way that is certainly faster makes use of cotton swabs. Unless large fluffy swabs are available, it may be necessary to fluff up the swab against the inside of the bacterial suspension tube in order to carry over enough volume to easily cover the plate with gentle strokes. A central streak is made, followed by overlapping perpendicular streaks. The ideal inoculum furnishes enough colony-forming units (CFU) to just provide confluent growth after 24 hr of incubation. The faintly turbid suspension referred to, to provide just confluent growth, is easily accomplished with experience.

Phages may be applied to the seeded lawns singly via pipette, loop, capillary tube, or No. 27 guage needle and 1-ml syringe. But if even a fair amount of work is to be done, a device should be obtained that applies all phages at once. A device (Steers et al., 1959) that applies phage via multiple metal posts dipped in a well template reservoir of phages is not recommended. Since phages in this case are applied to the agar first, followed by flooding of the plate with bacterial suspension, there is loss of phage particles into the agar, as well as various amounts of phage scatter by the flood of inoculum. A multiple-loop apparatus (Lidwell, 1959) uses two sets of suspended loops with horizontal bends at the ends, one set at each end of a rotating boom. Sterilized by flame, one set of loops is dipped into a reservoir block of phage suspensions and lowered into a typing plate while the other set is cooling. This machine, as all of those described, may be used for phage or pyocin typing.

A stainless steel and multiple-glass syringe device (Zierdt et al., 1960) has been most successful in laboratories around the world (Figs. 1—4). The slidable sleeves made of Teflon spaghetti tubing (0.018 in. inner diameter) must be wet with a silicone spray before use so that they continue to hang the drops from the tips only rather than creep up the lower extremity of the tubing. This maintains about a 2-mm air gap when the drops touch the agar and fall away, preventing needle contact with the inoculated agar surface. A practice plate or two is used to adjust the needle sleeves to the same height and to get rid of excess silicone oil. Silicone oil has no adverse effect on bacterial growth or on developing phage reactions. The multiple-syringe applicator, to function optimally, requires some mechanical dexterity and thoughtful attention to assembly and adjustments. Granted these prerequisites, the instrument delivers superb performance year after year. It is not available commercially and must be constructed by a machine shop from blueprints requested from the author. It remains the most rugged, accurate, and completely autoclavable dispenser of phage drops to seeded typing plates. Syringe

Fig. 1. Multiple-syringe dropper for phage typing. It is also used for bacteriocin typing and replica plating for Fisk and other techniques. (From Zierdt et al., 1960.)

Fig. 2. View of needles with drops ready to be touched off. The adjustable Teflon sleeves can just be seen.

and adaptor plates holding 26, 32, and 52 syringes are interchangable, to be used with larger dishes. Phages may be left in the syringes and the machine stored at 4°C. Then the only action required for use is to remove the machine from its case. Modifications of this machine that are more economical have been described by Farmer (1970a) and Farmer et al. (1975).

Fortunately *P. aeruginosa* phages are more stable in suspension at 4°C than *S. aureus* phages. Titers may be checked at monthly intervals, but Lindberg and Latta (1974) suggested annual testing for quality control. Sutter et al. (1965) found phage stocks to be

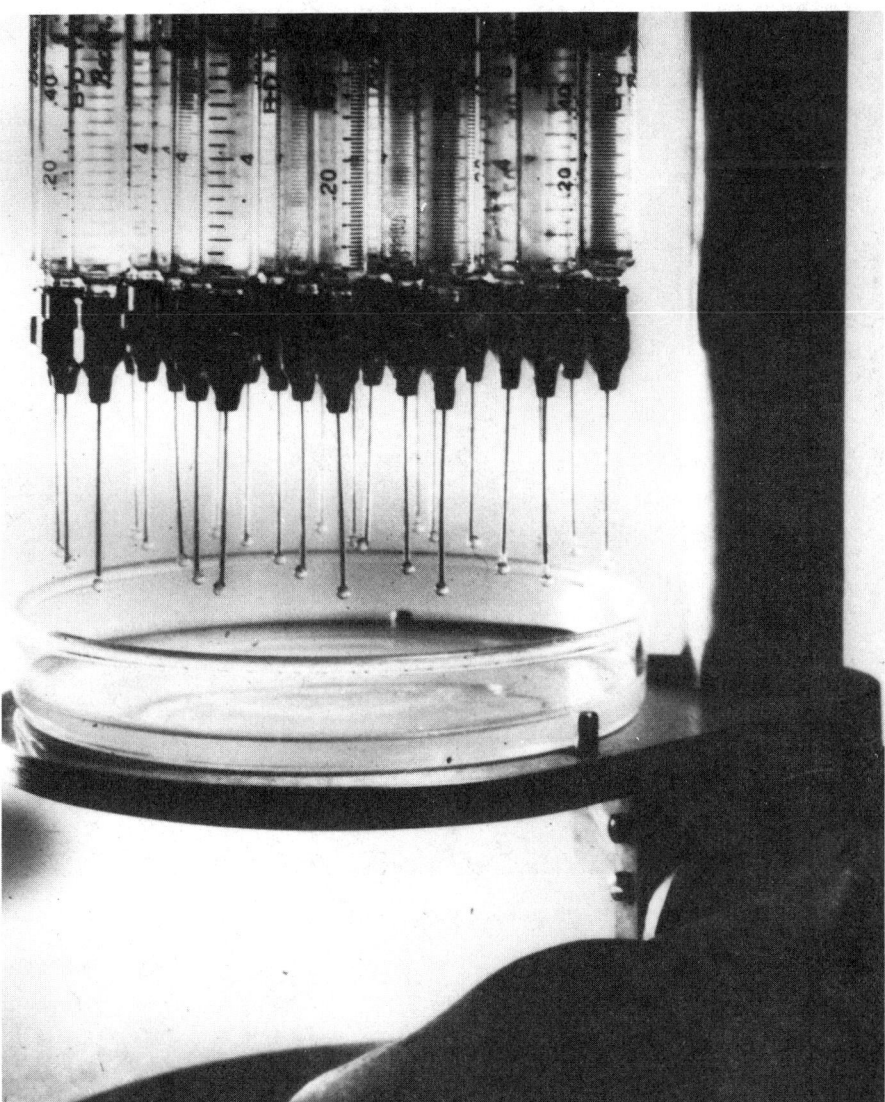

Fig. 3. Plate with fresh lawn of organisms to be typed is raised to touch off drops of phage suspension.

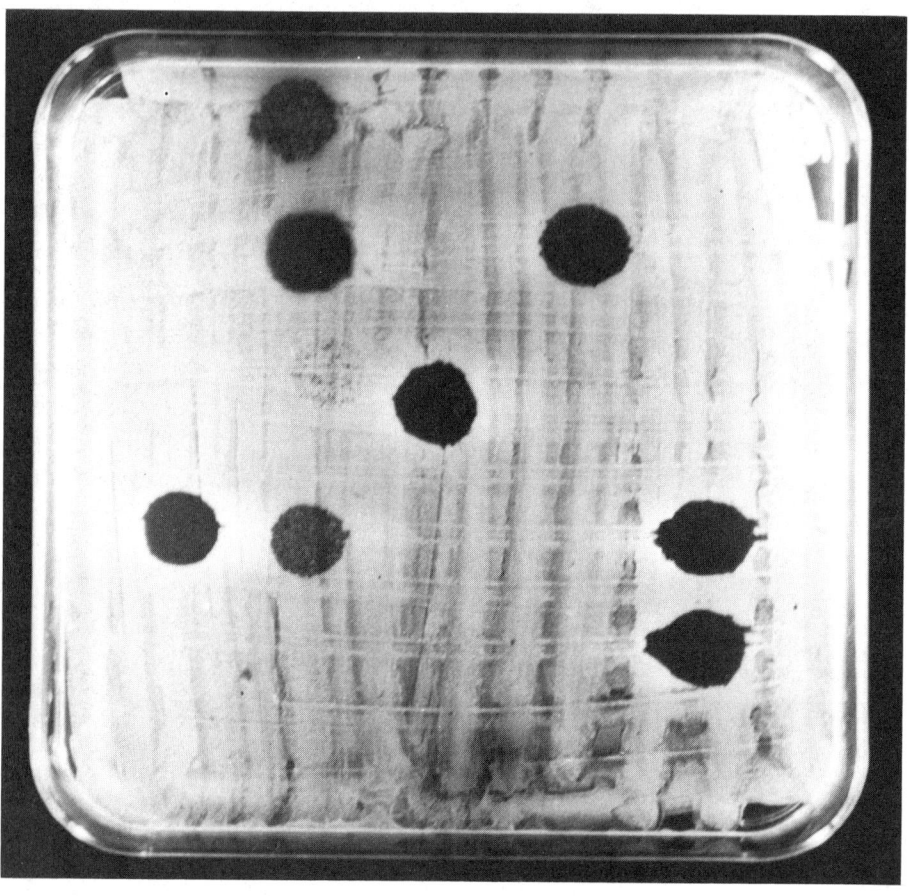

Fig. 4. Typing plate with mostly confluent lysis reactions. Plaques can be seen in three areas.

stable for 1 year, but reported that suspensions at routine test dose (RTD) required weekly checks. An RTD is the highest lysate dilution achieving confluent lysis of the host strain. The phage concentration for typing is usually at RTD, but some workers found that 100 RTD provided clearer and stronger lysis. There are two factors that favor the latter view. Inhibition reactions, which may be so strong as to bare the agar, thus mimicking true phage confluent lysis, are also specific and reproducible and may be used in typing. Inhibition reactions are only moderately increased at 100 RTD. Also

Pseudomonas aeruginosa Typing

100 RTD provides stronger lysis that prevents secondary growth of phage-resistant lawn organisms. This view is supported by the work of Bergan (1972).

Reading Phage Reactions

After overnight incubation, typing plates may be held at 4°C if desired before reading the reactions. During a single study it is of primary importance that all of the phage reactions be read by the same worker. Also, it is well known that phages vary in their activity dependent on the manufacturers lot number of Tryptic Sor Agar (Difco, Detroit). A single epidemiological study should be completed if possible on a single lot number of Tryptic Sor Agar.

Unlike a lytic area on a lawn of S. aureus which is clear-cut and unequivocal, baring the agar in its stark revelation of destroyed bacterial cells, many phage reactions on P. aeruginosa lawns are of partial lysis. This partial lysis may be caused by one or more of a variety of mitigating interactions, some having to do with less-than-perfect phage attack and lysis and some with resistant mutants within the lawn population. These resistant colonies may grow rapidly enough to coalesce and obscure the initial lytic reaction, leaving only a depression on the lawn. This depression can vary from just visible to being deep enough to almost bare the agar surface. All of these factors are also at work when the lytic reaction is not confluent but presents as plaques. Plaques may initially be open and "punched out" and then close as the lawn becomes luxuriant.

There is evidence that colonial variants or dissociants of a single strain can have markedly different typing patterns (Zierdt and Schmidt, 1964). In some strains this may be a real concern, although it is difficult in most cases to differentiate phage pattern differences due to phenotypic colony changes and differences that occur from unknown causes during repetitive typing of successive strain subcultures. There is a discussion in the pyocin section of colony selection for typing that is pertinent to this question.

Mucoid strains or slime strains producing an excess of mucopolysaccharide may inhibit phage attachment of some phages in the set, but usually not all of them; however, the lytic pattern achieved is obviously unreliable if part of the phage set is rendered ineffective by the slime coating. Also, mucoid overgrowth may follow lysis, resulting in a lawn with circular depressions rather than agar-baring lysis. These depressions should be read as lysis.

The iridescent phenomenon is evident in a significant number of phage lysis reactions. The phenomenon occurs naturally in perhaps 25% of strains, but also is inducible by phage action. It thereby occurs more frequently and must be differentiated from true phage action. There are two outstanding differential criteria. Iridescent plaques always have an easily seen metallic sheen, and they do not lyse through the lawn to bare the agar. Their lysis is always partial. In the presence of diffuse iridescence, or iridescence without plaques, true phage plaques may be masked with a thin overlay of iridescence. True phage plaques, free of the complicating intrusions discussed, are never iridescent.

The number of plaques in a phage drop area that are required for a typing reaction is a nebulous figure dependent on the criteria of the individual laboratory. Some workers adhere more or less to the criteria established for *S. aureus* phage typing (Blair and Williams, 1961) as follows:

50 plaques to confluent lysis	++
20–50 plaques	+
Less than 20 plaques	±
No lysis	Nontypable

According to this scheme, all reactions of 50 plaques to confluent lysis are called strong reactions. Lesser reactions may be entered in the laboratory record, but the phage pattern, as reported, includes only major (strong) reactions.

Many workers, including the author, believe that fewer clear plaques, on rare occasions even one, constitute a typing reaction. The statistical verification in the author's laboratory is better, in fact, for this concept than for the 50-plaque miminum for typing reactions (Zierdt and Marsh, 1962). The arbitrary cutoff number of 50 plaques, if strictly practiced, is an artificial standard. Too often this results in inclusion of phages in a pattern during one typing, while requiring that they be deleted in a subsequent typing because they are under the required 50-plaque minimum.

A few more general rules apply. One of these is that it is much easier to relate phage patterns of strains with great certainty when they are derived from the same eipidemic and the same ward in a closely knit time frame. There is somewhat less confidence when strains are related between wards, and somewhat less again when the entire hospital is considered. When long time intervals are encountered in strain isolation and typing, phage pattern relatedness decisions are more tenuous. Again invoking the *S. aureus* phage typing analogy, the present-day 80/81 strain is not the same as the 80/81 pandemic strain of 1957–1965.

The difference of two phage reactions criterion (Blair and Williams, 1961) is not a useful tool in interpreting *P. aeruginosa* or *S. aureus*

phage patterns (Smith, 1972). This rule states that a difference of
more than two phages between two strain patterns indicates nonrelatedness. The rule is a great oversimplification and in practice soon
founders in a sea of dichotomies, except for the shorter, strong patterns, where it is neither necessary nor expedient to invoke the rule.
For that matter, the shorter, strong patterns also are exceptions to
the influence of space and time on relating patterns with certainty.
Parker et al. (1975), speaking for a symposium panel, recommends
that a difference of three phages be permitted in *P. aeruginosa*
phage typing, for identity of two or more strains. A better recommendation is not to impose a limitation at all.

Bacteriophage patterns of a single *P. aeruginosa* strain are subject to considerable variation (Bergen, 1972; Beumer et al., 1972;
Pavlatou and Kaklamani, 1962; Zierdt and Schmidt, 1964). Thus
different typing patterns are possibly indicative of only one epidemiological strain. When there is rigid acceptance of differing typing
patterns as separate bacterial strains, serious errors can be made.
Intelligent inspection of phage patterns from repeated tests avoids
errors and leads to accurate interpretation. It is usually seen that
there are key phages in the longer patterns. These provide a line
of identity during repetitive typing that permits the strain to be recognized and avoids the pitfall of assigning strain status to each different pattern. Actually, single strains are the expected finding at
one site or even any site of one patient, so that if apparently different strains are evidenced by phage typing, the burden of proof is
to confirm their separateness. It is usual that a patient carry his
P. aeruginosa strain for lengthy periods, up to many years, or often for life in the case of cystic fibrosis patients. Obviously, long
phage-patterned strains of *S. aureus*, or *P. aeruginosa* in the rarer
situation, might delete most of those phages on a repeated typing and
regain them on the next. More important is the presence of phages
in one pattern that are not represented on repeated typing in another
strain otherwise having the same pattern. Relating phage patterns
is done with most confidence on sequential *P. aeruginosa* isolates
from the same patient, even if the patterns vary considerably. The
best way of determining relatedness of strains through phage patterns is to view them all side by side, with all of each patient's patterns in chronological order and the patient's strains separated by
ward. There may be a real hesitancy in decision making as to whether
two *P. aeruginosa* strains are the same based on a set of repeated
typing reactions. A difficult decision whether to lump as one a number of strains having similar but different phage patterns may influence workers to avoid that doubt by assigning separate strain status
to each of them. The latter interpretation seems to place the responsibility on a higher power.

With phage typing of *S. aureus*, long-term close reproducibility
of the shorter patterns is close to 100% in a laboratory repeatedly

using the same lots of lysate, following the same technique, and utilizing a lone worker. However, when separate laboratories type the same set of strains, this rule is not applicable. This should be somewhat of an embarrassment to workers who proposed that standardized phage typing techniques would result in the same phage patterns for selected strains typed in distant laboratories. This has not been borne out in practice, with the exception of a few very strong and clear-cut strain reactions, for example, 3B/3C/55/71, 187, and perhaps 80/81. The complexities of lawn density, medium, subtle phage mutation, variation within the strain, and human foibles in other aspects of an intricate technique preclude this laudable hope. Relating this to the situation in P. aeruginosa typing and to what is known of the difficulties there, it is abundantly evident that the benefits of phage typing patterns as derived in a particular hospital are applicable only to that hospital, and to no other, even if the same set of *ostensibly* identical phages is used in different hospitals. However, results obtained from a typing center may be applied to all hospitals using that center's service.

Variation in phage type was studied after culture transfer, lyophilization, and selection of single colonies after primary typing (Beumer et al., 1972). Of 100 strains, 23 underwent type changes after one transfer, 39 of 100 after lyophilization, and 13 of 29 (45%) strains changed phage type after selection of single colony types. However, in experiments of this type, one must not assume stability of the phage type (or pyocin type) of the strain before subculture, lyophilization, or colony type selection. These manipulations are performed over a base of variation, so that to measure the variation-enhancing effect of lyophilization, for example, one should be able to subtract base-line variation from imposed variation. This was not done by Beumer et al. Experiments could be designed with selected strains to better establish natural variation so that the effect of lyophilization could be measured. If lyophilization indeed has a genetic effect expressed as alteration in phage type, it would raise the number of typing changes to a new level and could be expressed as the difference between lyophilization effects and base variation assuming adequate statistical analysis.

Sjoberg and Lindberg (1968) reported reproducible phage patterns from typing of 667 strains from their hospital environment, although only 64% of the cultures returned the same pattern after storage for 4 months. Successful use of phage typing in hospitalized burn patients has been reported (Gould and McLeod, 1960; Graber et al., 1962).

Wretling et al. (1973) attempted to correlate P. aeruginosa serotyping, antibiograms, and phage typing with synthesis of protease, elastase, DNase, RNase, lecithinase, egg yolk factor, staphylolytic enzyme, and hemolysins to different red cells. No correlation between antibiogram and any enzyme or toxin existed. Wretlind et al. noted that most strains from any body site produced the substances studied: "No statistical difference in the qualitative production of these various metabolites was found between the two groups or between individual serotypes or phage-typing patterns."

Tejedor et al. (1982) reported that phage active on *P. aeruginosa* was present in the sputum of eight patients with *P. aeruginosa* bronchopulmonary colonization. Since the organism was also present in the sputum, this would be the expected finding. Early on, *P. aeruginosa* phage was detected in pus from *P. aeruginosa*-caused abcesses.

PYOCIN TYPING

History

Bacteriocin is the generic name for the soluble or particulate proteins directly related to bacteriophage that are incapable of reproduction but have lethal action on strains of the host species, with a specificity similar to the whole virus. Pyocin is the bacteriocin of *P. aeruginosa*. *Pseudomonas aeruginosa* bacteriophages and pyocins are often produced together, free in the culture broth.

Pyocins were first described and were named by Jacob (1954). Pyocin typing of *P. aeruginosa* was first done by Holloway (1960), who employed techniques first described by the American bacteriologist Fisk (1942a,b), who innovated isolation of *S. aureus* phages and their use in typing. The Fisk technique consists of superposing drops of the unkown, possibly phage- or bacteriocin-producing strains on just prepared lawns of an indicator set of distinctive strains. Phage or "cine" production (spontaneous) is read after incubation as a clear zone about the growth spot of the producing (superposed) strain. Phage plaques may be distinguished on the periphery of this clear, lysed zone, but further "plaquing" must be done to be sure. It is inevitable that some reactions read as pyocin activity are actually caused by phage. Assuming that a questionable suspension of phage or pyocin is pure, it is easy to distinguish the two by dilution and dropping onto a lawn of a susceptible host. Phage eventually appears as viral plaques, while pyocin activity more quickly disappears with dilution, never forms plaques, and dissipates as a diffuse partial inhibition.

Within recent years the particulate nature of many *P. aeruginosa* pyocins (also called aeruginocins) has been established. Most of these are rodlike (rhapidosomes) and have been related to phage tail material. They are lethal to strains of *P. aeruginosa* to which they are specifically attached at pili or other receptor sites. Thus these parts of phages have the attachment characteristics of the whole phage and are lethal to the cell without any possibility of an injection and replication sequence. A collection of pyocins selected on individuality of action has much in common with a set of typing phages. At least 12% of pyocin reactions during typing are actually phage reactions (Bobo et al., 1973; Jones et al., 1974a).

Exciting discoveries of the nature and the mode of action of pyocins have been made in recent years. Takeya et al. (1967) described a small rod-shaped pyocin. Kageyama and Egami (1962) and Kageyama (1964) described pyocin R, inducible by ultraviolet light or mitomycin C, as a simple protein. By sedimentation analysis and electron microscopy, they showed pyocin R to be a rodlike particle with a similarity to the bacteriophage tail. The rod is a double hollow cylinder 120 μm in diameter consisting of a sheath and a core (Ishii et al., 1965). The sheath is capable of contraction. The mode of action is described as inactivation of ribosomes, with subsequent cessation of RNA, DNA, and protein synthesis (Kaziro et al., 1964). Pyocin R is unaffected by trypsin, even though it is a pure protein. Higerd et al. (1967) and Govan (1974a,b) found that one pyocin resembled headless contractile phage tails. It was necessary for contraction to occur before lethal action (disruption) on P. aeruginosa. When lipopolysaccharide receptors for contractile pyocins are extracted from the bacterial cells and added to pyocin suspensions, they attach themselves to and thereby neutralize the pyocins. The receptor-attached contracted pyocin particles have no action when added to sensitive bacteria. Attachment, but not contraction, occurs at 4°C and lethality is prevented.

Pyocin Sensitivity (Passive Typing)

A set of pyocin-producing strains is selected that produces as little phage as possible. It is not possible, of course, to find strains that are not mixed with phage, so strains are selected whose free phage is least likely to furnish overt lysis on the strains to be typed. An indicator set must be maintained for quality control of the pyocin stock suspensions, by cross-lytic reactions. Osman (1965) used four pyocins for typing and Dasomson et al. (1970) used a typing set of 24 pyocins. Jacob et al. (1973) also typed by passive typing. Bobo et al. (1973) reported that typing by pyocin sensitivity (passive) was more prone to changes in pattern and lack of correlation with serotyping and phage typing than was typing by pyocin production (active).

Rampling et al. (1975) described a set of "phage-free" pyocins for testing by passive or sensitivity typing. Pyocin suspensions as thin layers in Petri dishes were treated for 10 min with ultraviolet light to inactivate phage, but no data were given in which tests were made for phage reduction after ultraviolet treatment or for the effect of ultraviolet light on the presumably all-protein pyocins. These authors reported that typing by 17 stock pyocin suspensions was equal to the other method.

Pyocins produced in broth may be increased with ultraviolet light, shaking (Williams and Govan, 1973), mitomycin C (Farmer and Herman, 1969; Jones et al., 1973), or in broth with 1% potassium nitrate. The addition of 1% potassium nitrate to Trypticase Soy Broth (Baltimore Biological Laboratories) without glucose is simple and produces high titers of pyocin when inoculated with the producing strain and incubated at 32°C for 18 hr. The organism is then killed by saturation of the medium with chloroform. Centrifugation to remove debris is usually not necessary. Filtration results in unnecessary loss of pyocin by absorption to the filter. This technique is useful for producing pyocins from a standard *P. aeruginosa* set for passive typing by sensitivity. It is also apparent to the reader that the broth technique is equally applicable to active pyocin typing (pyocin production by unknown strains).

Reproducible pyocin typing patterns are greatly dependent on pyocin concentration (Jacob, 1954; Jacob et al., 1973; Osman, 1965; Wahba, 1963; and others). It would be too labor intensive to type by pyocin *production* with a *standardized* dose of pyocin. Since the unknown strain is the producer, the routine test dose (RTD) of the produced lysate would have to be determined on member strains of the indicator set. This RTD would then be applied to each culture of the indicator set. A strong case could be made for pyocin typing by sensitivity using a titrated and stored standard pyocin set, where all that would be required for typing would be to drop standardized pyocin suspensions onto lawns of the *P. aeruginosa* strains to be typed. According to Merrikin and Terry (1972), typing by sensitivity required pure, titered pyocin suspensions induced by mitomycin.

In an attempt to partially automate pyocin typing, Farkas-Himsley and Pagel (1977) used pyocin typing in the test tube and measured leakage of ultraviolet absorbing material from pyocin-lysed *P. aeruginosa*. They reported lysis patterns corresponding to those obtained by the conventional method.

Pyocin Production (Active Typing)

Since pyocin production by the unknown strains to be typed cannot be well controlled in content or strength, it would seem that the controlled system of typing by pyocin sensitivity (passive typing) is superior where the pyocins can be titered and quality controlled. Such has not proven to be the case, at least when world use of the two systems is measured.

The "scrape-and-streak" method of pyocin typing by production (Wahba, 1963; Gillies and Govan, 1966) involves a central streak of the unknown strain on horse blood agar. During the 24-hr incubation pyocins are produced and diffuse into the agar. The bacterial growth is removed by adroit scraping with a glass slide, and chloroform is added to kill remaining bacteria. The chloroform is allowed

to evaporate and indicator strains are streaked at a right angle to
the former central streak area. Incubation continues for 18—24 hr,
during which time pyocins from the original streak may kill some of
the indicator strains (Fig. 5). The killings are recorded as consec-
utive pyocin numbers which then may be shortened, according to
a conversion scheme, to a single-numbered pyocin type. Because
of lack of sensitivity in the original Gillies-Govan scheme, five more
indicator strains were added (Rose et al., 1971; Govan and Gillies,

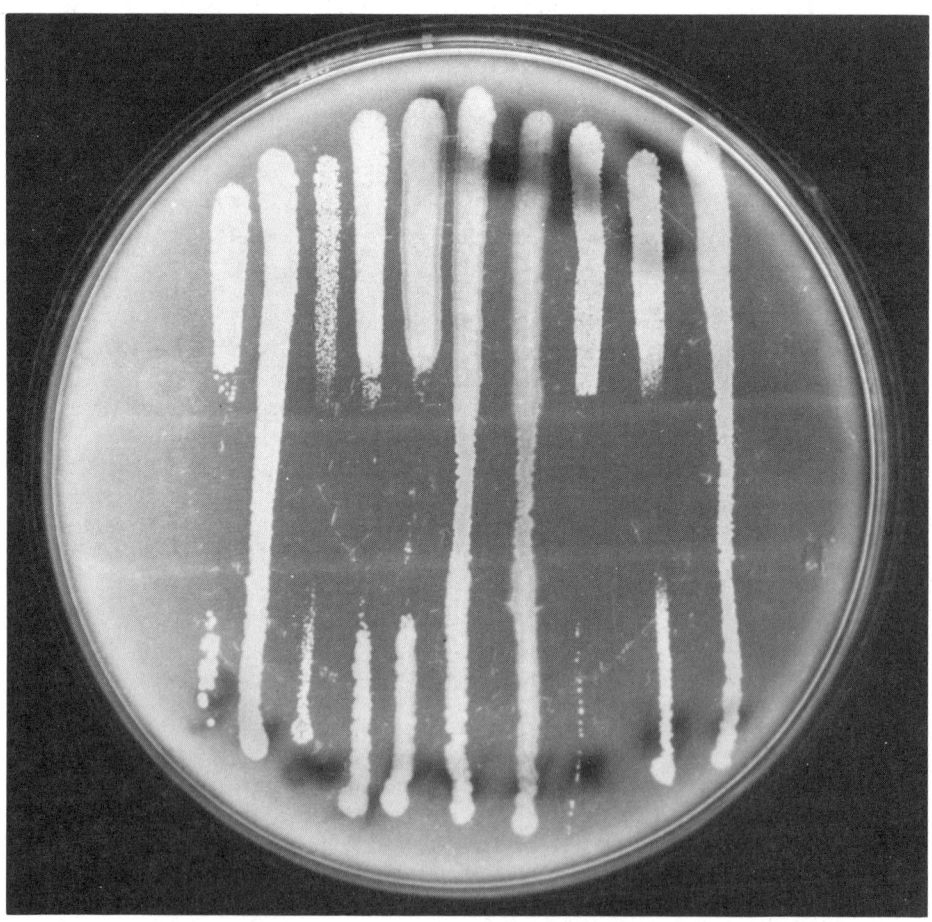

Fig. 5. Pyocin typing by pyocin production. The visible area is
where the center growth of P. *aeruginosa* strain to be typed has been
removed. The agar was cross-streaked with 10 strains of the indicator set, 6 of which were killed.

1969). Bobo et al. (1973) favor typing by pyocin production, as they state that pyocin production is a plasmid-determined character, more stable than the refluxing state of cell wall receptors. On the other hand, the evidence seems to indicate that the produced pyocins are also fluctuant in nature.

To avoid the chores of scraping the bacterial growth from the agar and chloroform-sterilizing the agar before applying indicator strains, Barnes and Gitta (1979) insert filter paper between an agar base layer and a top layer on which the pyocin producers grow. Removal of the top layer is done by lifting the filter paper, leaving the lower layer with diffused pyocins ready for inoculation with the indicator strains.

Active typing may also be done by a broth method (Farmer and Herman, 1969; Jones et al., 1973; Williams and Govan, 1973). Major advantages of the broth method are (a) increased sensitivity, (b) a phage applicator (multiple-syringe dropping device) device may be used to apply pyocin suspensions in broth to the indicator set, and (c) a much larger indicator set can be used because of the labor saved by the phage applicator. In spite of these advantages, the technological ease of the scrape-and-streak technique seems to assure its continued dominance.

Bergan (1968) described techniques of pyocin typing by production. To achieve more clear-cut pyocin clearing reactions, Bergan removed all of the bacterial growth strip, including the agar, before cross-streaking with indicator strains. Küchler (1976) and Küchler and Gunther (1976) had better success in relating strains by pyocin production using more than one indicator set and comparing reactions on one set to those on the other.

For typing by pyocin production, the 18-strain Alabama (ALA) indicator strain set developed at the University of Alabame (Jones et al., 1974a,b) is widely used. A strain is grown overnight in nitrate broth, diluted 1:400 in 4 ml of the same medium, and then poured onto a very dry trypticase soy agar plate. Excess liquid is removed, and the plate is allowed to adsorb residual inoculum suspension for 1 hr before the pyocin preparations are applied. Some workers find shaken cultures advantageous (yielding higher-titered lysates and better-quality lysates), as they produce a larger maturation and burst of the desired pyocin particles. Less hazard was thereby ensured from the action of possible accompanying pyocins of different *P. aeruginosa* strain specificity.

Better results were reported by typing only colonies of certain morphology, following Shionoya and Homma (1968), who, in a study of phenotypic colony differences, reported two general types in most cultures. The first, termed la, is a large, flat, moist colony, while the second, sm, is a small, round, convex colony. Applying phage, pyocin, and serotyping to these two major colonial types, they concluded that distinct differences in phage and pyocin patterns are usual, but that serotypes of the two colonial morphologies (from the same strain) are the same. If phage and pyocin typing are restricted to the 1a colony growth, results are more reproducible.

When grown in broth the sm colony forms a heavy pellicle. It also tends to agglutinate spontaneously during serotyping. The 1a colony grows in even suspension in broth and does not spontaneously agglutinate. Therefore, even though the serotype of the two colonies is the same when it is possible to type both of them, the 1a type is selected because of its superior growth characteristics. Naito et al. (1972a) conclude that the 1a type colony should always be selected for pyocin typing. No differences in antibiotic susceptibility were noted between any colony types. Zierdt and Schmidt (1964) earlier described marked phage pattern differences between colonial types of *P. aeruginosa*.

Colony selection is essential to the best technique, whether it be the pyocin, phage, or serological typing technique. The special difficulties posed by mucoid strains make it most important that these strains be streaked out to achieve well-isolated colonies, so that nonmucoid 1a type dissociant colonies may be selected for subculture. Even the most strongly mucoid strains usually throw off nonmucoid colonies on first plating. This small convex colony (sm) is the most usual first dissociant. It is suggested that the sm colony be used for typing (be it pyocin, phage, or serotyping) if the 1a colony is not present. Rather than attempt typing the pure mucoid form of *P. aeruginosa*, it may be better not to type the organism at that time, but to continue to look for the 1a colony type. It is preferable to select an inoculum by harvesting single colonies of known morphology, but if subculture of the 1a colony is made so as to achieve mass growth, there is little chance that dissociation may occur within the mass growth sufficient to alter the typing reaction.

A low incubation temperature, 30 or 31°C, was used by Shriniwas (1974) in preference to the usual 36–37°C. It has been stated that it provides more reactions that are more clear-cut in appearance, with less tendency to develop secondary overgrowth.

Wahba and Lidwell (1963) have devised a simple mechanical apparatus that assists in pyocin typing by stamping the indicator strains onto the agar. A "broomette" handle of aluminum, with 10 parallel stainless steel prongs inserted, is used by Tagg and Mushin (1971) to dip into an aluminum base with 10 wells holding indicator strains of *P. aeruginosa*. The prong tips are then drawn lightly across the typing plate to simultaneously inoculate the indicator strains. For dropping pyocin suspensions, for either active or passive pyocin typing, one of the multiple-syringe applicators described in the phage section can be used.

Pseudomonas aeruginosa Typing

Reading Reactions

Much of the description of phage reactions and the handling of phage patterns applies to pyocin typing. In most laboratories the phage reactions encountered during pyocin typing are included with the pyocin reactions. This is an operational compromise, but to differentiate between phage and pyocin would require a plating step, burdensome to a routine procedure. The phage reactions have sensitivity at least as good as those from pyocin.

Reading of pyocin lytic reactions requires interpretation and judgment. Expertise is acquired with experience. As with phage typing, pyocin typing plates are best read by one person for an entire study. There are no plaques to indicate sensitivity to pyocin. The presence of one plaque might be termed positive in phage typing. Such fineness is not available in pyocin typing. Luckily the majority of pyocin lethal actions are complete, baring the agar surface. The degree to which lesser reactions are perceived is up to the reader. This large gray area is best approached with caution. The simplest scheme is probably the best. A decision could be made to read all 100% lysis reactions as positive, and all lesser reactions, where there is bacterial growth in the inhibition area, as negative. Isolated, resistant colony growth in the area may be disregarded and the reaction called positive. Pyocin effect on organism growth is seen also as a variety of surface changes vis-à-vis the growth outside the pyocin area. Unless growth is severely diminished, these changes should be regarded as negative typing.

Jones et al. (1974a) devised a system of symbols such as t_w, while retaining the positive reaction. While the weak designation should not be reported, it may be helpful, when retyping the same strain, either repetitively or as sequential isolates from environment or patients.

The longer pyocin patterns may be abbreviated to a single digit according to a conversion code for sequences of patterns of pyocin designations. Farmer (1970a,b) uses a mnemonic based on assigning one number to a known order of three numbers in longer pyocin patterns, thus reducing their length by a factor of 3. Chadwick (1976a) devised a code of nomenclature for pyocin patterns to simplify reporting. These systems are particularly helpful if the pyocin pattern is long. Since pyocin typing sets include as many as 27 pyocins, a significant percentage of patterns are indeed long.

Typing reproducibility varies greatly among reports on the subject. There are many reports citing constancy in repeated pyocin typing of *P. aeruginosa* strains and in the typing of new isolates from the same patients. These reports are a source of puzzlement to workers whose notebooks are full of data indicating less than satisfactory reproducibility. It has often been recommended that the relative stability of serotyping be combined with the greater sensitivity of pyocin or phage typing. Tripathy and Chadwick (1971) cited many possible reasons for change in pyocin patterns. They advocated inclusion of 0.5 µg/ml of mitomycin C in the typing agar to induce more pyocin from producing strains and to reduce residual growth in the lysing area. They found the pyocin pattern to be dependent on the method used. Csiszár and Lányi (1970) reported typing of 543 strains by Gillies and Govan's scrape-and-streak pyocin production technique, and by Lányi's serotype scheme. They found that strains within each serotype belong to one or two pyocin types, revealing an important relationship between O antigens and pyocin types of P. aeruginosa.

In a study of pyocin pattern variations Chadwick (1972) described the predominant pattern of a strain, with variant patterns of this strain still recognizable as the same strain. He believed that variation in the pattern "may reflect instability in the organisms themselves and may have to be accepted as normal phenomenon."

Only the fluorescent pseudomonads are pyocin typable, according to Jones et al. (1974a). These are *Pseudomonas fluorescens, Pseudomonas putida,* and *P. aeruginosa*. A much lower percentage of *P. putida* typed (13 of 58 strains) than *P. aeruginosa* (97 of 100 strains). Apyocyanogenic strains of *P. aeruginosa* typed poorly (4 of 26 strains), and only 18 of 38 *P. fluorescens* strains were pyocin typable. These results are similar to those that Yabuuchi et al. (1971) reported based on serological typing.

Naito et al. (1972a) reported on the reproducibility of patterns by both pyocin production and pyocin sensitivity. In a report (Naito et al., 1972b) on typing by sensitivity using the method of Darrell and Wahba (1964), they concluded that there are frequent changes in pyocin patterns on repetitive typing. Only 19 of 95 strains retained the same pattern after 5 months in cultivation.

Use in Epidemiology

Development of pyocin typing methods was followed by its use in studying the epidemiology of *P. aeruginosa* infections in hospitals. This is evidenced by a number of reports in addition to those already discussed (Bruun et al., 1976; Dayton et al., 1974; Deighton et al., 1971; Farmer and Herman, 1969; Heckman et al., 1971; Shriniwas, 1974; Tagg and Mushin, 1971; Zabranski and Day, 1969). These were studies generally confirming the usefulness of the technique. Tinne (1977) reported a single strain of *P. aeruginosa* causing severe infections in patients in a large general hospital. This strain, pyocin type 10, had the unusual characteristics of ability to colonize patients and virulence in causing disease. It was not seen in cultures of the environment and presumably was transmitted from person to person. Chitkara et al. (1977) reported pyocin typing (active) in a hospital in India, with reported relative constancy of pyocin types.

Falkiner and Keane (1977) compared typing by "active" pyocin production and "passive" pyocin set sensitivity in nine outbreaks of infection. They found acceptable correlation in six of the nine. Of the remaining episodes typing by pyocin production in the unknown strains provided the better evidence. These results confirmed those of Rampling et al. (1975). Even so, the best correlation achieved between active and passive typing was about 75%.

Active pyocin typing (pyocin production by unknown strains) was used by Baltch et al. (1979) to look for correlation between pyocin type and levels of lecithinase, protease, and elastase among strains isolated from patients succumbing to bacteremia versus those surviving bacteremia. These strains were also contrasted with *P. aeruginosa* strains isolated from sputum, urine, and skin. No clear-cut strain differences were detected, except that the highest enzyme producers came from septicemia. One-third of all strains produced minimal or no extracellular enzymes regardless of their source. There was no particular pyocin strain incriminated in this study.

Leg ulcers on cardiology patients in a Florence, Italy, hospital (Urbano et al., 1983) were usually caused by *P. aeruginosa*. Most of the ulcers were caused by a single serotype and pyocin type strain. Infecting strains in this study could not be related to environmental strains, but cross-infection between patients was demonstrated.

Wahba (1965b) reported that pyocin typing by sensitivity using the 10-strain set gave insufficient information, but that pyocin plus serotyping provided 22 groups and was useful.

BIOTYPING

Biotyping might logically include antibiotic testing, pyocin and phage testing, and serotyping, but in this section is restricted to the classic biochemical reactions, odor, pigmentation, colony form, and motility. Perhaps in this sense biotyping is most useful as an alarm signal to the possible presence of an infectious outbreak, to be followed by more certain typing procedures.

A classic example of biotyping *P. aeruginosa* is of course monitoring the mucoid strains that are a flag to cystic fibrosis, which are discussed separately. To these tests may be added description of colony morphology. Except in unusual circumstances, colony morphology differences cannot be trusted for *P. aeruginosa*, for often colonial dissociation continues from any selected colony form (Zierdt and Schmidt, 1964).

In the absence of more sensitive systems, a carbohydrate profile can be worked up of acid production from glucose, arabinose, xylose, mannose, galactose, and ribose, in oxidative–fermentative medium base. These reactions are 96–100% positive for *P. aeruginosa*. In addition, tests are in order for motility, arginine dihydrolase, nitrate use, and citrate use. Asence of pigment in *P. aeruginosa* is a particularly useful marker, as is presence of the brown-red pigment pyorubin. Motility is rarely absent in *P. aeruginosa*, so that when this occurs it is a very nice marker.

From this phalanx of tests a useful profile or fingerprint may arise that can be used to identify an epidemic strain, under the constraints to be discussed. The cautious investigator can make use of unequivocal biochemical differences in strains of a species.

Smith (1983) expressed the difficulties inherent in the use of biotyping in epidemiology. He cited studies that reported inadequate reproducibility during replicate testing. For example, 110 Enterobacteriaceae isolates retested on the API 20E system yielded identical results on only 55.5% of the tests. "The clinical microbiologist should be alware that it is risky to identify various strains of epidemiologically significant organisms solely on the basis of one or two differences in their biotypes unless he or she is willing to invest additional time and effort to define the normal variation to be expected in these same cultures."

The sharp observer can make use of any distinguishing bacterial characteristic in tracing a single strain epidemiologically, but it would be rare indeed for two or more strains to exhibit separate identifying characteristics in the same outbreak. Recently a series of cultures were sent to the author for typing, with the information that the imputed epidemic strain had a distinguishing musty odor. Indeed it did, and it was confirmed as a single strain by serotyping. Thus odor judged by sophisticated noses is useful in biotyping. This type of characterization is more likely to be taken seriously by medical staff if confirmed by more accepted typing methods.

ANTIBIOGRAMS AND MISCELLANEOUS TYPING TECHNIQUES

The same precautions given in the section on biotyping apply fully to this section. Perhaps the key requisite to the use of antibiotic susceptibility differences in strain separation would be repetition of the antibiogram on suspected epidemic strains. Subculture of itself can change the antibiogram, as resistance may change to sensitivity. In any case, subculture is an unfortunate necessity before further testing. The possible loss resistance is a risk that cannot be avoided.

Antibiogram differences have value in strain tracing of *P. aeruginosa* only where there are outstanding differences. Resistance to carbenicillin, colistin, gentamicin, amikacin, and tobramycin can be used as epidemiological markers. Sensitivity to kanamycin, streptomycin, chloramphenicol, and tetracycline can also be used as epidemiological markers for *P. aeruginosa*, where a difference from the expected is encountered.

It should always be kept in mind that during therapy, especially multiple therapy, resistance to many antibiotics via R factors or by other mechanism can develop rapidly. Antibiogram differences in sequential *P. aeruginosa* isolates from a patient may be reflecting this. An assumption that the differences indicate distinct strains would be incorrect. Appropriate controls must be included in the use of antibiograms for typing purposes. If not, conclusions reached from the data can be completely at odds with the facts. In no case should results be reported that have not been repeated to check reproducibility. The literature on antibiotic susceptibility testing is full of conclusions drawn from data gathered without appropriate controls. Demko and Thomassen (1980) stated that degrees of heterogeneity of antibiotic susceptibilities were common among isolates of the same serotype of *P. aeruginosa*, even among colonies chosen from a single sputum culture. As discussed here and in other sections, strains may be characterized by phagocytosis rates; serocidal assay; quantitation of exotoxin A, protease, elastase, and lecithinase; virulence in various infectious models; mucoid quality; and pigment production. *Pseudomonas aeruginosa* strains may be characterized by their stable rate of phagocytic uptake following Michaelis−Menten kinetics, according to Winnie et al. (1983). These may hardly be considered standard typing methods, but they have value as strain markers in the epidemiology of infections, and certainly in research studies of *P. aeruginosa* infections.

Differences in minimum inhibitory concentration of one doubling dilution in the tube dilution test form no basis for epidemiological characterization of strains. A two-tube difference (fourfold) in minimum inhibitory concentration would also seem to have doubtful value in this regard, unless the distinction is carefully documented by repeated testing. Differences of three tubes (eightfold) or greater are acceptable evidence for epidemiological surveillance.

In the fortunate case where the *P. aeruginosa* strain involved has a unique antibiogram, such as one with a marked resistance to colistin, then this antibiogram provides a legitimate marker for epidemiological purposes, but the laboratory that seriously plans epidemiological surveillance of *P. aeruginosa*, should have another method at hand. Bobo et al. (1973) were of the opinion that antibiotic susceptibility alone is unreliable as an epidemiological tool, especially for strains with colonial dissociation, where different colony forms tended to vary in antibiogram.

Tagg and Mushin (1971) studied strains from 3 New South Wales hospitals and 10 Victoria hospitals, particularly by pyocin typing but also by reaction to 10 antibiotics and sulfonamide. Over 10% of 372 strains were susceptible to all drugs, and 5% were resistant to all drugs. Some endemic strains of pyocin type UC were susceptible only to chloramphenicol and tetracycline, while some UCA strains were additionally susceptible to streptomycin and occasionally to sulfonamide and kanamycin. It was noted that the antibiograms changed rather capriciously on retesting. Neomycin, colistin, polymyxin, and gentamicin were unsuitable for typing purposes, as 364 of 372 strains were susceptible to all. A useful division of strains was sometimes possible with designations of either resistance or susceptibility to streptomycin, chloramphenicol, tetracycline, sulfonamide, and kanamycin.

In an outbreak of surgical infections caused by P. aeruginosa, resistance to gentamicin and tobramycin was noted (Falkiner et al., 1977). This resistance was transferred in vitro to a recipient strain of P. aeruginosa. Resistant strains with the same serological, phage, and pyocin type were cultured from urine bottles, bedpans, and the hands of attendant staff. Inadequate disinfection played a major role in cross-infection.

Typing by antibiogram alone was practiced in a general ward where five patients died in a year (Williams et al., 1960) from P. aeruginosa septicemia. Using streptomycin, tetracycline, chloramphenicol, and sulfonamide, 5 strains were detected among 16 isolated cultures. No attempt was made to repeat the antibiograms. It is evident that the epidemiology would have been strengthened by adding at least another typing system.

Bobo et al. (1973) compared five different typing techniques: pyocin production, pyocin sensitivity, serological agglutination, antibiotic susceptibilities, and phenotypic properties such as pigmentation and colony form. The nursery epidemic studied by the five techniques was caused by a single strain of P. aeruginosa which spread from one infant with P. aeruginosa pneumonia to resuscitation equipment and on to eight other infants. Invoking the five typing techniques separately, it was concluded from pyocin production and serotyping that a single strain was causing the epidemic. Typing by pyocin sensitivity indicated that two strains were involved. Typing by antibiogram or by miscellaneous markers, such as colony form and pigmentation, indicated that four different strains were causing the infections. Thus typing by pyocin production or serotyping gave correct results, while each of the other methods by itself gave incorrect results. Antibiogram patterns varied in sulfadiazine and sulfathiazole reaction from resistant to susceptible. Pigmentation varied from subculture to subculture, according to which colony type was chosen for subculture. The colonial types themselves further dissociated on subculture. Pigmentation was entirely due to pyocyanin, and differentiation was on the basis of shades of green or blue-green. Melanin-(pyorubrin) producing strains are unique and rare. Because of its rarity, melanin production is a more reliable marker than pyocyanin, but, also because of its rarity, it is of little use in epidemiology.

Serotyping and antibiogram typing were combined to study an epidemic of P. aeruginosa infections in a pediatric burn unit (Dayton et al., 1974). A strain of P. aeruginosa resistant to 250 µg of gentamicin and carbenicillin infected a 4-year-old girl and then spread to 18 other patients. There were 51 patients in the ward at this time.

All 19 strains were immunotype 1 of Fisher (Parke-Davis). The other strains of different serotypes present in the ward at this time were susceptible to gentamicin and carbenicillin. The child first presenting infection by this strain had no antibiotics up to that point. It can be concluded that the exceptional strain showing extraordinary resistance to a drug or drugs to which most strains are susceptible can be considered a unique and stable strain fingerprinted by its antibiogram so that it can be followed epidemiologically with the control methods described.

In general, the methods discussed in this section are best combined with serotyping, phage typing, or pyocin typing to obtain reliable information for use in epidemiology. Granted that the rarer strains with unusually strong and stable deviance from the mode may be identified by these markers, the rarity of their occurrence calls for the availability of at least one other technique. These were also the conclusions of Brokopp and Farmer (1979).

MUCOID STRAINS

Mucoid strains of bacterial species other than *P. aeruginosa* have been reported in cystic fibrosis (mucovisicidosis) patients (Macone et al., 1981). The condition is far more frequent for *P. aeruginosa*. The mucoid factor is transducible. It may be transiently induced in random strains through growth on *P. aeruginosa*-depleted agar media and may be perpetuated by mineral additions to media. Although the mucoid quality is rapidly lost in subculture, rare strains are mucoid stable. It may be maintained by colonial selection in subculture. The polysaccharide itself is a copolymer of mannuronic and guluronic acids in various ratios. It is described as an alginic acid (Evans and Linker, 1973) similar to that found in brown seaweed. It is produced in large quantities in shaken cultures grown in synthetic broth. No such benign assignation as alginic acid is granted to *P. aeruginosa* exopolysaccharide by Sensakovic and Bartell (1974), who found it to contain fatty acid moieties similar to those in *P. aeruginosa* lipopolysaccharide and to be two to three times more toxic than LPS to mice. These materials may have been contaminants in the alginic acid. Bartell and Krikszens (1980) isolated a glycolipoprotein fraction from the exopolysaccharide slime of *P. aeruginosa* to which they assigned the toxic properties.

Markowitz et al. (1978) concluded that polysaccharide from mucoid *P. aeruginosa* was probably not plasmid controlled, but was controlled by chromosomal DNA. Govan (1975) was also of this opinion.

There has been extensive research into mucoid *P. aeruginosa* strains causing colonization of the airways of cystic fibrosis patients. Since the metabolic defect underlying this disease is unknown, the possibility of a basic relationship between cystic fibrosis and the mucoid condition in *P. aeruginosa* has inspired workers to investigate all possible aspects.

So far, no basic connection between cystic fibrosis and mucoid P. aeruginosa has been reported.

Mucoid presence is of itself a form of typing of P. aeruginosa. Mucoidicity is a marker for cystic fibrosis-derived P. aeruginosa and a reliable flag for the disease itself (Reynolds et al., 1976). There is a disclaimer of these facts by Rivera and Nicotra (1982). The latter work was based on retrospective examination of their laboratory results. But much description of mucoid P. aeruginosa in disease other than cystic fibrosis results from confusion of luxuriant, raised, moist growth of P. aeruginosa as mucoid growth. These two conditions are pheotypically completely separate.

In the author's laboratory, of thousands of P. aeruginosa strains isolated over a period of many years, only six were overtly mucoid that were not isolated from cystic fibrosis patients. Of the six patients, four were possibly cases of adult-type cystic fibrosis. They could not be finally diagnosed because of terminal unrelated diseases such as Wegener's granulomatosis and chronic lymphocytic leukemia. Isolation of overtly mucoid P. aeruginosa from any patient, whether or not another disease is present, can be an indicator of cystic fibrosis (Reynolds et al., 1976).

At this point it would seem that not much basic information about cystic fibrosis has come of the great expenditure of research on these mucoid strains of P. aeruginosa associated with the disease. Some workers have reported increased resistance to phagocytosis of the slime-coated P. aeruginosa strains (Schwarzman and Boring, 1971; Roe and Jones, 1974), while others cannot confirm this (Le Blanc et al., 1982; Borowski and Schiller, 1983). The latter reported that a nonmucoid variant was more resistant to phagocytosis and serum killing. They believed that altered, absent, or inaccessible cell surface components of the nonmucoid cells were responsible. Similar results had been published by Hoiby and Olling (1977) as regards serum bactericidal effect.

Isolated IgC antibodies from cystic fibrosis patients' sera were found to have about half of the opsonin activity of IgG from normal sera (Fick et al., 1981). That was stated to be a result of low immunoglobulin affinity for alveolar macrophage membrane Fc γ receptors. These authors believe that the use of rabbit alveolar macrophages are not fully appropriate for experiments with human serum opsonins against P. aeruginosa. Nguyen et al. (1982) compared alveolar macrophages from humans, rabbits, rats, and hamsters for differences in phagocytosis and killing. They concluded that experiments related to human disease should only be done with human alveolar macrophages.

Winnie et al. (1982) induced chronic P. aeruginosa pulmonary infection in cats with repeated intrabronchial injections of agarose bead–bacterium slurries. Serum from these cats inhibited phagocytosis of P. aeruginosa by normal cat alveolar macrophages by 30—79%. This indicates that similar inhibitions by the serum of cystic fibrosis patients is not a phenomenon unique to cystic fibrosis.

One of the benefits from immunological research into P. *aeruginosa* and cystic fibrosis has been extended knowledge by Reynolds and colleagues of specific immunoglobulins and their properties in respiratory secretions from diseased and healthy individuals and their interactions with phagocytosis and phagocytic killing (Reynolds and Thompson, 1973; Fick et al., 1981).

A surprisingly low P. *aeruginosa* colonization rate (50%) in 160 cystic fibrosis patients was reported by Kulczycki et al. (1978), who attributed this highly desirable state specifically to the reduced use of antibiotics, but also to minimal hospital visits and reduced use of mists and aerosols, citing these last as possible sources of P. *aeruginosa* infection.

Henry et al. (1982) determined Schwachman scores (a complex estimation of severity of cystic fibrosis) and forced expiratory volume for 206 age-matched children and adolescents with cystic fibrosis. Sputum cultures were also done to identify those patients infected with mucoid P. *aeruginosa* strains. Schwachman scores and forced expiratory volume measurements were much lower for patients infected with mucoid P. *aeruginosa*. These values were roughly the same for patients infected with nonmucoid P. *aeruginosa* and those with no P. *aeruginosa* colonization at all.

There are reports that mucoid strains are more sensitive to antibiotics (Demko and Thomassen, 1980; Thomassen et al., 1979). A very large data bank (unpublished) in the author's laboratory of agar diffusion as well as microtiter determinations indicates no remarkable difference in antibiograms of mucoid and nonmucoid variants of the same strains. It is possible to relate lower bacterial counts (viable) in suspensions of mucoid growth used as inocula for drug sensitivities to an apparent but false marginal sensitivity increase.

There is much confusion in reporting mucoid P. *aeruginosa* from patients' samples. A very large, convex, moist colony may be interpreted as mucoid when such is not the case, at least not by the definition of mucoid colonies from cases of cystic fibrosis. Reports are often not critically reexamined, and once accepted are transmitted, further strengthening the misdiagnosis. When the definition of mucoid is restricted to those colonies producing overt slime, the confusion may be obviated. Unless the cystic fibrosis patient is studied most carefully and the bacteriologist has clear access to the patient before his colonization with P. *aeruginosa*, it is impossible to say whether it is transferred to him by another cystic fibrosis patient as the fully mucoid organism. The proportion of mucoid to nonmucoid P. *aeruginosa* colonies on the primary culture plates of a patient's sputum varies capriciously from day to day. In the extreme case there may be only nonmucoid colonies on one day and only mucoid on the next (Zierdt and Schmidt, 1964). These different colonial forms are of the same strain, as demonstrated by phage, pyocin, and serotyping (Bergan, 1973; Bergan and Hoiby, 1975; Diaz et al., 1970; Hoiby et al., 1975; Naito et al., 1972a; Shionoya and Homma, 1968; Williams and Govan, 1973; Zierdt and Schmidt, 1964; Zierdt and Williams, 1975). If the lungs of a good percentage of cystic fibrosis patients are infected with one serological or phage type of a mucoid strain, then it is reasonable to assume that most of those patients were infected initially with a mucoid strain of the organism and it did not develop from rough to mucoid after infection. Doggett et al. (1971) believe that the latter situation prevails. If this is the case, most cystic fibrosis-associated strains should be of random serotypes or phage types; however, they are not if a sufficiently sensitive typing system is applied, such as the original serotyping set of Homma (unfortunately disbanded) or an appropriate phage typing schema (Zierdt and Williams, 1975; Hirao et al., 1977). Even typing with the current IATS seroset often uncovers a predominance of serotype 6 in cystic fibrosis patients, corresponding to the old Homma serotype 8, the cystic fibrosis serotype. Other mucoid strain serotypes are involved, as well as nonmucoid strains, and these seem to be randomly distributed.

Most workers have reported more nontypability in typing mucoid strains. The slime has been cited as interfering with attachment by phage and pyocin and masking reaction sites for agglutination. Still, even when fully mucoid most strains are typable by any system, it is usually possible to quickly isolate nonmucoid strains for typing. Even nonmucoid isolates from cystic fibrosis patients have a high rate of nontypability, between 30 and 40%. Penketh et al. (1983) reported that *P. aeruginosa* strains from cystic fibrosis patients who were "relatively fit" had typical cultural and serological characteristics. Strains from cystic fibrosis patients with chronic bronchopulmonary infection had lost their O serotype and acquired a new one and had become sensitive to normal human serum.

DISCUSSION AND RECOMMENDATIONS

The very real problems associated with phage typing and pyocin typing of *P. aeruginosa* have not been squarely faced by many authors. At risk of sermonizing, some of the pitfalls will be reviewed once again. The greatest of these is the erratic reproducibility of some phage and pyocin patterns. Some authors have expressed confidence in this parameter, but an examination of their typing tables may not support such confidence. Also, if the best data are selected for publication, a false picture of reproducibility may be drawn. In the author's experience, repeated phage typing of strains showed marked variation in at least 50% of patterns.

While reliable phage typing of *S. aureus* is not easy, phage or pyocin typing of *P. aeruginosa* must be classed as extremely difficult. This is not meant to imply that phage and pyocin systems should be abandoned for lack of usefulness. Still, too often there is a tendency toward blithe acceptance of typing patterns. The multinumbered patterns applied to a series of strains acquire a certain respectability and credibility simply because they are figures on paper. There is a strong tendency, finally, to accept typing patterns as numbers etched in bronze. This can lead to an interpretation that the patient is changing his colonizing strain almost every week, when in fact the phage pattern has varied from week to week on each new isolate of the same strain. Much of the typing data in the literature on *P. aeruginosa* are based on a single typing per strain. Repetitive typing controls are seldom applied rigidly and objectively. Equally rare is the typing of sequential isolates from the same patient. Not enough attention is given to relating *P. aeruginosa* strains by similarity patterns or, conversely, to requiring distinctive different patterns for strain separation. A strong common cluster of phages in phage patterns tends to repeat and form a nucleus of identity for similarity patterns. Judgmental decisions are necessary to relate these patterns.

The refluxing state of the phages and, to a lesser extent, of the pyocins in *P. aeruginosa* strains could explain type or pattern variation, citing only the bacterial variation, but other factors are certainly involved in a system where the biological reagents are subject to change with each propagation. Granted a research need for the much greater fingerprinting sensitivity of pyocin and particularly phage typing, the techniques are warranted, assuming that the commensurate technical backup is there. There is a very high level of technical expertise and dedication required for the construction, maintenance, and use of a phage set for *P. aeruginosa*. Phage propagation and titration of *P. aeruginosa* phages require more than following directions. It requires ingenuity, discipline, and perseverance. In the average busy hospital laboratory, there may be no time to develop this system. The on-call typing system must be immediately available for use. The gearing-up process for serological typing is fast, but to activate pyocin and bacteriophage systems might require many weeks. The expertise and effort required are far less to set up pyocin typing than phage typing. At the present time there is no distinct established superiority of one pyocin technique over another, except that active typing is probably better than passive typing. Most important is the skilled, critical application of whatever method is chosen, with adequate control cultures in each typing run. There is no substitute for experience, but experience only develops expertise when the worker is enthusiastic, dedicated, and talented.

Differing colony forms of *P. aeruginosa* in a primary culture usually do not represent different strains of *P. aeruginosa*. The reporting of *P. aeruginosa* 1, *P. aeruginosa* 2, and so on, from a clinical specimen because of colony differences is misleading and meaningless unless typing is done to establish that they are of the same or different strains. The observation of multiple colonial forms on single culture plates should be handled carefully. This seldom represents more than one *P. aeruginosa* strain; it almost always is the expression of colonial dissociation in *P. aeruginosa*. The false report of multiple strains in a patient's specimen may seem to be of minor importance, but nevertheless it is a false report and can be damaging to an otherwise well-run *P. aeruginosa* surveillance program.

A picture of the relative usefulness of *P. aeruginosa* typing techniques may be gotten from the volume of use to which they are put. Serotyping is used at least 10 times more often than any other typing system. Pyocin typing is a distant second. Various forms of "typing" by antibiogram, physiological reactions, and colony morphology constitute a more distant third, with bacteriophage typing in use by only a very few laboratories in the world. This is in spite of the fact that the last mentioned is easily the most sensitive in delineating differences between closely related strains. For genetic research and research that requires the added sensitivity to distinguish key strains, phage typing is a necessity. As an example of the latter, very good evidence for a worldwide strain of cystic fibrosis-related *P. aeruginosa* was obtained through phage typing. This evidence was supported by serotyping with the original large serotyping set devised by Homma, a set that was no longer available after Homma's new, shortened set was developed.

To conduct most epidemiology studies, serotyping has sufficient sensitivity. It is very unlikely that more than one unique strain of the same serotype will occur, for example, in the limited ecological niche of a single hospital's environment. Nevertheless, it must be understood that this possibility exists. A single serotype obviously includes, in an expanded environment, strains with distinguishing features in physiological, morphological, pathogenic, and antigenic makeup. Realizing this, some people prefer the term serogroup.

REFERENCES

Ádám, M. M., Kontrohr, T., and Horváth, E. (1971). *Acta Microbiol. Acad. Sci. Hung. 18*:307.

Adams, M. H. (1959). *Bacteriophages*, Interscience, New York, p. 592.

Al-Dujaili, A. H., and Harris, D. M. (1974). *J. Clin. Pathol. 27*:569.

Ansorg, R. (1978). *Zentralbl. Bakteriol. Hyg. I Abt. Orig. A 242*: 228.

Ansorg, R., Schmitt, W., and Schwerk, V. (1978). *Med. Microbiol. Immunol. 165*:181.

Aoki, K. (1926). *Zentralbl. Bakteriol. Parasitenkd. Infektionskr. Hyg. Abt. 1 Orig.* 98:186.
Asheshov, E. H. (1974). *Proceedings of the 6th National Congress on Bacteriology,* Athens (A. Arseni, ed.) Central Public Health Laboratory, London, p. 9.
Ayliffe, G. A. J., Lowbury, E. J. L., Hamilton, J. G., Small, J. M., Asheshov, E. A., and Parker, M. T. (1965). *Lancet* 2:365.
Ayliffe, G. A. J., Barry, D. R., Lowbury, E. J. L., Roper-Hall, M. J., and Walker, W. M. (1966). *Lancet* 1:1113.
Baltch, A., Griffin, P. E., and Hammer, M. (1979). *J. Lab. Clin. Med.* 93:600.
Baltimore, R. S., Dobek, A. S., Stark, F. R., and Artenstein, M. S. (1974). *J. Infect. Dis. Suppl.* 130:553.
Barnes, W. G., and Gitta, P. S. (1979). *Am. J. Med. Technol.* 45: 688.
Bartell, P. F., and Krikszens, A. (1980). *Infect. Immun.* 27: 777.
Bébéar, C., and Dulong de Rosnay, E. (1973). *Bordeaux Med.* 8: 1143.
Bergan, T. (1968). *Acta Pathol. Microbiol. Scand.* 72:401.
Bergan, T. (1972). *Acta Pathol. Microbiol. Scand. Sect. B* 80:177.
Bergan, T. (1973). *Acta Pathol. Microbiol. Scand. Sect. B* 81:70.
Bergan, T., and Hoiby, N. (1975). *Acta Pathol. Microbiol. Scand. Sect. B* 83:553.
Beumer, J., Cotton, E., Delmotte, A., Millet, M., vonGrunigen, W., and Yourassowsky, E. (1972). *Ann. Inst. Pasteur* 122:415.
Blair, J. E., and Williams, R. E. O. (1961). *Bull. WHO* 24:771.
Bobo, R. A., Newton, E. J., Jones, L. F., Farmer, L. H., and Farmer III, J. J. (1973). *Appl. Microbio.* 25:414.
Borowski, R. S., and Schiller, N. L. (1983). *Curr. Microbiol.* 9:25.
Bradley, E. (1972). *Genet. Res.* 19:39.
Bradley, D. E., and Pitt, T. L. (1975). *J. Hyg.* 74:419.
Brokopp, C. D., and Farmer III, J. J. (1979). In *Pseudomonas aeruginosa—Clinical Manifestations of Infection and Current Therapy* (R. G. Doggett, ed.), Academic, New York, pp. 90—133.
Brokopp, C. D., Gomez-Lus, R., and Farmer, J. J. (1977). *J. Clin. Microbiol.* 5:640.
Brutsaert, P. (1924). *C. R. Soc. Biol.* 90:1290.
Bruun, J. N., McGarrity, G. J., Blakemore, W. S., and Coriell, L. L. (1976). *J. Clin. Microbiol.* 3:264.
Burkhart, C., and Shapiro, R. (1980). *Cutis* 25:642.
Chadwick, P. (1972). *Can. J. Microbiol.* 18:1153.
Chadwick, P. (1976a). *Can. J. Public Health* 67:321.
Chadwick, P. (1976b). *Can. J. Public Health* 67:323.
Chester, I. R., and Meadow, P. M. (1973). *J. Gen. Microbiol.* 78:305.
Chitkara, Y. K., King, S. D., and French, G. L. (1977). *West Indian Med. J.* 26:12.
Christie, R. (1948). *Aust. J. Exp. Biol. Med. Sci.* 26:425.

Cross, A. S., Zierdt, C. H., Roup, B., Almazan, R.a, and Swan, J. C. (1983). *Am. J. Clin. Pathol.* 79:598.
Csiszár, K., and Lányi, B. (1970). *Acta Microbiol. Acad. Sci. Hung.* 17:361.
Darrell, J. H., and Wahba, A. H. (1964). *J. Clin. Pathol.* 17:236.
Dasomson, T., Roberts, Jr., C. E., and Panas-Ampol, K. (1970). *Southeast Asian J. Trop. Med. Public Health* 1:391.
Dayton, S. L., Blasi, D., Chipps, D., and Smith, R. F. (1974). *Appl. Microbiol.* 27:1167.
Deighton, M. A., Tagg, J. R., and Mushin, R. (1971). *Med. J. Aust.* 1:892.
Demko, C. A., and Thomassen, M. J. (1980). *Curr. Microbiol.* 4:69.
D'Herelle, F. (1926). *The Bacteriophage and Its Behaviour*, Bailliere, Tindall and Cox, London.
Diaz, F., Mosovich, L., and Neter, E. (1970). *J. Infect. Dis.* 121:269.
Dickinson, L. (1948). *J. Gen. Microbiol.* 2:154.
Dickinson, L., and Codd, L., (1952) *J. Gen Microbiol.* 6:1.
Doggett, R. G., Harrison, G. M., and Carter, R. E. (1971). *Lancet* 1:236.
Don, P. A., and van den Ende, M. (1950). *J. Hyg.* 48:196.
Drake, C. H. (1966). *Health Lab. Sci.* 3:10.
Duncan, N. H., Hinton, N. A., Penner, J. L., and Duncan, I. B. R. (1976). *J. Clin. Microbiol.* 4:124.
Edmonds, P., Suskind, R. R., Macmillan, B. G., and Holder, I. A. (1972). *Appl. Microbiol.* 24:213.
Evans, L. R., and Linker, A. (1973). *J. Bacteriol.* 116:915.
Falcoa, D. P., Mendoca, C. P., Scrassolo, A., de Almeida, B. B., Hart, L., Farmer, L. H., and Farmer III, J. J. (1972). *Lancet* 2:38.
Falkiner, F. R., and Keane, C. T. (1977). *J. Med. Microbiol.* 10:447.
Falkiner, F. R., Keane, C. T., Dalton, M., Clancy, M. T., and Jacoby, G. A. (1977). *J. Clin. Pathol.* 30:731.
Farkas-Himsley, H., and Pagel, A. (1977). *Infect. Immun.* 16:12.
Farmer, J. J. (1970a). *Appl. Microbiol.* 20:517.
Farmer III., J. J. (1970b). *Lancet* 2:96.
Farmer, J. J., and Herman, L. G. (1969). *Appl. Microbiol.* 18:760.
Farmer, J. J., Hickman, F. W., and Sikes, J. V. (1975). *Lancet* 2:787.
Farmer III., J. J., Weinstein, R. A., Zierdt, C. H., and Brokopp, C. D. (1982). *J. Clin. Microbiol.* 16:266.
Fastier, L. B. (1945). *J. Bacteriol.* 50:301.
Favero, M. S., Carson, L. A., Bond, W. W., and Peterson, N. J. (1971). *Science* 173:836.
Feary, T. W., Fisher, Jr., E., and Fisher, T. N. (1963). *Soc. Exp. Biol. Med.* 113:426.

Fick, Jr., R. B., Naegel, G. P., Matthay, R. A., and Reynolds, H. Y. (1981). *J. Clin. Invest.* 68:899.
Fierer, J., Taylor, P. M., and Gezon, H. M. (1967). *N. Engl. J. Med.* 276:991.
Fisher, M. W., Devlin, H. B., and Gnabasik, F. J. (1969). *J. Bacteriol.* 98:835.
Fisk, R. T. (1942a). *J. Infect. Dis.* 71:153.
Fisk, R. T. (1942b). *J. Infect. Dis.* 71:161.
Gaby, W. L. (1946). *J. Bacteriol.* 51:217.
Gillies, R. R., and Govan, J. R. W. (1966). *J. Pathol. Bacteriol.* 91:339.
Gould, J. C., and McLeod, J. W. (1960). *J. Pathol. Bacteriol.* 79:295.
Govan, J. R. W. (1974a). *J. Gen. Microbiol.* 80:1.
Govan, J. R. W. (1974b). *J. Gen. Microbiol.* 80:17.
Govan, J. R. W. (1975). *J. Med. Microbiol.* 8:513.
Govan, J. R. W., and Gillies, R. R. (1969). *J. Med. Microbiol.* 2:17.
Graber, C. D., Latta, R., Vogel, E. H., and Brame, R. (1962). *Am. J. Clin. Pathol.* 37:54.
Grün, V. L., Pillich, J., and Heyn, K. (1967). *Arch. Hyg. Bakteriol.* 151:640.
Habs, I. (1957). *Z. Hyg. Infektionskr.* 144:218.
Hadley, P. (1924). *J. Infect. Dis.* 40:1.
Hanessian, S., Regan, W., Watson, D., and Hadkell, T. H. (1971). *Nature London New Biol.* 229:209.
Hassan, E. M. (1979). Thesis, University of Cairo, p. 282.
Havelaar, A. H., Bosman, M., and Borst, J. (1983). *J. Hyg. Camb.* 90:489.
Heckman, M. G., Babcock, J. B., and Rose, H. D. (1971). *Am. J. Clin. Pathol.* 57:35.
Henry, R. L., Dorman, D. C., Brown, J., and Mellis, C. (1982). *Aust. Paediatr. J.* 18:43.
Higerd, T. B., Baechler, C. A., and Berk, R. S. (1967). *J. Bacteriol.* 93:1976.
Hirao, Y., Homma, J. Y., and Zierdt, C. H. (1977). *Jpn. J. Exp. Med.* 47:249.
Hoadley, A. W. (1977). In *Pseudomonas aeruginosa: Ecological Aspects and Patient Colonization* (V. M. Young, ed.). Raven Press, New York, p. 31.
Hobbs, G., Cann, D. C., Gowland, G., and Byers, H. D. (1964). *J. Appl. Bacteriol.* 27:83.
Hoff, J. C., and Drake, C. H. (1961). *Am. J. Public Health* 51:918.
Hohmann, A., and Wilson, M. R. (1975). *Infect. Immun.* 12:866.
Hoiby, N. (1975a). *Scand. J. Immunol. Suppl.* 21:187.
Hoiby, N. (1975b). *Acta Pathol. Microbiol. Scand. Sect. B* 83:321.

Hoiby, N. (1975c). *Acta Pathol. Microbiol. Scand. Sect. B 83*:328.
Hoiby, N. (1977). *Acts Pathol. Microbiol Scand. Sect. C Suppl. 262*:1.
Hoiby, N., and Axelsen, N. H. (1973). *Acta. Pathol. Microbiol. Scand. Sect. B 81*:298.
Hoiby, N., and Olling, S. (1977). *Acta Pathol. Microbiol. Scand. Sect. C 85*:107.
Hoiby, N., Andersen, V., and Bendixen, G. (1975). *Acta Pathol. Microbiol. Scand. Sect. C 83*:459.
Hojyo-Tomoka, M. T., Marples, R. R., and Kligman, A. M. (1973). *Arch. Dermatol. 107*:723.
Holder, I. A. (1977). In *Pseudomonas aeruginosa: Ecological Aspects and Patient Colonization* (V. M. Young, ed.), Raven Press, New York, p. 77.
Holloway, B. W. (1960). *J. Pathol. Bacteriol. 80*:448.
Holloway, B. W., and Krishnapillai, V. (1975). In *Genetics and Biochemistry of Pseudomonas* (P. H. Clarke and M. H. Richmond, eds.), Wiley, New York, pp. 99–132.
Homma, J. Y. (1974). *Jpn. J. Exp. Med. 44*:1.
Homma, J. Y. (1976). *Jpn. J. Exp. Med. 46*:329.
Homma, J. Y., Kim, K. S., Yamada, H., Ito, M., Shionoya, H., and Kawabe, Y. (1970). *Jpn. J. Exp. Med. 40*:347.
Homma, J. Y., Shionoya, H., Yamada, H., and Kawabe, Y. (1971). *Jpn. J. Exp. Med. 41*:89.
Homma, J. Y., Shionoya, H., Yamada, H., Enomoto, M., and Miyao, K. (1972). *Jpn. J. Exp. Med. 42*:171.
Homma, J. Y., Hirao, Y., Saku, K., Terada, Y., and Sugiyama, J. (1977). *Jpn. J. Exp. Med. 47*:195.
Hopkins, R. S., Abbott, D. O., and Wallace, L. E. (1981). *Public Health. Rep. 96*:264.
Ishii, S., Nishi, Y., and Egami, F. (1965). *J. Mol. Biol. 13*:428.
Jacob, F. (1952). *Ann. Inst. Pasteur 83*:671.
Jacob, F. (1954). *Ann. Inst. Pasteur 86*:149.
Jacob, F., Blobel, H., and Scharmann, W. (1973). *Zentralbl. Bakteriol. Parasitenkd. Infekt. Hyg. Abt. 1 Orig. Reihe A 224*:472.
Jacobson, J. A., Hoadley, A. W., and Farmer, J. J. (1976). *Am. J. Public Health 66*:1092.
Jacobsthal, E. (1912). *Muenchen Med. Wochenschr. 59*:1247.
Jadin, J. (1932). *C. R. Soc. Biol. Paris 109*:556.
Jones, L. F., Pinto, B. V., Thomas, E. T., and Farmer, J. J. (1973). *Appl. Microbiol. 26*:120.
Jones, L. F., Thomas, E. T., Stinnett, J. D., Gilardi, G. L., and Farmer, III., J. J., (1974a). *Appl. Microbiol. 27*:288.
Jones, L. F., Zakanycz, J. P., Thomas, E. T., and Farmer, J. J. (1974b). *Appl. Microbiol. 27*:400.
Kageyama, M. (1964). *J. Biochem. 55*:49.
Kageyama, M., and Egami, F. (1962). *Life Sci. 1*:471.

Kanzaki, K. (1934). *Zentralbl. Bakteriol. Parasitenkd. Infektionskr. Hyg. Abt. 1 Orig. 133*:89.
Kawaharajo, K. (1973). *Jpn. J. Exp. Med. 43*:225.
Kaziro, Y., Tanaka, M., and Shimazono, N. (1964). *Biochem. Biophys. Rs. Commun. 17*:624.
Khabbaz, R. F., McKinley, T. W., Goodman, R. A., Hightower, A. W., Highsmith, A. K., Tait, K. A., and Band, J. D. (1983). *Am. J. Med. 74*:73.
Kleinmaier, H. (1957). *Zentralbl. Bakteriol. Parasitenkd. Infektionskr. Hyg. Abt. 1 Orig. 170*:570.
Kleinmaier, H., and Muller, H. (1958). *Zentralbl. Bakteriol. Parasitenkd. Infektionskr. Hyg. Abt. 1 Orig. 172*:54.
Kleinmaier, H., Schreiner, E., and Graeff, H. (1958). *Z. Immunitaetsforsch. Exp. Ther. 115*:492.
Klieneberger, C. (1907). *Muench. Med. Wochenschr. 54*:1330.
Kodama, H., and Ishimoto, M. (1976). *Jpn. J. Exp. Med. 46*:383.
Köhler, W. (1957). *Z. Immunitaetforsch. Exp. Ther. 114*:282.
Kominos, S. D., Copeland, C. E., and Grosiak, B. (1972). *Appl. Microbiol. 23*:309.
Kominos, S. D., Copeland, C. E., and Delenko, C. A. (1977). In *Pseudomonas aeruginosa: Ecological Aspects and Patient Colonization* (V. M. Young, ed.), Raven Press, New York, p. 59.
Kono, M., and Sei, S. (1977). *Jpn. J. Exp. Med. 47*:1.
Koval, S. F., and Meadow, P. M. (1975). *J. Gen. Microbiol. 91*: 437.
Küchler, R. (1976). *Zentralbl. Bakteriol. Parasitenkd. Infektionskr. Hyg. Abt. 1 Orig. Reihe A 234*:202.
Küchler, R., and Gunther, D. (1976). *Zentralbl. Bakteriol. Parasitenkd. Infektionskr. Hyg. Abt. 1 Orig. 235*:413.
Kulczycki, L. L., Murphy, T. M., and Bellanti, J. A. (1978). *Am. Med. Assoc. 240*:30.
Kurup, V. P., and Sheth, N. K. (1976). *Am. J. Clin. Pathol. 65*: 557.
Kush, B. J., and Hoadley, A. W. (1980). *Am. J. Public Health 70*: 279.
Lányi, B. (1966—1967). *Acta Microbiol. Acad. Sci. Hung. 13*:295.
Lányi, B. (1970). *Acta Microbiol Hung. 17*:35.
Lányi, B., Adam, M. M., and Szentmihalyi, A. (1975). *J. Med. Microbiol. 8*:225.
LeBlanc, C. M. A., Bortolussi, R., Issekutz, A. C., and Gillespi, T. (1982). *Clin. Invest. Med. 5*:125.
Leigh, D. A. (1971). *J. Clin. Pathol. 24*:295.
Lidwell, O. M. (1959). *Bull. Ministry Health Public Health Lab. Serv. 18*:49.
Lindberg, R. B., and Latta, R. L. (1974). *J. Infect. Dis. 130*:33.
Liu, P. V., Matsumoto, H., Kusama, H., and Bergan, T. (1983). *Int. J. Syst. Bacteriol. 33*:256.

Loiseau-Marolleau, M. L. (1973). *Pathol. Biol. 21*:163.
Lowbury, E. J. L., Thom, B. T., Lilly, H. A., Babb, J. R., and Whittall, K. (1970). *J. Med. Microbiol. 3*:39.
Lüderitz, O., Staub, A. M., and Westphal, O. (1966). *Bacteriol. Rev. 30*:192.
Lüderitz, O., Jann, K., and Wheat, R. (1968). In *Comprehensive Biochemistry*, Vol. 26 (M. Florkin and E. H. Stotz, eds.), Elsevier, Amsterdam, P. 105.
Lüderitz, O., Westphal, O., Staub, A. M., and Nikaido, H. (1971). In *Microbiol Toxins*, Vol. 4 (G. Weinbaum, F. Kadis, and S. J. Ajl, eds.), Academic, New York, p. 145.
McCausland, W. J., and Cox, P. J. (1975). *J. Environ. Health 37*: 455.
Macone, A. B., Pier, G. B., Pennington, J. E., Matthews, W. J., and Goldmann, D. A. (1981). *N. Engl. J. Med. 304*:1445.
MacMillan, B. G., Edmonds, P., Hummel, R. P., and Maley, B. A. (1973). *J. Trauma 13*:627.
Markowitz, S. M., Macrina, F. L., and Phibbs, Jr., P. V. (1978). *Infect. Immun. 22*:530.
Matsumoto, H., and Tazaki, T. (1969). *Jpn. J. Microbiol. 13*:209.
Mayr-Harting, A. (1948). *J. Gen. Microbiol. 2*:31.
Meitert, E., Meitert, T., Sima, F., Savulian, C., and Mihalache, V. (1976). *Arch. Roum. Pathol. Exp. Microbiol. 35*:83.
Merrikin, D. J., and Terry, C. S. (1972). *J. Appl. Bacteriol. 35*: 667.
Mikkelsen, O. S. (1970). *Acta Pathol. Microbiol. Scand. Sect. B 78*: 163.
Moody, M. R. (1977). In *Pseudomonas aeruginosa: Ecological Aspects and Patient Colonization* (V. M. Young, ed.), Raven Press, New York.
Moody, M., Young, V. M., Schimpff, S., and Kenton, D. M. (1971). *Antimicrob. Agents Chemother.* 249.
Munoz, J., Scherago, M., and Waever, R. H. (1949). *J. Bacteriol. 57*:269.
Muraschi, T. F., Bolles, D. M., Moczulski, C., and Lindsay, M. (1966). *J. Infect. Dis. 116*:84.
Mutharia, L. M., Nicas, T. I., and Hancock, R. E. W. (1982). *J. Infect. Dis. 146*:770.
Naito, T., Iwanaga, Y., and Koura, M. (1972a). *Trop. Med. 14*:1.
Naito, T., Koura, M., and Iwanaga, Y. (1972b). *Trop. Med. 14*:71.
Nguyen, B. T., Peterson, P. K., Verbrugh, H. A., Quie, P. G., and Hoidal, J. R. (1982). *Infect. Immun. 36*:504.
Noone, M. R. (1983). *Br. Med. J. 286*:341.
Osman, M. A. M. (1965). *J. Clin. Pathol. 18*:200.
Parker, M. T., Pitt, T. L., Asheshow, E., and Martin, D. R. (1975). *Proceedings of the International Colloquium on Phage Typing and*

Other Laboratory Methods for Epidemiological Surveillance, 6th Wernigerode, Germany, p. 468.
Pavlatou, M., and Kaklamani, E. (1962). Ann. Inst. Pasteur 102: 300.
Penketh, A., Pitt, T., and Roberts, D. (1983). Am. Rev. Respir. Dis. 127:605.
Pitt, T. L. (1980). J. Hosp. Infect. 1:193.
Pitt, T. L., and Bradley, D. E. (1975). J. Med. Microbiol. 8:97.
Pitt, T. L., and Erdman, Y. J. (1977). J. Med. Microbiol. 11:15.
Pollack, M., Taylor, N. S., and Callahan, III., L. T. (1977). Infect. Immun. 15:776.
Pollack, M., Longfield, R. N., and Karney, W. W. (1983). Am. J. Med. 74:980.
Postic, B., and Finland, M. (1961). J. Clin. Invest. 40:2064.
Rabinowitz, G. (1934). J. Bacteriol. 28:221.
Rajashekaraiah, K. R., Rice, T. W., and Kallick, C. A. (1981). J. Infect. Dis. 144:482.
Rampling, A., Whitby, J. L., and Wildy, P. (1975). J. Med. Microbiol. 8:531.
Rasmussin, J. E., and Graves, W. H. (1982). Am. J. Dis. Child. 136:553.
Reid, T. M. S., and Porter, I. A. (1981). J. Hyg. Camb. 86:357.
Reyes, M. P., and Lerner, A. M. (1983). Rev. Infect. Dis. 5:314.
Reynolds, H. Y., and Thompson, R. E. (1973). J. Immunol. 111: 358.
Reynolds, H. Y., Levine, A. S., Wood, R. E., Zierdt, C. H., Dale, D. C., and Pennington, J. E. (1975). Ann. Intern. Med. 82: 819.
Reynolds, H. Y., Di Sant Agnese, P. A., and Zierdt, C. H. (1976). J. Am. Med. Assoc. 236:2190.
Rinke, C. M. (1983). J. Am. Med. Assoc. 250:2031.
Rivera, M., and Nicotra, M. B. (1982). Am. Rev. Respir. Dis. 126:833.
Roe, E. A., and Jones, R. J. (1974). Brit. J. Exper. Pathol. 55: 336.
Rose, H. B., Babcock, J. B., and Heckman, M. G. (1971). Appl. Microbiol. 22:475.
Rose, H. D., Franson, T. R., Sheth, N. K., Chusid, M. J., Macher, A. M., and Zierdt, C. H. (1983). J. Am. Med. Assoc. 250:2027.
Salmen, P., Dwyer, D. M., Vorse, H., and Kruse, W. (1983). J. Am. Med. Assoc. 250:2025.
Sandvik, O. (1960). Acta Vet. Scand. 1:221.
Sausker, W. F., Aeling, J. L., and Fitzpatrick, J. E. (1978). J. Am. Med. Assoc. 239:2362.
Schiotz, P. O., and Hoiby, N. (1975). Acta Pathol. Microbiol. Scand. Sect. C 83:469.

Schroth, M. N., Cho, J. J., Green, S. K., and Kominos, S. D. (1977). In *Pseudomonas aeruginosa: Ecological Aspects and Patient Colonization* (V. M. Young, ed.) Raven Press, New York, p. 1.
Schwartzmann, S., and Boring, J. R., III (1971). *Infect. Immun.* 3:762.
Sensakovic, J. W., and Bartell, P. F. (1974). *J. Infect. Dis. 129*: 101.
Sherertz, R. J., Sarubbi, F. A. (1983). *J. Clin. Microbiol. 18*:160.
Shionoya, H., and Homma, J. Y. (1968). *Jpn. J. Exp. Med. 38*:81.
Shionoya, H., Arai, H., and Ohtake, S. (1977). *Jpn. J. Exp. Med. 47*:185.
Shionoya, H., Goto, S., Tsukamoto, N., and Homma, J. Y. (1967). *Jpn. J. Exp. Med. 37*:359.
Shooter, R. A., Walker, K. A., Williams, V. R., Horgan, G. M., Parker, M. T., Asheshov, E. H., and Bullimore, J. F. (1966). *Lancet 2*:1331.
Shriniwas (1974). *J. Clin. Pathol. 27*:92.
Sjoberg, L., and Lindberg, A. A. (1968). *Acta Pathol. Microbiol. Scand. 74*:61.
Smirnova, N. E. (1976). *Zh. Mikrobiol. Epidemiol. Immunobiol. 53*: 126.
Smith, P. B. (1983). *Clin. Microbiol. Newslett. 5*:165.
Smith, P. B. (1972). In *The Staphylococci* (J. O. Cohen, ed.), Interscience, New York, p. 431.
Stafford, S. R., Nixon, M., and Brieseman, M. A. (1982). *N.Z. Med. J. 95*:179.
Steers, E., Foltz, E. L., and Graves, B. S. (1959). *Antibiot. Chemother. 9*:307.
Strand, C. L., Bryant, J. K., Morgan, J. W., Foster, J. G., McDonald, H. P., and Morganstern, S. L. (1982). *J. Am. Med. Assoc. 248*:1615.
Sutter, V. L., Hurst, V., and Fennell, J. (1963). *J. Bacteriol. 86*:1354.
Sutter, V. L., Hurst, V., and Fennell, J. (1965). *Health Lab. Sci. 2*:7.
Suzuki, N., Tsunematsu, Y., Matsumoto, H., and Tazaki, T. (1977). *Infect. Immun. 15*:692.
Swanson, J. (1975). *Vie Med. Can. Fr. 4*:1456.
Tagg, J. R., and Mushin, R. (1971). *Med. J. Aust. 1*:847.
Takeya, K., Minamishima, Y., Amako, K., and Ohnishi, Y. (1967). *Virology 31*:166.
Tejedor, C., Foulds, J., and Zasloff, M. (1982). *Infect. Immun. 36*:440.
Tereda, Y., Sugiyama, J., and Orikasa, M. (1977). *Jpn. J. Exp. Med. 47*:203.

Thomassen, M. J., Boxerbaum, B., Demko, C. A., Kuchenbrod, P. J., Dearborn, D. G., and Wood, R. E. (1979). Pediatr. Res. 13:1085.
Tinne, J. E. (1977). Scott. Med. J. 22:16.
Tripathy, G. S., and Chadwick, P. (1971). Can. J. Microbiol. 17: 829.
Trommsdorff, R. (1916). Zentralbl. Bakteriol. Parasitenkd. Infektionskr. Hyg. Abt. 1 Orig. 78:493.
Tunstall, A. M., and Gowland, G. (1975). J. Appl. Bacteriol. 38: 159.
Urbano, P., Dei, R., Tarantelli, F., and Mazza, A. (1983). Med. Microbiol. Immunol. 172:41.
Van Den Ende, M. (1952). J. Hyg. 50:405.
Verder, E., and Evans, J. (1961). J. Infect. Dis. 109:183.
Veron, M. (1961). Ann. Inst. Pasteur 101:456.
Vieu, J. F. (1969). Bull. Inst. Pasteur 67:1231.
Vogt, R., LaRue, D., Parry, M. F., Brokopp, C. D., Klaucke, D., and Allen, J. (1982). J. Clin. Microbiol. 15:571.
Wahba, A. H. (1963). J. Hyg. 61:431.
Wahba, A. H. (1965a). Br. Med. J. 1:86.
Wahba, A. H. (1965b). Zentralbl. Parasitenkd. Infekt. Hyg. Abt. 1 Orig. 196:389.
Wahba, A. H., and Lidwell, O. M. (1963). Appl. Bacteriol. 26:246.
Warner, P. T. J. (1950). Br. J. Exp. Pathol. 31:242.
Washburn, J., Jacobson, J. A., and Marston, E. (1976). J. Am. Med. Assoc. 235:2205.
Williams, R. J., and Govan, J. R. W. (1973). J. Med. Microbiol. 6: 409.
Williams, R. W., Williams, E. D., and Hyams, D. E. (1960). Lancet 13:376.
Winnie, G. B., Klinger, J. D., Sherman, J. M., and Thomassen, M. J. (1982). Infect. Immun. 38:1088.
Winnie, G. B., Dearborn, D. G., Klinger, J. D., and Thomassen, M. J. (1983). Curr. Microbiol. 9:63.
Wokatsch, R. (1964). Zentralbl. Bakteriol. Parasitenkd. Infektionskr. Hyg. Abt. 1 Orig. 192:468.
Wretling, B., Heden, L., Syobert, L., and Wadstrom, T. (1973). J. Med. Microbiol. 6:91.
Yabuuchi, E., Miyajima, N., Hotta, H., and Furu, Y. (1971). Appl. Microbiol. 22:530.
Yamamoto, T., and Chow, C. T. (1968). Can. J. Microbiol. 14:667.
Young, V. M., and Moody, M. R. (1974). J. Infect. Dis. 130:47.
Zabranski, R. J., and Day, F. E. (1969). Appl. Microbiol. 17:293.
Zierdt, C. H. (1959). Am. J. Clin. Pathol. 31:326.
Zierdt, C. H. (1971). Antonie van Leeuwenhoek J. Microbiol. Serol. 31:319.
Zierdt, C. H. (1979). Appl. Environ. Microbiol. 38:1166.

Zierdt, C. H. (1982). *J. Clin. Microbiol. 16*:517.
Zierdt, C. H., and Marsh, H. H. (1962). *Am. J. Clin. Pathol. 38*: 104.
Zierdt, C. H., and Schmidt, P. J. (1964). *J. Bacteriol. 87*:1003.
Zierdt, C. H., and Williams, R. L. (1975). *J. Clin. Microbiol. 1*:521.
Zierdt, C. H., Fox, F. A., and Norris, G. F. (1960). *Am. J. Clin. Pathol. 33*:233.
Zierdt, C. H., Robertson, E. A., Williams, R. L., and MacLowry, J. D. (1980). *Appl. Environ. Microbiol. 39*:623.

10
Transformation Assays for *Acinetobacter* and *Moraxella*

ELLIOT JUNI   *The University of Michigan, Ann Arbor, Michigan*

INTRODUCTION

The statement that a particular bacterial strain has been identified implies that this strain is closely related to another strain that has been designated as the type or neotype strain of a certain species. The conventional general approach for strain identification is based upon a comparison of some of the properties of the strain under investigation with the corresponding properties of the type strain, or the properties of a group of strains considered to be representative of typical strains of a certain species. Because of their lack of unique distinguishing phenotypic characteristics, it has frequently been difficult to identify strains of the gram-negative non-sugar-fermenting bacteria using conventional methodology.

Newer methods based upon nucleic acid homologies have been most useful in microbial taxonomy. When single stranded DNA samples from two closely related bacteria are mixed, under appropriate conditions, they can associate to form stable complexes (1,2). The closer the relationship of the strains the greater is the stability of the complex. The ability to demonstrate such DNA:DNA homology provides clear evidence for relatedness of the strains being compared. The ultimate criterion for relatedness of two independently isolated bacteria is the ability of their respective chromosomes to undergo ready genetic recombination. This process can take place only when the base sequences of the two DNA molecules are very nearly identical for the genes being recombined.

It has been known for approximately 30 years that DNA from a streptomycin-resistant strain of *Haemophilus influenzae* can transform different streptomycin-sensitive strains of *H. influenzae* to streptomycin resistance (3). Furthermore, it was also demonstrated that in addition to interstrain transformation, interspecies transformation of the streptomycin resistance marker can occur in *Haemophilus, Rhizobium, Neisseria, Bacillus,* and *Moraxella* (4). Interspecies transformation takes place at considerably lower frequencies than interstrain or intrastrain transformation within a particular species. The ratio of the frequency of interspecies transformation to the frequency of intrastrain transformation is taken as a measure of species relatedness; the closer the ratio approaches unity, the closer the relationship (5). The ratio of interstrain to intrastrain transformation frequencies has been shown to be fairly close to unity within species of *Moraxella* (5).

QUALITATIVE TRANSFORMATION ASSAY

Determination of strain relatedness by quantitative transformation of the high-level streptomycin resistance marker requires preliminary isolation of a streptomycin-resistant mutant of each strain to be tested (5). The isolation of these mutants, preparation of DNA from such streptomycin-resistant strains, and performance of quantitative transformation assays are not operations normally carried out in a diagnostic laboratory. When it was first discovered that a particular strain of *Acinetobacter calcoaceticus* is highly competent for genetic transformation, it was also shown that DNA samples from other acinetobacters readily transformed auxotrophs of the competent strain to prototrophy, as evidenced by growth of colonies derived from transformed cells on a simple medium devoid of vitamins, amino acids, or other growth factors (6). During these studies a simple method was devised for preparation of crude transforming DNA that makes possible the analysis of large numbers of strains within 24 hr (6). Qualitative transformation assays for identification of strains of other bacterial species were developed subsequently (7—10). The qualitative transformation assay is essentially an "all or nothing" test. The DNA from an unidentified strain will either transform the auxotrophic mutant strain of a particular species or it will fail to do so. In a few cases there may be weak interspecies transformation for strains belonging to species that are especially closely related. The specificity of transformation assays is discussed below.

ORGANISMS THAT CAN BE IDENTIFIED BY THE QUALITATIVE TRANSFORMATION ASSAY

In order to prepare a transformation assay for identification of strains of a given species, it is essential that at least one strain of that species be

Transformation Assays

Table 1 Assay Mutant Strains

Organism	Assay strain	Required nutrient
Moraxella osloensis	ATCC 29721	Tryptophan
Moraxella (Branhamella) catarrhalis	B12[a]	Hypoxanthine
Moraxella nonliquefaciens	MN64[a]	Tryptophan
Moraxella bovis	MB13[a]	Tryptophan
Moraxella urethralis	MU5[a]	Hypoxanthine
Acinetobacter calcoaceticus	ATCC 33308	Tryptophan
Psychrotrophic achromobacter	A74[a]	Hypoxanthine

[a]This assay mutant strain will be deposited in the American Type Culture Collection in the near future, but is available upon request from the author.

stably competent for genetic transformation. Fortunately, competent strains have been found for many of the gram-negative nonfermentative species. For some organisms, such as *Acinetobacter*, competent strains are relatively rare (11,12). Competent strains of certain moraxellae have not yet been discovered (5). The organisms of this group for which qualitative transformation assays have been developed are listed in Table 1 together with the assay mutant strains found to be most suitable in these tests. In addition to the assays for strains of the species listed in Table 1, a transformation assay for identification of strains of *Neisseria gonorrhoeae* has been reported (8), as well as several modifications of the original procedure (13–15). Another description of the transformation assay for strains of *Acinetobacter* has also been reported (16).

SPECIFICITY OF QUALITATIVE TRANSFORMATION ASSAYS

The transformation assay has been shown to be completely specific for strains of *Acinetobacter*, since DNA samples from a wide variety of gram-negative bacteria failed to give any transformation of the assay mutant strain (6). Similar results have been obtained in the assay for psychrotrophic achromobacters (10). The studies of Bøvre and Henriksen have resulted in the recognition of several distinct species of *Moraxella* (5). All strains of each species have ratios of interstrain to intrastrain transformation of the streptomycin resistance marker close to unity, whereas such ratios for interspecies transformation are usually less than 10^{-3} and most frequently range from 10^{-4} to less than 10^{-6} (5). In the qualitative transformation assay for strains of Moraxella osloensis there was no interference from other species of Moraxella (7). Strains of Moraxella urethralis all show close relatedness in the transformation assay for strains of this organism (9). Since M. urethralis is not actually a species of Moraxella (5), it is not surprising to find that there is no transformation of M. urethralis auxotrophs with DNA from any Moraxella species (9).

In quantitative transformations of the streptomycin resistance marker in Moraxella bovis and Moraxella nonliquefaciens, ratios of interspecies to intraspecies transformation as high as 3×10^{-3} have been observed (5). In the qualitative transformation assays for these organsims there is a small interspecies interaction, as illustrated below. Because DNA samples from all strains of M. bovis and M. nonliquefaciens transform their respective auxotrophs at similar high frequencies, a particular DNA sample showing a considerably lower frequency of transformation can be discerned readily in the assay, as demonstrated below.

Bøvre (5) has demonstrated that the ratio of interspecies to intraspecies transformation for strains of Moraxella lacunata, M. bovis, and M. nonliquefaciens (the so-called Moraxella lacunata group) can vary from 2×10^{-3} to 10^{-2}. It is possible, therefore, that DNA from a strain of M. lacunata will be found to give low frequencies of transformation in the qualitative transformation assay for both M. bovis and M. nonliquefaciens. It has been observed that the isolation of strains of M. lacunata from human sources is relatively rare (5).

Transformation of the Moraxella (Branhamella) catarrhalis assay strain has been found to be specific for strains of this species (E. Juni and G. A. Heym, unpublished results).

TRANSFORMATION ASSAY PROCEDURE

Principle

Studies of the nutritional requirements of the various competent organisms for which qualitative transformation assays have been developed have made it possible to devise a single-assay medium capable of supporting excellent growth of all the wild-type strains. This medium lacks the amino acid tryptophan as well as any purine or pyrimidine base. Mutants of each species that require one of the missing components in order to grow have been isolated following chemical mutagenesis and are consequently unable to grow when plated on this medium. Following exposure to wild-type parental DNA or DNA from strains of the same species, these mutants are transformed to grow on the assay medium.

Transformation Assays

The assay procedure starts with the preparation of crude transforming DNA from cells of the strain to be identified. This is accomplished by suspending cell paste of the unknown strain in a detergent-based lysing solution and heating the suspension to lyse the cells with release of DNA, as well as other cellular components. This process also serves to sterilize the suspension. A loopful of this crude DNA preparation is then mixed with cell paste of an appropriate mutant strain on the surface of a Heart Infusion plate. During this mixing the small amount of detergent present in the loopful of DNA solution is absorbed into the agar and does not usually lyse the mutant cells to any significant extent. The high molecular weight DNA remains on the agar surface, where it can be taken up by the competent mutant cells during growth on the plate.

Following incubation some of the cell–DNA mixture is streaked on a sector of the assay medium plate, which is then incubated. If the DNA tested is derived from a strain of the same species as the assay mutant strain, the DNA taken into the competent cells will recombine with the chromosomal DNA of the recipient bacteria. For some of these cells this recombination will serve to replace some or all the defective gene in the assay mutant strain that is responsible for its mutant characteristic with DNA that now permits expression of the originally mutated gene. Such transformed cells will be able to grow on the assay medium and will give rise to visible colonies, in contrast to the large number of nongrowing cells in which transformation of the mutated gene did not take place. Although only a small proportion of the plated cells will be able to grow on the assay medium, a sufficiently large number of them will give rise to colonies in the streaked area so that a positive transformation can be recognized readily. By contrast, a control streaking on another sector of the assay medium of the mutant unexposed to DNA will fail to show visible colonies after suitable incubation. Similarly, a streaking of the mixture of the assay mutant and DNA from an unrelated strain will also fail to show colonies on the streaked area of the assay medium plate.

Materials and Bacterial Strains Required for the Transformation Assay

1. Assay mutant strains (see Table 1)
2. Lysing solution (see Table 2)
3. Assay medium plates (see Table 3)
4. Water bath or heating block at 60–70°C
5. Incubator at 33–36°C
6. Heart Infusion (Difco) plates

Detailed Steps of the Transformation Assay Procedure

1. The strain to be assayed should be grown overnight on a suitable semisolid medium. Using a sterile loop, remove a small amount of cell paste of this strain, an amount just visible to the naked eye, and suspend it thoroughly in 0.5 ml of sterile lysing solution (Table 2). Heat the suspension at 60—70°C for 30—60 min. The amount of cell paste used is not critical, but the suspension should be slightly turbid when first prepared. Heating periods longer than 1 hr are permissible. In most cases lysis will be evident after heating by disappearance of the initial turbidity. Any number of crude DNA samples may be prepared simultaneously and stored at room temperature or refrigeration temperature until ready for use.

If too much cell paste is used in the preparation of crude DNA, this may produce a viscous solution which will result in adherence of excess detergent to the recipient cells and cause some lysis on the plate.

2. Any of the assay mutant strains to be used (Table 1) should be grown on Heart Infusion plates for 8—24 hr at 33—36°C, with the exception of the psychrotrophic achromobacter, which should be grown at room temperature. Small amounts of cell paste of the assay mutant strain should be placed in the centers of squares marked A—F of another Heart Infusion plate, using a sterile loop, as shown in Figure 1. The arrangement shown is suitable for the simultaneous assay of five strains. The amount of cell paste used is not critical.

3. Using a sterile 2-mm-diameter loop, remove a loopful of crude DNA, as prepared in step 1 above, and use this to suspend and spread the cell paste in one of the squares marked B—F (Fig. 1)

Table 2 Lysing Solution[a]

Trisodium citrate · $2H_2O$	0.44 g
Sodium chloride	0.88 g
Sodium dodecylsulfate	0.05 g
Distilled water	100 ml

[a]This solution is sterilized by autoclaving for 20 min or by passage through a 0.22-μm filter and 0.5-ml aliquots are dispensed into sterile 13 X 100 mm screw-capped test tubes. These tubes may be stored indefinitely at room temperature if the caps are screwed on tightly to prevent evaporation.

Transformation Assays

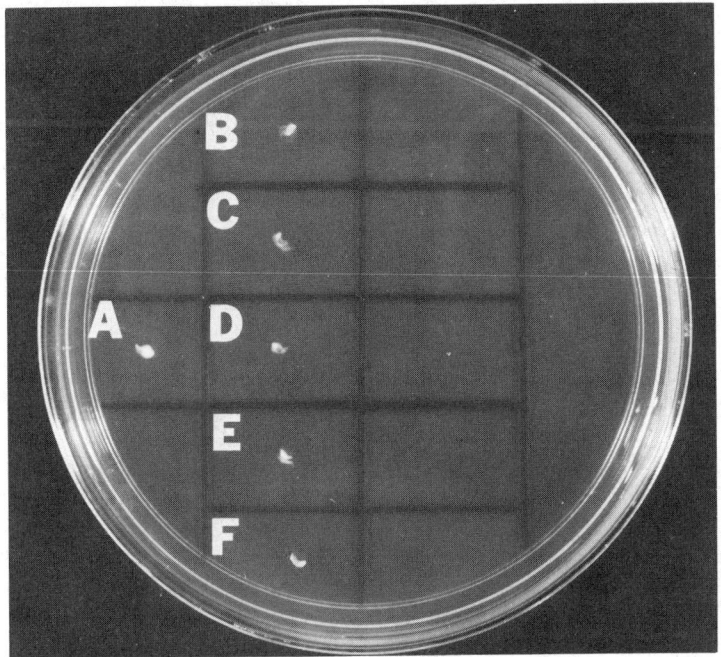

Fig. 1 Preparation of a Heart Infusion plate for assay of five DNA samples for relatedness to the assay mutant strain of *Moraxella non-liquefaciens*. A small but visible amount of cell paste of the assay mutant strain was placed in squares A–F.

over an area approximately 5–8 mm in diameter. A loopful of the same DNA preparation is also placed on the surface of the square to the right of the cell–DNA mixture just prepared to test for the sterility of this material. Other DNA samples should make use of the remaining squares containing cell paste of the assay mutant strain in a similar manner. The cell paste in square A of Figure 1 is spread in this area but without addition of DNA, and the subsequent growth will be used to check the behavior of the non-DNA-treated assay mutant strain. The plate is then incubated at 33–36°C for 8–24 hr. (When the assay mutant strain of the psychrotrophic achromobacter is used, incubation should take place at room temperature.)

4. After incubation for 8–24 hr the plate prepared in step 3 should appear as shown in Figure 2. There should be no growth in any of the squares immediately to the right of growth areas B–F, since the DNA samples placed here should all have been sterile. The appearance of even a single colony in any of these areas implies that the par-

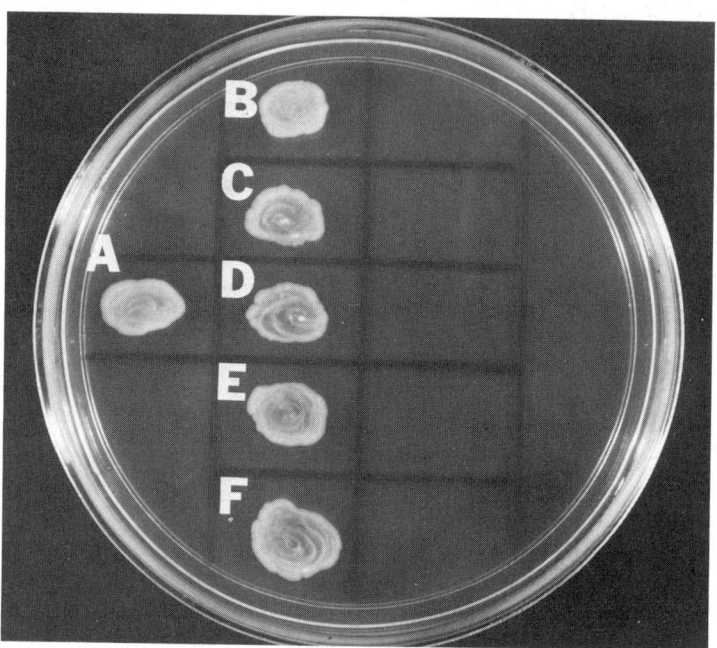

Fig. 2 Appearance of the heart infusion plate shown in Figure 1 after overnight incubation following mixing of the assay mutant strain in squares B—F with crude DNA preparations from five different bacteria. The cell paste in square A was spread without DNA to serve as a non-DNA-treated control for stability of the assay mutant strain. A loopful of each DNA preparation was placed in the appropriate square immediately to the right of squares B—F prior to incubation. The lack of growth in these squares demonstrates that all five DNA samples were sterile.

ticular DNA sample used was not sterile. In such a case either a new DNA preparation should be made or the original DNA preparation should be reheated and step 3 repeated. If there is no growth in the squares upon which the DNA samples were placed, some of the adjacent cell paste should be removed with a sterile loop and streaked to a sector of an assay medium plate. The amount of cell paste removed from growth areas of the plate in Figure 2 for this purpose is illustrated in Figure 3. The cell paste removed should be streaked so as to cover the entire sector of the assay medium plate. Growth of the assay strain in area A of Figure 2 should also be streaked to a separate sector of the assay medium plate in a similar manner. The

Table 3 Assay Medium

One liter of double-strength assay medium is prepared by adding the following components, one at a time, to 750 ml of distilled water until completely dissolved.

Na_2HPO_4	5.6 g
KH_2PO_4	2.0 g
NaCl	10.0 g
NH_4Cl	1.0 g
$MgSO_4 \cdot 7H_2O$	0.9 g
Casein hydrolysate[a]	16.0 g
L-Proline	1.0 g
L-Serine	1.0 g
Glutathione (reduced)	0.8 g
Nicotinic acid	5 mg
Calcium pantothenate	2 mg
Folic acid (saturated solution)	20 ml
Biotin	100 µg
Thiamine HCl	100 µg
Monosodium L-glutamate	10 g
Sodium lactate (60%)	10 ml

Following adjustment of the volume to 1 liter with distilled water the medium is sterilized by passage through a 0.22-µm filter. Since the filter tends to clog before passage of the entire volume, because of material in the casein hydrolysate, it is suggested that the medium be passed through a prefilter (Millipore filter AP15) prior to final filter sterilization. Assay medium plates are prepared by pouring 200 ml of the above medium into 200 ml of recently melted sterile 3% agar, mixing, adding 4 ml of filter sterilized ferric ammonium citrate (Sigma F-5879), mixing once again, and pouring approximately 20 ml per plate. After drying in the inverted position, plates can be stored in double plastic bags at room temperature.

[a] Vitamin-free, salt-free (ICN Nutritional Biochemicals, catalog no. 104778).

350   Juni

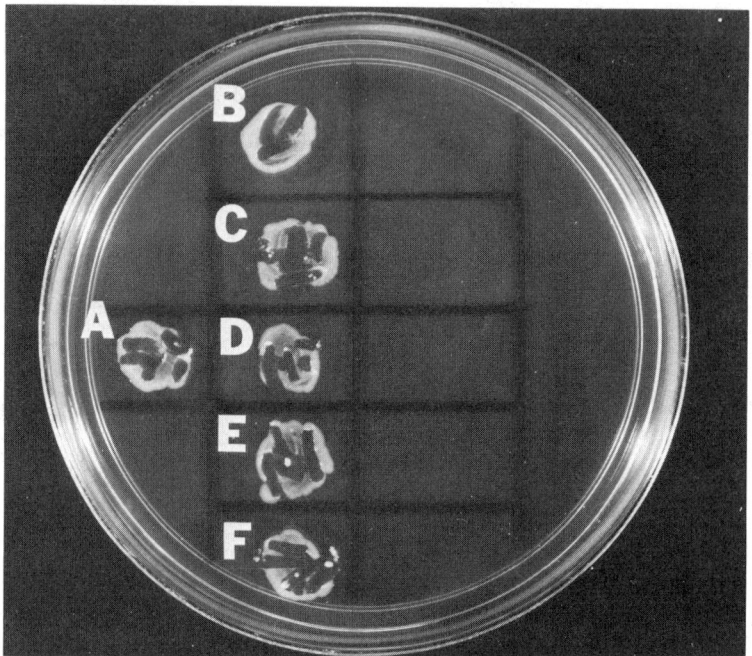

Fig. 3 Illustration of the amount of cell paste removed from each growth area of the Heart Infusion plate of Figure 2 for spreading on sectors of an assay medium plate.

assay medium plate is then incubated at 33–36°C. (Room temperature incubation is required for the psychrotrophic achromobacter strain.)

Usually incubation for 10–20 hr of the streaked assay medium plate is sufficiently long for the appearance of colonies derived from transformed cells, as shown in Figure 4. For some of the assay mutant strains (particularly the assay mutant strains of M. urethralis and the psychrotrophic achromobacter) it may be necessary to observe the streaked areas of the assay medium plate with a dissecting microscope (5- to 10-fold magnification) in order to observe colonies derived from transformed cells. Upon further incubation for another day all colonies derived from transformed cells will be visible to the naked eye. There should be no colonies in the area of the assay medium on which the cell paste of sector A (Fig. 4) was streaked, since this area contains the assay mutant strain that was not exposed to DNA and this represents an essential control of the assay.

Transformation Assays

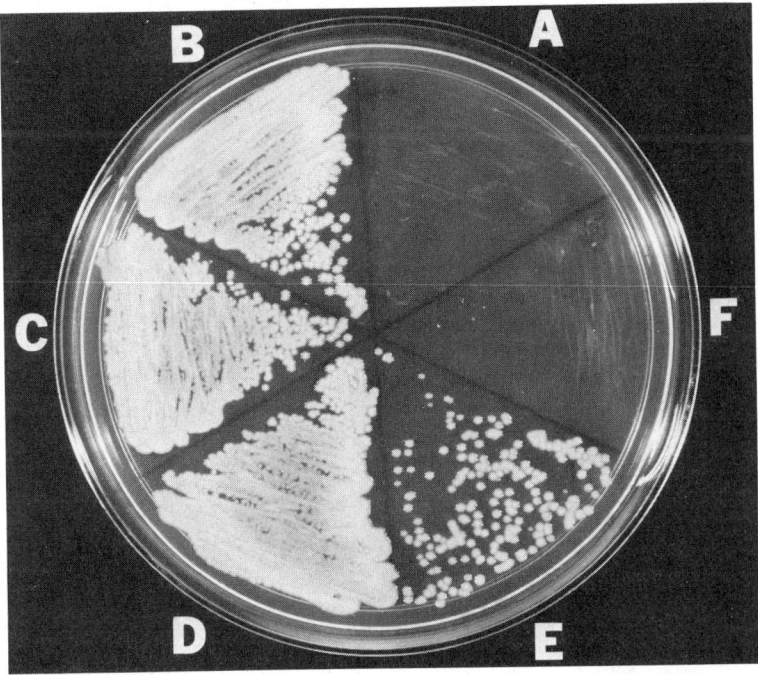

Fig. 4 Assay medium plate streaked with DNA-assay mutant strain mixtures (from the plate of Fig. 3) after incubation for 24 hr. The assay mutant strain of *M. nonliquefaciens* was used to assay DNA samples from five bacteria. Sector A contains a streaking of the non-DNA-treated assay mutant strain. Only the original amount of cell paste used to streak sector A can be seen, no growth having taken place on this medium. The following sources of DNA were used to prepare the DNA-assay mutant strain mixtures streaked on the remaining sectors: (B) *M. nonliquefacines,* ATCC 19975, the parent of the assay mutant strain; (C) *M. nonliquefaciens,* ATCC 17953; (D) *M. nonliquefaciens,* MN58, isolated as the causative agent of endophthalmitis (20); (E) *M. bovis,* ATCC 10900, a strain of a species closely related to *M. nonliquefaciens;* and (F) *M. osloensis,* ATCC 19976, a strain of a species distantly related to *M. nonliquefaciens.*

INFORMATION RELATING TO THE TRANSFORMATION ASSAYS

Stability of the Assay Mutant Strains

The assay mutant strains used in the assays described above are relatively stable and can be maintained during testing periods by daily transfer to Heart Infusion plates. The strains can be maintained indefinitely by suspension in broth and adding an aliquot to sterile glycerol to give a final concentration of 67% glycerol and storage at -65°C. When streaked on control sectors (sector A, Fig. 4) of assay medium plates these mutants will fail to grow. Rare revertants having the ability to grow on the assay medium are occasionally observed as single colonies growing on sector A. Revertants such as these will in no way invalidate the results of the assay, since a positive result for DNA derived from a strain being tested will almost invariably show massive transformation (see sectors B—D, Fig. 4).

Should the culture of an assay mutant strain appear to contain cells able to form colonies on the assay medium, arising from previous spontaneous mutation, it is a simple matter to streak the culture for colony isolation on a Heart Infusion plate, select a few colonies, and check these for ability to grow on the assay medium. Colonies which are unable to grow on the assay medium may then be grown on Heart Infusion plates and maintained as the stock assay mutant strain.

Discussion of Individual Transformation Assays

Moraxella osloensis

The assay mutant strain for this species (Table 1) has been tested with more than 30 independently isolated strains of the species and shown to be transformed massively by DNA from each strain in the qualitative transformation assay procedure (7). The mutant strain is very stable and is not transformed by DNA samples from other species of Moraxella or from a variety of other gram-negative bacteria (7). Quantitative studies of interspecies transformation of the streptomycin resistance marker have demonstrated that M. osloensis is quite distantly related to the other moraxellae (5). Colonies of the assay mutant strain of M. osloensis on the assay medium arising from transformed cells are visible to the naked eye after 18—20 hr of incubation.

Moraxella (Branhamella) catarrhalis

DNA samples from more than two dozen strains of this species transform the assay mutant strain with high efficiency (E. Juni and G. A. Heym, unpublished results). Although there are extremely weak interactions with other moraxellae using the streptomycin resistance marker (5), no transformation of the assay mutant strain of

M. catarrhalis has been observed with DNA samples from other species of *Moraxella* (E. Juni and G. A. Heym, unpublished results).

Moraxella nonliquefaciens and Moraxella bovis

These two species and *M. lacunata* make up what is known as the "*Moraxella lacunata* group" because quantitative interspecies to intraspecies transformation ratios, making use of the streptomycin resistance marker, are relatively low (10^{-2}–10^{-3}) compared to such ratios for other species of *Moraxella* (4×10^{-5} to less than 10^{-6}) (5). Each of the assay mutant strains of *M. nonliquefaciens* and *M. bovis* is highly specific for DNA samples of the same species (E. Juni and G. A. Heym, unpublished results). The DNA from a strain of another species of the "*M. lacunata* group" may give a weak interaction in the transformation assay (see Fig. 4, sector E) but such poor transformation is readily distinguished from the massive transformation observed when DNA samples from strains of the same species are used (E. Juni and G. A. Heym, unpublished results). When a suspected strain of *M. nonliquefaciens* or *M. bovis* is tested in the transformation assay, it is suggested that a control DNA sample from an authentic strain of that species be used together with DNA samples from the strains being identified in order to compare the intensity of the transformation. Should a relatively weak transformation be observed with either assay mutant strain of *M. nonliquefaciens* or *M. bovis*, the DNA sample giving this result should also be tested with the other assay mutant strain. An assay mutant strain for *M. lacunata* is not currently available. In any case, a weak transformation with either of the available assay mutant strains serves to identify the unknown organism as a member of the "*M. lacunata* group." DNA samples from other species of *Moraxella* as well as from several other gram-negative bacteria failed to transform either of the assay mutant strains for *M. nonliquefaciens* and *M. bovis* (E. Juni and G. A. Heym, unpublished results).

Moraxella urethralis

Although tentatively classified as a species of *Moraxella*, it is recognized that this organism is, in fact, unrelated to the other moraxellae (5). DNA samples from more than two dozen independently isolated strains of *M. urethralis* have been shown to give massive transformation of the assay mutant strain for this species (Table 1) (9). No transformation was observed with DNA samples from any of the moraxellae as well as from a variety of gram-negative bacteria.

Acinetobacter calcoaceticus

In a study of several hundred acinetobacters it was shown that DNA from each of these strains was capable of transforming the assay mutant strain listed in Table 1 (6). Unlike the situation for the moraxellae, where all strains of a given species transform an auxotrophic mutant strain of the same species with high frequency, transformation of a single auxotrophic mutant of *Acinetobacter* by DNA samples from different acinetobacters takes place with a wide variation in the proportion of recipient cells that are transformed (6). In a few cases, where strains were derived from soil or water, extremely low efficiencies of transformation have been observed (E. Juni and G. A. Heym, unpublished results). This finding suggests that different species of *Acinetobacter* may exist in nature. Although attempts have been made to discover natural groupings of acinetobacters, no valid basis has been found for speciation (12). Because of the wide variety of environments in which acinetobacters reside, from the human body to soil and water (11), it seems likely that continuous mutational changes occurred to enable different strains to adapt to a particular environment. Perhaps it is a consequence of all these mutational changes that interstrain transformation frequencies can vary so widely in *Acinetobacter*. Another manifestation of this variation in DNA composition is the wide range of efficiency of DNA:DNA hybridization demonstrated for strains of *Acinetobacter* (17).

The Psychrotrophic Achromobacters

The psychrotrophic achromobacters also represent a group of bacteria that are found in many different environments; from water as commensals on the skins of fish, and from processed foods. These organisms are gram-negative, nonmotile, oxidase-positive coccobacilli and were at one time considered to be members of the genus *Acinetobacter* (18). Currently only the oxidase-negative strains are classified as strains of *Acinetobacter* (19). Although most psychrotrophic achromobacters grow best at 20—25°C and do not grow at all at 37°C, strains having temperature optima near 37°C have been isolated (10) and some have been implicated in human infection, these organisms having been tentatively characterized as strains of *Moraxella paraphenylpyruvica* (E. Falsen, personal communication).

When it was found that many psychrotrophic achromobacters were competent for genetic transformation, a transformation assay for identification of related strains was developed (10). The psychrotrophic achromobacters are very similar to the acinetobacters in that they are found in nature in a variety of ecological niches and appear to have evolved considerably. DNA samples from different strains of psychrotrophic achromobacters transform the assay mutant strain of this organism with a large variation in frequency of transformation (10), unlike the uniformly high frequency of transformation of an assay mutant strain for any *Moraxella* species by DNA samples from all strains of that particular species. DNA samples from moraxellae fail to transform the assay mutant strain for psychrotrophic achromobacters (10). Low-level but clearly positive transformation of this strain can be taken as proof of the identity of the strain from which the assayed DNA sample was derived (10).

The psychrotrophic achromobacters do not appear to be closely related to any other previously described group of bacteria. We plan to propose shortly that all strains of psychrotrophic achromobacter be classified as strains of *Psychrobacter immobilis*.

ACKNOWLEDGMENT

The investigations by the author, described in this chapter, were supported by Public Health Service grant AI-10107 from the National Institute of Allergy and Infectious Diseases.

REFERENCES

1. B. J. McCarthy and E. T. Bolton, *Proc. Nat. Acad. Sci. U.S.A.* 50:156 (1963).
2. J. H. Crosa, D. J. Brenner, and S. Falkow, *J. Bacteriol.* 115: 904 (1973).
3. G. Leidy, E. Hahn, and H. E. Alexander, *J. Exp. Med.* 104: 305 (1956).
4. P. Schaeffer, in *The Bacteria,* Vol. 5 (I.C. Gunsalus and R. Y. Stanier, eds.), Academic, New York, 1964, p. 87.
5. K. Bøvre and N. Hagen, in *The Prokaryotes* (M. P. Starr, H. Stolp, H. G. Truper, A. Balows, and H. G. Schlegel, eds.), Springer-Verlag, Berlin, 1981, p. 1506.
6. E. Juni, *J. Bacteriol.*, 112:917 (1972).
7. E. Juni, *Appl. Microbiol.* 27:16 (1974).
8. A. Janik, E. Juni, and G. A. Heym, *J. Clin. Microbiol.* 4:71 (1976).
9. E. Juni, *J. Clin. Microbiol.* 5:227 (1977).
10. E. Juni and G. A. Heym, *Appl. Environ. Microbiol.* 40:1106 (1980).
11. E. Juni, *Annu. Rev. Microbiol.* 32:349 (1978).
12. E. Juni, in *Bergey's Manual of Systematic Bacteriology,* Vol. 1 (N. R. Krieg and J. G. Holt, eds.), Williams and Wilkins, Baltimore, 1984, p. 303.
13. S. K. Sarafian and H. Young, *J. Med. Micriobiol.* 13:291 (1980).
14. L. Zubrzycki and S. S. Weinberger, *Sex. Transm. Dis.* 7:183 (1980).
15. L. O. Butler and R. D. J. Knight, *J. Clin. Microbiol.* 15:810 (1982).
16. K. Brooks and T. Sodeman, *Appl. Microbiol.* 27:1023 (1974).

17. J. L. Johnson, R. S. Anderson, and E. J. Ordal, *J. Bacteriol.* *101*:568 (1970).
18. M. J. Thornley, *J. Gen. Microbiol.* *49*:211 (1967).
19. E. F. Lessel, *Int. J. Syst. Bacteriol.* *21*:213 (1971).
20. J. E. Ebright, J. R. Lentino, and E. Juni, *J. Clin. Pathol.* *77*:362 (1982).

Index

Achromobacter group Vd
  clinical role of, 205
  identification of, 78–82
Achromobacter xylosoxidans
  clinical role of, 204
  identification of, 78–82
Acidovorans group (see
  Pseudomonas acidovorans)
Acinetobacter calcoaceticus
  antimicrobial susceptibility of, 172–173
  assay mutant strains, 353–354
  community-acquired adult pneumonia by, 162–166
  ecology of, 159–160
  endocarditis by, 167
  exoenzymes in, 251
  foundry pneumonia by, 164–166
  identification of, 52–54
  meningitis by, 167–168, 171
  nosocomial pneumonia by, 160–162
  pediatric respiratory infections by, 166
  peritonitis by, 170

[Acinetobacter calcoaceticus]
  skin and wound infections by, 168–169
  urinary tract infections by, 169–170, 170–172
Agrobacterium radiobacter
  clinical role of, 212
  identification of, 82
Alcaligenes
  clinical role of, 202–204
Alcaligenes denitrificans, 70–72
Alcaligenes faecalis, 70–72
Alcaligenes group (see Pseudomonas alcaligenes)
Alcaligenes odorans, 70–72
Antimicrobial agents
  aminoglycosides, 127, 172–173, 185, 191–193, 196–207, 211–212
  beta-lactam antibiotics, 173
  cephalosporins, 126, 173, 185, 191–193, 196–207, 211–212
  chloramphenicol, 133, 185, 191–193, 196–207, 211–212
  minocycline, 172, 196, 199
  nalidixic acid, 185, 191–193, 196–207, 211–212

[Antimicrobial agents]
  novobiocin, 185, 191–193,
    196–207, 211–212
  penicillins, 185, 191–193,
    196–207, 211–212
  phosphonic acids, 135
  polymyxins, 185, 191–193,
    196–207, 211–212
  tetracyclines, 134, 185,
    191–193, 196–207, 211–
    212
  trimethoprim-sulfamethoxa-
    zole, 173, 185, 191–193,
    196–207, 211–212
  quinolones, 134
  ureidopenicillins, 126
Automated identification systems
  Autobac ID, 93–95
  Automicrobic, 88–93

Bacteriophage typing (see
  Pseudomonas aeruginiosa)
Bordetella bronchicanis
  clinical role of, 205–206
  identification of, 72–75
Branhamella catarrhalis (see
  Moraxella catarrhalis)

Comamonas (see *Pseudomonas
  acidovorans*)

DNA:DNA homology, 341,
  354
Diminuta group (see *Pseudomonas diminuta*)

Farcey *(see Pseudomonas
  mallei)*

Flavobacterium
  animal pathogenicity of,
    206–207
  ecology of, 206–207
  identification of, 62–70
  occurrence in humans of,
    206–207, 209–210
Flavobacterium breve, 63
Flavobacterium group IIe, 63
Flavobacterium group IIf, 63
Flavobacterium group IIh, 63
Flavobacterium group IIi, 63
Flavobacterium group IIj, 70
Flavobacterium indologenes,
  62–63
*Flavobacterium meningosepti-
  cum*
  adult meningitis due to, 208
  antimicrobial susceptibility
    of, 211–212
  identification of, 62
  neonatal meningitis due to,
    207–208
  pneumonitis due to, 208
Flavobacterium odoratum, 63
Flavobacterium multivorum
  (see Sphingobacterium)
Fluorescent group *(see Pseudomonas aeruginosa)*

Glanders *(see Pseudomonas
  mallei)*
Glucose-nonfermenting rods
  identification methods, 4–15
    acidification of carbohy-
      drates, 4–5
    alkalinization of organic
      salts and amines, 9
    cytochrome oxidase, 5–6,
      12
    decarboxylase activity,
      6, 13

Index

[Glucose-nonfermenting rods]
  dihydrolase activity, 6
  fluorescence, 8-9, 14
  fluorescence-indole-denitrification medium, 9
  fluorescence-lactose-denitrification medium, 8
  motility, 7, 14
  motility-nitrate medium, 7-8
  nitrate reduction, 7-8, 14
  oxidative low-peptone medium, 3-5, 9, 11
  oxidative-fermentative medium, 3-5, 11
  incubation temperature, 1
  isolation media, 2-3
  isolation precedures, 1-2
  presumptive identification, 3-4
    L-alanine-4-nitroanilide test in, 4, 10
    potassium hydroxide (3%) test in, 4
  rapid identification systems, 85-115
Group EO-2, 54
Group IVc-2, 72-75
Group IVe, 72-75, 204
Group Va (see *Pseudomonas pickettii*)
Group Ve (see *Pseudomonas* group Ve)

Human melioidosis (see *Pseudomonas pseudomallei*)

Kingella, 173-176

Moraxella
  antimicrobial susceptibility of, 175
  assay mutant strains, 352-353
  conjunctivities by, 174

[*Moraxella*]
  ecology of, 174
  endocarditis by, 174
  identification of, 54-62
  pericarditis by, 174
  septic arthritis by, 175
Moraxella atlantae, 61
Moraxella bovis, 174, 353
Moraxella catarrhalis, 352-353
Moraxella lacunata, 61
Moraxella nonliquefaciens, 61, 175, 353
Moraxella osloensis, 61, 175, 352
Moraxella phenylpyruvica, 61
Moraxella urethralis, 62, 353

Nonautomated identification systems
  API 20E, 98-103
  Flow N/F, 110
  MicroScan, 114
  Minitek, 103-114
  Oxi/Ferm, 96-98
  Rapid NFT, 115

Pseudomallei group (see *Pseudomonas pseudomallei*)
Pseudomonas acidovorans
  clinical role of, 201
  ecology of, 200-201
  identification of, 41
Pseudomonas aeruginosa
  antibiogram typing in, 322-324
  antimicrobial susceptibility of, 125-135
  bacteremia by, 136-138
  bacteriophage typing, 300-313
    establishing a phage typing set, 310-313
    reading phage reactions, 309-311

[*Pseudomonas aeruginosa*]
  use of a phage typing set, 303–313
  variation in phage type, 312–313
  central nervous system infections by, 141–142
  in cystic fibrosis, 140–141
  ecology of, 118–119
  elastase in, 246, 236, 238–239
  endocarditis by, 138–139
  exoenzyme assays in, 238–240
  exotoxin A in, 236, 238–244, 259–262
  extracellular enzymes in, 235–237
  frequency of exoenzyme production in, 237–248
  hemolysin in, 246–248, 236, 238–239
  identification of, 21–25
  in vitro virulence, determination of, 255–262
    agar plate methods, 255–258
    enzyme profiling, 259
    immunological methods, 259
    inhibition of protein synthesis, 259–261
    spectrophotometric methods, 258–259
    tissue culture, 262
  in vivo virulence, determination of, 262–275
    burn-compromised models, 263–268
    corneal ulcer models, 271–274
    eye infection models, 273–274
    intact animal models, 262–263
    lung infection models, 269–271
    neutropenic animal models, 274–275

[*Pseudomonas aeruginosa*]
  mucoid strains of, 325–328
  ophthamological infections by, 148
  osteomyelitis by, 144–147
  otolaryngological infections by, 142–144
  pathogenic properties of, 121–124
  peritonitis by, 149
  phospholipase C in, 246–248, 236, 238–239
  proteases in, 236, 238–239, 244–246, 255–259
  pulmonary infections by, 139–141
  pyocin typing, 313–321
    active typing, 315–317
    epidemiological use of, 320–321
    passive typing, 314–315
    pyocin production, 315–317
    pyocin sensitivity, 314–315
    reading pyocin reactions, 319–320
  pyocyanin in, 248–249
  septic arthritis by, 144–147
  serotyping, 287–300
    agglutination reactions in, 292–293
    dominance of type O:11, 298–300
    flagellar antibodies in, 290–291
    lipopolysaccharide in, 289–290
    multitypes, 295
    pilus antibodies in, 290–291
    standardization of, 297–298
  skin infections by, 148–149
  thermal injuries by, 135–136
  treatment of infections by, 150–151
  urinary tract infections by, 147
  vaccines, 151

Pseudomonas alcaligenes
  clinical significance of, 200
  identification of, 40
Pseudomonas cepacia
  antimicrobial susceptibility of, 196
  clinical significance of, 194–196
  ecology of, 193–194
  exoenzymes in, 250–251
  identification of, 25–29
Pseudomonas diminuta
  clinical role of, 200
  identification of, 41–44
Pseudomonas extorquens, 49–52
Pseudomonas fluorescens
  antimicrobial susceptibility of, 185
  clinical significance of, 184–185
  ecology of, 183–184
  identification of, 25
Pseudomonas gladioli, 29
Pseudomonas group Ve-1, 47–49, 202
Pseudomonas group Ve-2, 47–49, 202
Pseudomonas mallei, 25, 193
Pseudomonas maltophilia
  antimicrobial susceptibility of, 199
  clinical significance of, 197–199
  ecology of, 197
  exoenzymes in, 249–250
  identification of, 32–34
Pseudomonas mendocina, 32
Pseudomonas paucimobilis
  clinical role of, 202
  exoenzymes in, 251
  identification of, 44–47
Pseudomonas pickettii
  clinical role of, 201
  identification of, 29
Pseudomonas pseudoalcaligenes
  clinical significance of, 200
  identification of, 40

Pseudomonas pseudomallei
  antimicrobial agents in human melioidosis, 188–190
  antimicrobial susceptibility of, 191–192
  clinical features in human melioidosis, 188–190
  ecology and transmission, 185–186
  identification of, 25
  infection in animals, 187–188
  pathology in human melioodosis, 191
  serodiagnosis in human melioidosis, 190–191
  toxin production, 187–188
Pseudomonas putida
  antimicrobial susceptibility of, 185
  clinical significance of, 184–185
  ecology of, 183–184
  identification of, 25
Pseudomonas putrefaciens
  antimicrobial susceptibility of, 199
  clinical significance of, 199
  ecology of, 199
  identification of, 34–37
Pseudomonas stutzeri
  antimicrobial susceptibility of, 197
  clinical significance of, 196–197
  ecology of, 196
  identification of, 29–32
Pseudomonas testosteroni
  clinical role of, 201
  ecology of, 200–201
  identification of, 41
Pseudomonas thomasii (see *Pseudomonas pickettii*)
Pseudomonas vesicularis
  clinical role of, 200
  identification of, 41–44
Psychrobacter immobilis, 355
Psychrotrophic achromobacters, 354–355

Pyocin typing (see *Pseudomonas aeruginosa*)

Serotyping (see *Pseudomonas aeruginosa*)
Sphingobacterium, 70
Stutzeri group (see *Pseudomonas stutzeri*)

Transformation assay, 341–355

Virulence determination (see *Pseudomonas aeruginosa*)

Xanthomonas (see *Pseudomonas maltophilia*)